Control of structures

MECHANICS OF ELASTIC STABILITY
Editors: H.H.E. Leipholz and G.Æ. Oravas

H.H.E. Leipholz, Theory of elasticity. 1974.
  ISBN 90-286-0193-7
L. Librescu, Elastostatics and kinetics of anisotropic and heterogeneous shell-type structures. 1975.
  ISBN 90-286-0035-3
C.L. Dym, Stability theory and its application to structural mechanics. 1974.
  ISBN 90-286-0094-9
K. Huseyin, Nonlinear theory of elastic stability. 1975.
  ISBN 90-286-0344-1
H.H.E. Leipholz, Direct variational methods and eigenvalue problems in engineering. 1977.
  ISBN 90-286-0106-6
K. Huseyin, Vibrations and stability of multiple parameter systems. 1978.
  ISBN 90-286-0136-8
H.H.E. Leipholz, Stability of elastic systems. 1980.
  ISBN 90-286-0050-7
V.V. Bolotin, Random vibrations of elastic systems. 1984.
  ISBN 90-247-2981-5
D. Bushnell, Computerized buckling analysis of shells. 1985.
  ISBN 90-247-3099-6
L.M. Kachanov, Introduction to continuum damage mechanics. 1986.
  ISBN 90-247-3319-7
H.H.E. Leipholz and M. Abdel-Rohman, Control of structures. 1986.
  ISBN 90-247-3321-9

Control of structures

by

H.H.E. Leipholz

University of Waterloo
Ontario, Canada

M. Abdel-Rohman

University of Kuwait
Kuwait

1986 **MARTINUS NIJHOFF PUBLISHERS** 
a member of the KLUWER ACADEMIC PUBLISHERS GROUP
DORDRECHT / BOSTON / LANCASTER

Distributors

for the United States and Canada: Kluwer Academic Publishers, 190 Old Derby Street, Hingham, MA 02043, USA
for the UK and Ireland: Kluwer Academic Publishers, MTP Press Limited, Falcon House, Queen Square, Lancaster LA1 1RN, UK
for all other countries: Kluwer Academic Publishers Group, Distribution Center, P.O. Box 322, 3300 AH Dordrecht, The Netherlands

Library of Congress Cataloging in Publication Data

```
Leipholz, H. H. E. (Horst H. E.), 1919-
   Control of structures.

   (Mechanics of elastic stability ; 11)
   Includes bibliographies and index.
   1. Structural dynamics.  2. Control theory.
I. Abel-Rohman, M.  II. Title.  III. Series.
TA654.L44  1986       624.1'71          86-5365
ISBN 90-247-3321-9
```

ISBN 90-247-3321-9 (this volume)
ISBN 90-247-2743-X (series)

Preface

This book is an outgrowth of several years of teaching and research
of the two authors in the field of structural dynamics and control.
The content of the book is based on structural dynamics, classical
and modern control theory and involves also recent developments
that took place with respect to the control of systems with distributed masses. It is hoped that the book will serve the researcher
and the practicing engineer in the areas of civil, mechanical and
aeronautical engineering. It may also be of interest to applied
mathematicians and to physicists. There is no question that the
book can be used as a reference book for advanced courses in the
above mentioned areas. The numerous examples will provide students
with the necessary material for exercising themselves and for self-
studying.

Thanks are due to Mrs. Cynthia Jones for preparing
patiently and competently the typescript of the book. The services
of Mrs. Linda Strouth, Solid Mechanics Division, University of
Waterloo, rendered in producing the camera-ready copy of the book
with great skill and devotion, are gratefully acknowledged.

Special thanks go to Mr. Ir. Ad. C. Plaizier, at
Martinus Nijhoff/Dr. W. Junk, who with much understanding and
enthusiasm supervised the production of the book as Publisher.

May the readers of it enjoy it, and may they have the
feeling of having gained something in turning to it and using it.

H.H.E. Leipholz
Waterloo, Ontario, Canada

M. Abdel-Rohman
Kuwait

Spring, 1985

Table of Contents

Contents

page

Chapter I

Introduction to Structural Control

1.1 INTRODUCTION

A safe design of a structure is ensured if three requirements are
being satisfied. The first one is that there is exact information
concerning the loads applied to the structure. The second one is
that there is exact information on behaviour and strength of the
building materials. The third one is that efficient methods are
being used to analyze and design the structure. In reality, these
requirements can hardly be met completely, and to take the uncertainties involved in the available information into account, various
factors of safety have been introduced into structural design.
Scientific progress has nowadays led to the application of probabilistic and structural identification concepts for determining (to a
certain extent) the structure's characteristics and the loads
applied to the structure. New materials have been introduced characterized by higher strength and predetermined behaviour, e.g.,
high strength steel and prestressed concrete. The methods of
analyzing and designing a structure have been continuously improved
for example by introducing the finite element method. All these
factors enable the structural engineers to venture on the erection

of large flexible structures such as high-rise buildings, long-span bridges, TV-towers, and off-shore platforms. However, the flexibility and low damping capacity of these structures have given rise to some problems as:

(a) easily possible failure of these structures due to uncertainty with respect to natural disturbances as wind and earthquakes;

(b) large vibrations of these structures causing discomfort and psychophysiological effects to the users of these structures;

(c) failure of some specific structural elements which leads for example to cracks in walls, and to cladding failure.

The Tacoma Bridge, which was designed to resist a steady wind speed of 100 mph, has for example failed by virtue of torsional instability caused by a wind speed of only 42 mph [1.1]. The Golden Gate Bridge has vibrated under a storm with unacceptable large amplitudes, and had, subsequently, to be stiffened [1.2]. Zuk has also reported that when Zetlin and his associates [1.3] had designed a 1,000 ft high tower of a diameter varying from 60 to 90 ft, they found that the maximum deflection would be 2.5 ft due to a wind speed of 150 mph. In order to restrict the lateral deflection to less than one foot, they also found that the structure has to be stiffened using sections of dimensions at least two times larger than those of the original design. It thus became obvious that large expenditures would occur when stiffening highly flexible structures to guarantee a reduction of their responses. In order to avoid such expenditures but still warrant the safety of the structure and human comfort, the concept of structural control was introduced to overcome such problems encountered in connection with modern civil engineering structures. The control concept has been used for a long time for airplanes, ships, and space structures. However, its application to civil engineering structures is only recently being taken into consideration.

1.2 STRUCTURAL CONTROL

Structural control involves basically the regulating of pertinent
structural characteristics as to ensure desirable response of the
structure under the effect of its loading. The control can be
exerted by using either active or passive control mechanisms or
both. These mechanisms provide in general a system of auxiliary
forces, the so-called control forces. These forces are designed
so that they regulate continuously the structural response.

Passive control mechanisms operate without using any
external energy supply. They use the potential energy generated by
the structure's response to supply the control forces. In this case,
the control forces are only able to control the structure's response
up to a certain limit imposed by the lack of energy needed to tackle
larger responses.

Active control mechanisms operate by using an external
energy supply. Therefore, they are more efficient than the passive
control mechanisms because they can control displacement, velocity,
and acceleration of the structure to any extent. Most of the active
control mechanisms can operate also as passive ones if the supplied
control energy were stopped.

One realizes that the cost of an active control is in
general greater than that of a passive control. However, the advantages involved in using an active mechanism for the control of
large flexible structures outweigh by far the economical disadvantages.

The interaction between the structure and the control
mechanism is analyzed by mainly using the principles of current
control theory. Most of the applications of the theory involve
active control mechanisms and are concerned with the question how
one may provide the energy needed to achieve a desirable control.
The theory has a wide range of applications in connection with
airplanes, space structures, ships, and general mechanical systems.

The aerodynamic and hydrodynamic actions to which airplanes and
ships, respectively, are subjected provide easily the active control
forces without too much power involved. However, for situations
involving heavy civil engineering structures, like long-span bridges
or high-rise buildings, the designer must carefully investigate the
energy which may be required to warrant a desirable control.

In this book, active control of simple-span bridges and
tall buildings are presented in detail. The control obtained by
various control mechanisms and various control methods is scrutinized.
In this way, the reader is supposed to become familiar with the
various techniques presently used. Some factors which are first
neglected in the mathematical modelling of the structures for the
sake of simplicity will be investigated later on in order to show
the feasibility of active control even in the presence of uncertainties in loading and structural parameters. The last chapter
of the book contains an introduction to the control of distributed
parameters system in accordance with some basic considerations on
morphology of control as presented in Chapter 2.

1.3 CLASSIFICATION OF STRUCTURAL CONTROL

In the previous section, structural control was classified into
passive control and active control. However, some confusion occurs
when researchers try to relate this classification with a classification of the control systems. Control systems are either of the
open-loop or of closed-loop type. Open-loop control systems are
sometimes called predetermined control systems. Closed-loop control
systems are usually called feedback control systems. In order to
eliminate any confusion with respect to the relationship between
the classification of structural control and that of a control
system, this section is devoted to a detailed exploration of such
a relationship.

An active control system is any such system equipped with an external energy supply. The energy can be supplied manually or automatically. The active control system may be operated in an open-loop or a closed-loop fashion. An open-loop active control system is a system in which the control action is determined from the initial state of the system. This means that the control action is previously known from information given by the configuration of the system, its initial state, and by the applied disturbance. A closed-loop active control system is a system in which the control action is depending on the current state of the system. This is the reason for it being usually called feedback control system. Feedback control systems are suitably used in the presence of uncertainties in the applied loads and the structural parameters.

Examples of manually and automatically open-loop active structural controls are shown in Figures 1.1(a) and 1.1(b). A simple span bridge is controlled by tendons, as shown for a King post truss. In order to control the mid-span deflection in an open-loop automatic active fashion, a sensor is placed somewhere so that it can give a signal when a load, which is known previously, approaches the bridge. The signal releases a trigger to an on-position thus causing the tensioning of the cable according to a predetermined scheme. One realizes that the control in this case may be dangerous and unsafe if the load is widely random or differs from the expected one. However, that type of control is frequently used for some problems in which the input is applied continuously as in chemical plants and electrical power plants. If the same bridge is manually controlled, then one may assume an operator using his muscles for the tensioning of the cable according to a predetermined scheme. The operator uses an appropriate device, e.g., a steering wheel, to tension the cable once he is being informed of or sees himself the approaching vehicle.

Examples for manually or automatically feedback active structural control are shown in Figures 1.2(a) and 1.2(b) for the

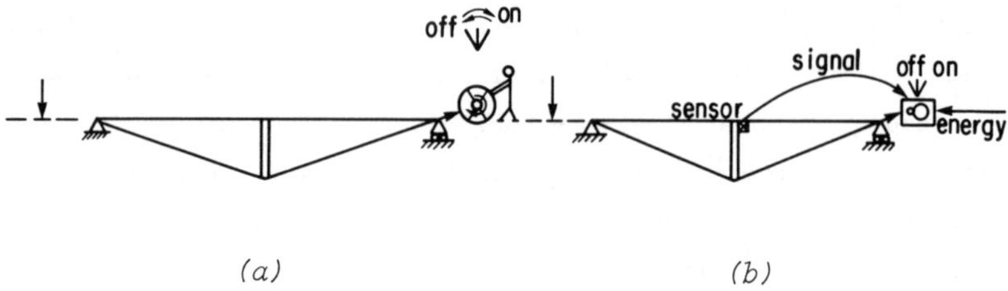

Figure 1.1 - (a) Manual Open-Loop Control
(b) Automatic Open-Loop Control

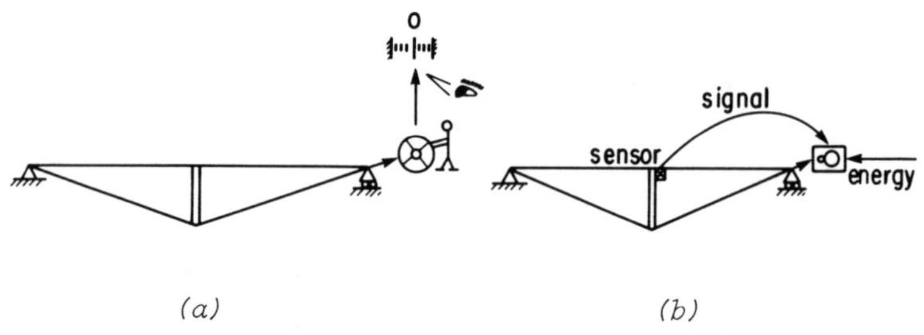

Figure 1.2 - (a) Manual Closed-Loop Control
(b) Automatic Closed-Loop Control

same King post truss. There, the control action is continuously
generated using information about the structural response. Automatically, it could be done if a sensor fed the structural response
into a device which would regulate the tension of the cable. In
this case, the control force is dependent on the measuring of the
current response of the bridge. Manually, it could be done if an
operator used his muscles for the tightening or releasing of the
cable, in this way adjusting the deflection at the mid-span within
a certain prescribed range. The operator would use the information
given by the sensor to keep the deflection within the proper range.
Several examples for automatic active control of civil engineering
structures are shown in Figures 1.3 to 1.8, [1.3-1.7].

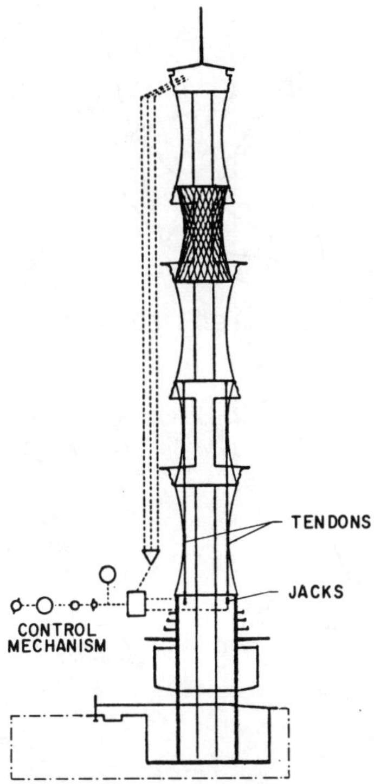

Figure 1.3 - Proposed Tower using Automatic Controlled Stressing Tendons. Taken from Reference [1.3]

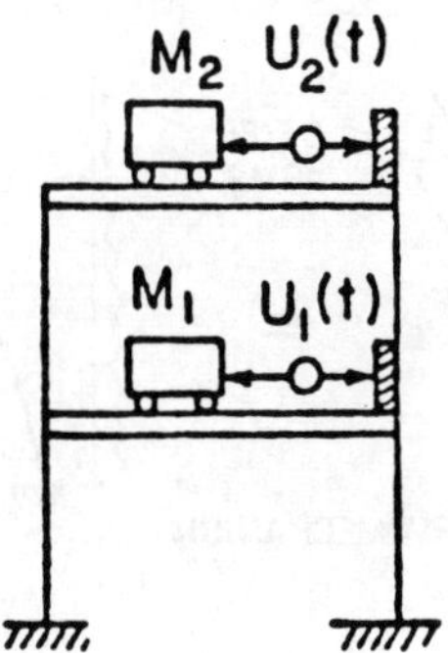

Figure 1.4 - Active Control by Mass Absorber Taken from Reference [1.4]

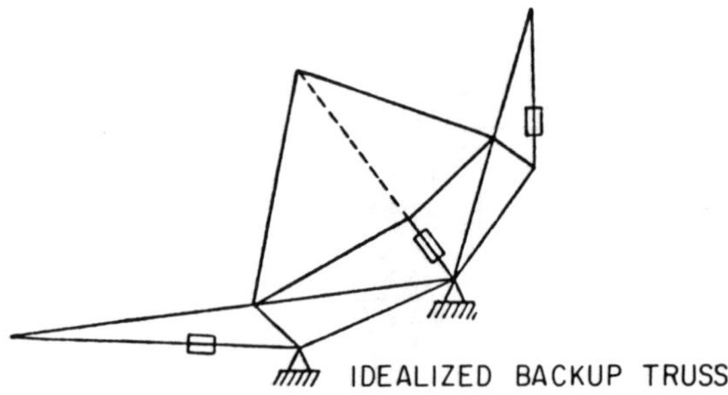

*Figure 1.5 - Active Control of Antenna by Tendons
Taken from Reference [1.5].*

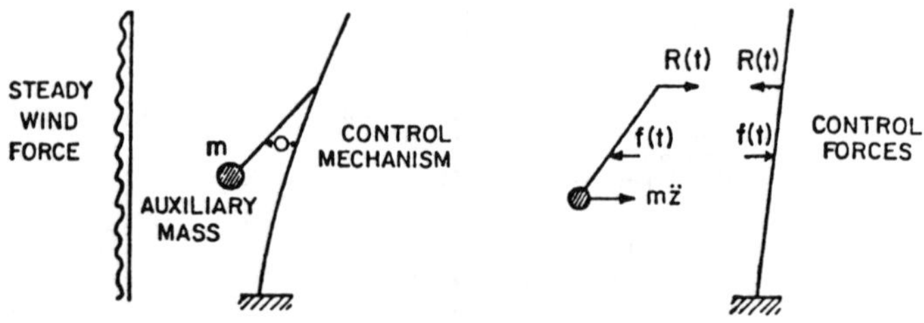

*Figure 1.6 - Active Control by Mass Absorber
Taken from Reference [1.6].*

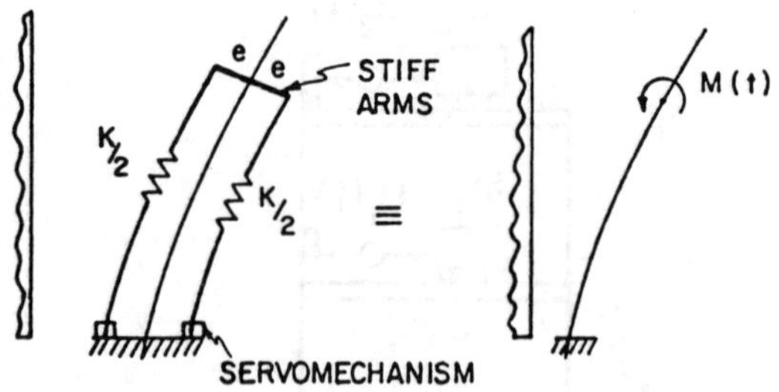

*Figure 1.7 - Active Control by Tendons
Taken from Reference [1.6].*

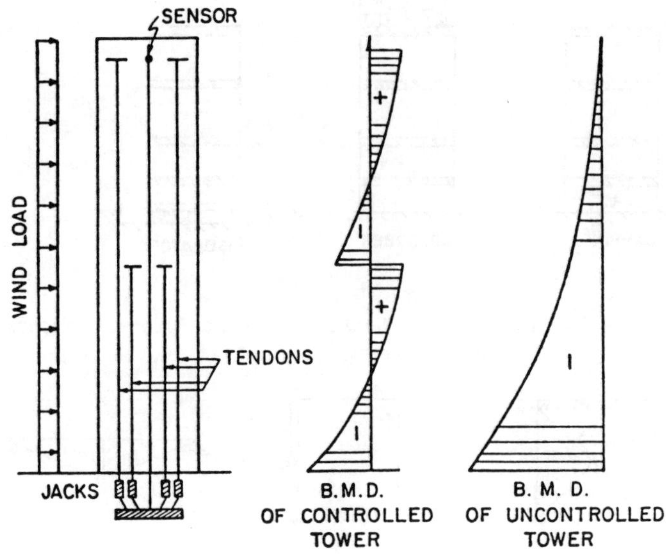

*Figure 1.8 - Active Control of a Tower
at Top and Mid-Points.
Taken from Reference [1.3]*

A passive control system does not need any external
energy to operate. Therefore, terms like automatically active or
manually active should not be used when talking about passive
structural control. Most of the passive control systems operate
in closed-loop fashion and there are only a few which operate in
open-loop fashion. Examples for closed-loop passive structural
control are shown in Figures 1.9 and 1.10 [1.8-1.11]. Each control
mechanism generates the necessary control forces when the structure
is being disturbed or its response exceeds certain limits. Therefore, the current states of the structure are the ones which force
the control mechanism to generate the forces needed to control the
structure's subsequent states. However, the control action is
limited, it depends on the momentary potential energy of the structure.

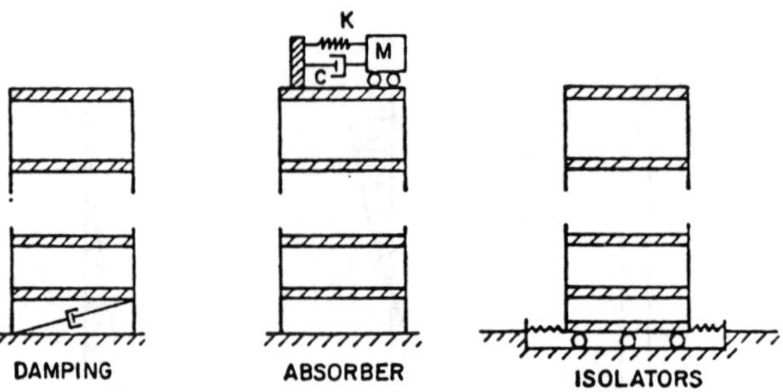

Figure 1.9 - Passive Control Methods against Seismic Effects Taken from Reference [1.5].

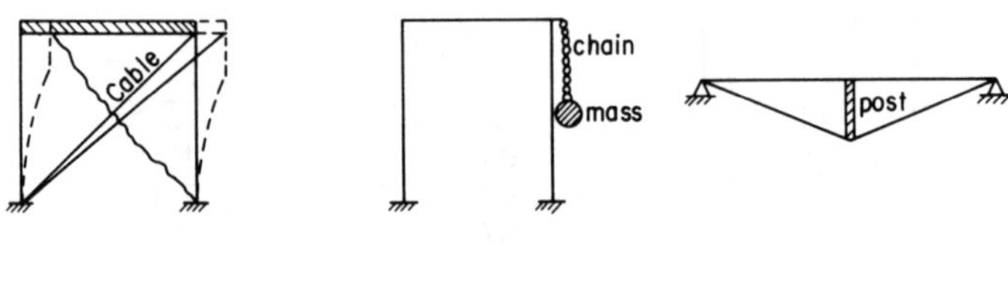

(a)   (b)   (c)

Figure 1.10 - (a) Frame with Cross Bracing Cables
(b) Frame with a Hanging Chain Damper
(c) Simple Span Bridge with Prestressed Cable
Taken from Reference [1.10].

An example for open-loop passive structural control is using prestressed concrete members. The degree of control is predetermined and does not depend on the current state of the member. The control action is potentially always present regardless whether the member is disturbed or not.

In Figure 1.11, the various classes of structural control are summarized.

1.4   ELABORATION OF VIBRATION CONTROL

The dynamic analysis of a structure consists primarily in determining the time variation of the deflection from which then the stresses can be calculated. Since the natural period of the structure depends upon its mass and stiffness, these two quantities are of great importance to the structural, controlled response. In this section, the parameters affecting the structural response of a one-degree-of-freedom system shall be considered.

1.4.1 *Response to a Constant Force with Finite Rise Time*

Consider an undamped one-degree-of-freedom system of mass M and stiffness K subjected to a constant force F_1 which is applied within the time t_r, as shown in Figure 1.12. This is the actual, practical situation when it is being claimed that the structure is subjected to a "sudden impact load": In reality, the load can never be applied instantaneously, and it takes some time, t_r, to affect the structure. Therefore, the equations of motion of the system are [1.14]

$$M\ddot{y} + Ky = \frac{t}{t_r}\, F_1 , \qquad t \leq t_r , \qquad (1.1)$$

$$M\ddot{y} + Ky = F_1 , \qquad t \geq t_r . \qquad (1.2)$$

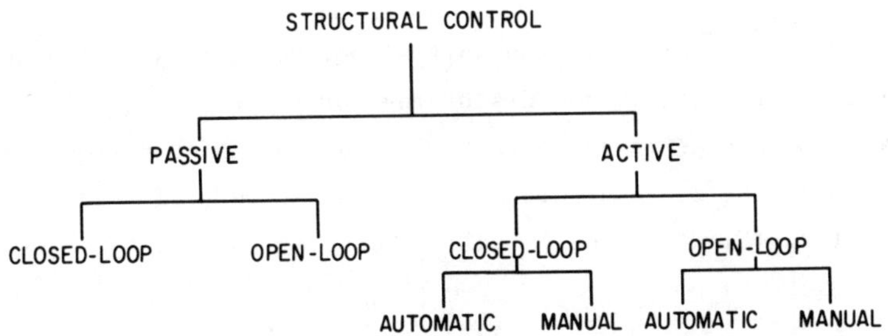

Figure 1.11 - Classification of Structural Control

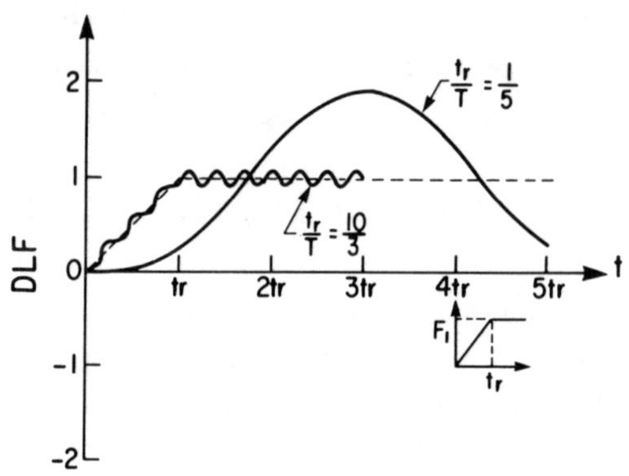

*Figure 1.12 - Response to Constant Force with
Finite Rise Time. Taken from Reference [1.14].*

Solutions of equations (1.1) and (1.2) can be written,
in general, as

$$y(t) = \frac{\dot{y}_0}{\omega} \sin\omega t + y_0 \cos\omega t + \int_0^t \frac{F(\tau)}{M} \frac{\sin\omega(t-\tau)}{\omega} d\tau , \quad (1.3)$$

in which $\omega = \sqrt{K/M}$; y_0 is the initial deflection; $\dot{y}_0$ is the initial
velocity; and $F(\tau)$ is the disturbing force.

Considering zero initial conditions, the ratio between
the dynamic deflection following from equation (1.3) and the static
deflection (F_1/K), which is called "dynamic load factor" (DLF), is
given by

$$DLF = \frac{1}{t_r} \left[1 - \frac{\sin\omega t}{\omega} \right] , \qquad t \leq t_r , \qquad (1.4)$$

$$DLF = 1 + \frac{1}{\omega t_r} [\sin\omega(t-t_r) - \sin\omega t] , \qquad t \geq t_r . \qquad (1.5)$$

The system's response to different t_r/T ratios is plotted
in Figure 1.12. The maximum DLF as a function of t_r/T is shown in

Figure 1.13. It is obvious from Figure 1.13 that the dynamic load factor becomes a minimum when the ratio t_r/T is an integer. But, the ratio t_r/T can be adjusted to be an integer if one is able to adapt the natural period of the system by changing its mass M and/ or its stiffness K because the natural period is related to these two quantities by

$$T = \frac{2\pi}{\omega} = 2\pi \sqrt{\frac{M}{K}} \ .$$

(1.6)

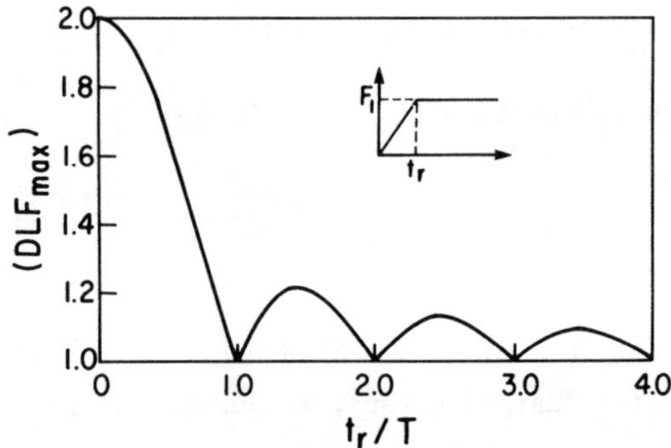

Figure 1.13 - Maximum DLF to Constant Force with
        Finite Rise Time. Taken from
        Reference [1.14].

Structural control can carry out these changes without really adding a mass or increasing the stiffness. If the control force is a function of the deflection or the acceleration of the system, one can execute these changes very easily.

1.4.2 *Effect of Damping*

Consider a damped one-degree-of-freedom system as shown in Figure 1.14. The equation of motion of the system is

$$M\ddot{y} + C\dot{y} + Ky = F(t) \ ,$$

(1.7)

in which C is the damping coefficient (force/velocity).

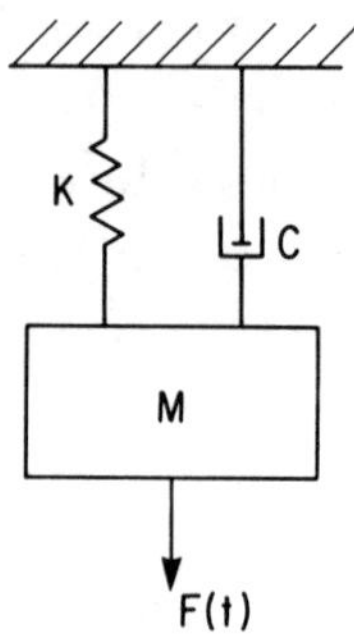

Figure 1.14 - Damped ODOF System

Solution to equation (1.7) is, in general, given by

$$y(t) = e^{-\beta t}\left[\frac{\dot{y}_0 + \beta y_0}{\omega_d} \sin\omega_d t + y_0 \cos\omega_d t\right] + \int_o^t \frac{F(\tau)}{M} e^{-\beta(t-\tau)} \frac{\sin\omega_d(t-\tau)}{\omega_d} d\tau ,$$

$$(1.8)$$

in which the quantity $\beta = C/2M$ is called "damping ratio", and $\omega_d = \sqrt{\omega^2 - \beta^2}$ is called "damped natural frequency" of the system.

It can be observed that the free vibrations decay gradually with time in the presence of damping. Damping has also an important effect on the forced vibration response as can be noticed from the integral term in equation (1.8). For example, if the applied force is pulsating, e.g., given by $F\sin\omega t$, the forced response of the system will be given by

$$y(t) = \frac{F}{K} \frac{\left[\left(1 - \frac{\Omega^2}{\omega^2}\right)\sin\omega t - 2\left(\frac{\beta\Omega}{\omega^2}\right)\cos\Omega t\right]}{\left[\left(1 - \frac{\Omega^2}{\omega^2}\right)^2 + 4\left(\frac{\beta\Omega}{\omega^2}\right)^2\right]} .$$

$$(1.9)$$

It is obvious from equation (1.9) that increasing damping will decrease the amplitude of the steady state response. The dynamic load factor as a function of the frequency ratio Ω/ω and the damping ratio β is plotted in Figure 1.15. It is of interest to notice that at resonance ($\Omega = \omega$), the maximum dynamic load factor is given by

$$DLF_{max} = \frac{\Omega}{2\beta} \,,$$

$$(1.10)$$

which indicates that for an undamped system, the maximum dynamic
load factor becomes infinite which means that the deflection will
be unbounded as shown in Figure 1.16. But with introducing any
amount of damping ratio, the deflection becomes bounded as shown
in Figure 1.17.

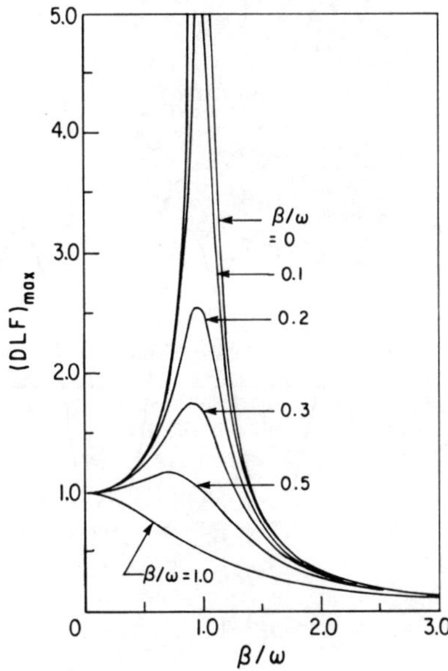

*Figure 1.15 - Maximum DLF for Damped One-Degree-of-Freedom System
subjected to Harmonic Load. Taken from Reference
[1.14].*

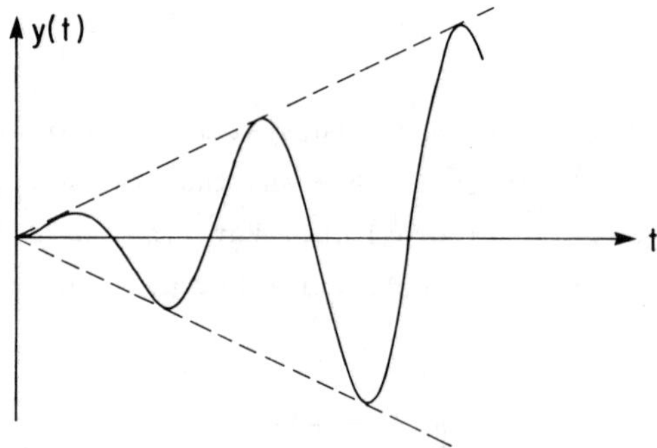

Figure 1.16 - Undamped System

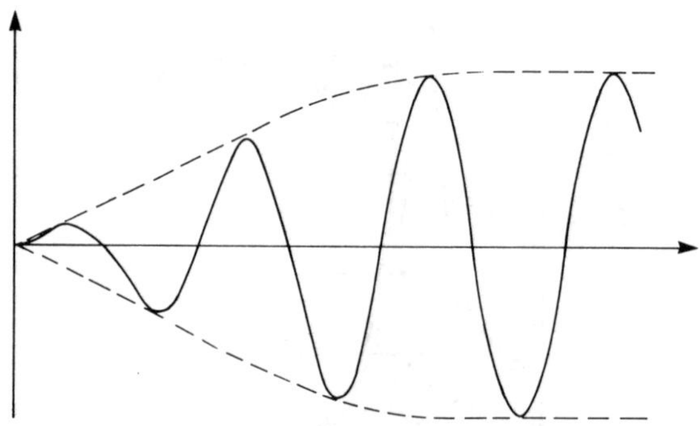

Figure 1.17 - Damped System

Therefore, in order to safeguard a structure against resonance, one should strive to introduce damping. Structural control can do this job if the control forces are functions of the structure's velocity response.

1.5 ACTIVE CONTROL OF ODOF SYSTEMS

It has been shown in the previous sections that controlling the dynamic response of the structure could be done by adjusting its mass and/or stiffness, by introducing damping to the structure, and/or by controlling the applied loads. Instantaneous changes in the structural parameters cannot really be achieved. Moreover, direct control of the applied loads is generally difficult and often impossible as far as natural loads like wind or earthquakes are concerned. Therefore, one would mainly revert to controlling the structural parameters. Structural control provides the techniques to implement such control through a set of control forces. These control forces must be properly located and designed so that no detrimental effects would occur through their action. To generate these control forces, one must use auxiliary structural elements which are meant to form a control mechanism. Such elements are for example auxiliary masses, auxiliary tendons, or auxiliary dampers. How to connect each of these elements, and how to develop the control forces is illustrated in the following sections for a one-degree-of-freedom system as an example.

1.5.1 *Control Using an Auxiliary Mass*

An auxiliary mass $\overline{M}$ could be mounted on the system of mass M such that the mass $\overline{M}$ can move freely and smoothly as shown in Figure 1.18. A device being servo-operated by external energy is used to push or pull the auxiliary mass $\overline{M}$. The control force generated by the servo shall be denoted by U. For this case, the equations of motion of the complete system read:

$$M\ddot{y} + C\dot{y} + Ky = F(t) - U(t) , \qquad (1.11)$$

$$\overline{M}(\ddot{z} + \ddot{y}) = U(t) , \qquad (1.12)$$

in which $\ddot{z}$ is the acceleration of the auxiliary mass.

Notice that if $\ddot{z} = 0$, i.e., if the auxiliary mass does not move independently, then one has a system of mass $(M + \overline{M})$ which reduces the natural frequency of the system and the load effect, $F(t)$. If $\ddot{z}$ is a function of $y(t)$, then the increase in the stiffness K will increase the natural frequency of the system.

Also damping C can be increased if $\ddot{z}$ is a function of $\dot{y}$. Furthermore, one can decrease the magnitude of the applied force F(t)
if $\ddot{z}(t)$ has the same time variation as F. Therefore, designing
the control device which actuates the mass $\overline{M}$ is the main task of
the designer when implementing a control mechanism.

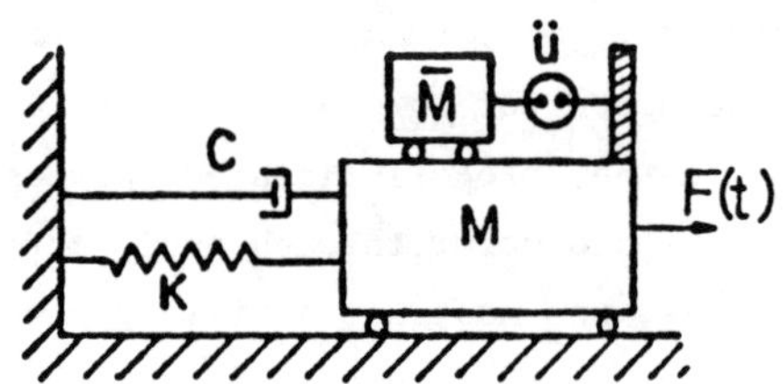

*Figure 1.18 - Active Control of a One-Degree-
of-Freedom through Auxiliary Mass*

    The control action $\ddot{z}(t)$ can be made a function of the
system response involving y, $\dot{y}$, $\ddot{y}$, if this response is being measured and fedback to the control device for generating the desirable control action $\ddot{z}(t)$. In this case, as mentioned previously,
one speaks of feedback control, since the control action is derived
from a comparison of the desired response with the current system's
response. One may also design the control action $\ddot{z}(t)$ as to follow
F(t) if this function is known in advance. Then, one has an open-
loop control. However, in the presence of uncertainties in loading
and structural parameters, feedback control has more relevance and
will therefore be of prime concern.

    In general, the relation between the control action $\ddot{z}(t)$
and the system response can be expressed by

$$\ddot{z}(t) = G_M y(t) \ , \tag{1.13}$$

where G_M is a gain operator on y(t) which may result in changes of
the deflection, velocity, or acceleration. Equations (1.11), (1.12)
and (1.13) can be represented by the block diagram of Figure 1.19

in which all quantities are expressed in terms of their Laplace
transforms. In order to determine the proper gain operator G_M,
certain specifications on the required controlled response must be
introduced. These specifications and a very brief summary of control system's theory is given in the Appendix A of this chapter.

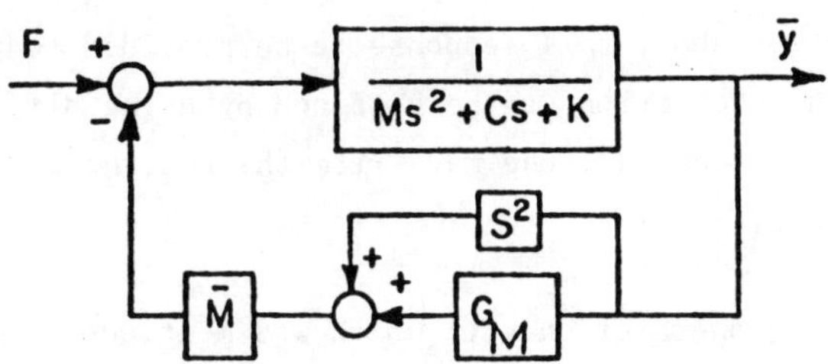

Figure 1.19 - Block Diagram

1.5.2 *Control using Auxiliary Tendons (Springs)*

An auxiliary spring or pretensioned tendon can be connected as shown
in Figure 1.20. The tendon is being tensioned or released by an
active control device. The equation of motion of a system controlled
in this manner is

$$M\ddot{y} + C\dot{y} + Ky + \bar{K}(y + z) = F(t) \; , \qquad (1.14)$$

where the active control force is given by $\bar{K}Z$, and $\bar{K}$ is the stiffness of the pretensioned tendon.

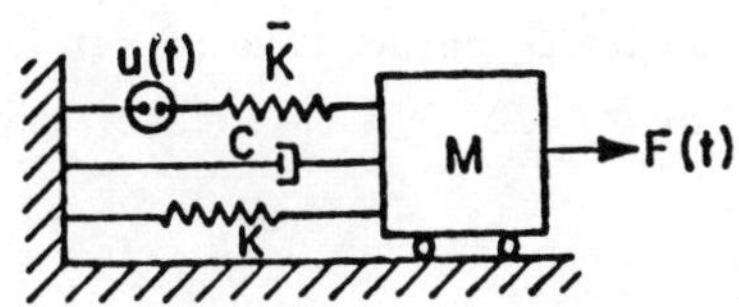

Figure 1.20 - Active Control of ODOF System
through Auxiliary Spring

The control action which is characterized by the behaviour
of z(t) must be designed as to provide a desirable response of the
system. For example, in order to introduce more damping to the
system, z(t) must be a function of $\dot{y}(t)$. In order to introduce more
stiffness, the control z(t) must be a function of y(t). The control
action can be made dependent on the system's response y(t) if there
is enough information about this response to be provided to the control device. Such information can be obtained by using different types
of sensors. Thus, in general, one may write the control action as

$$z(t) = G_s y(t) \, , \tag{1.15}$$

where G_s is the gain operator on y(t) which was mentioned already
in the preceding paragraph. Equations (1.14) and (1.15) can be
represented by the block diagram shown in Figure 1.21.

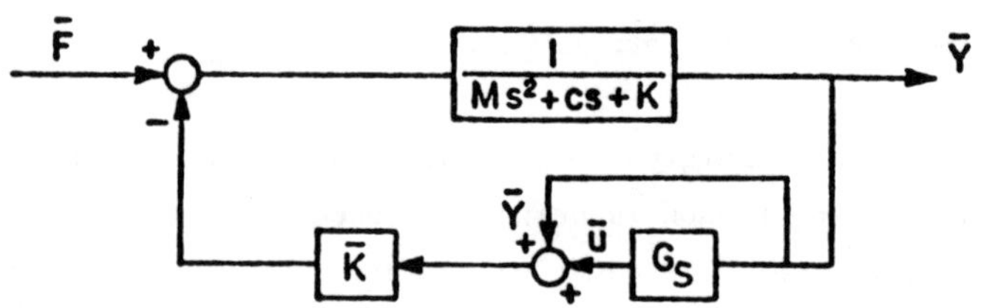

Figure 1.21 - Block Diagram

1.5.3 *Control Using an Auxiliary Damper*

Similarly, an auxiliary dashpot can be used to set up a control
mechanism as shown in Figure 1.22. In this case, the control
force is given by

$$U = \bar{C}\dot{z} \, , \tag{1.16}$$

in which $\bar{C}$ is the damping of the auxiliary damper; and $\dot{z}(t)$ is the
control action generated by the control device.

The equation of motion of the system now reads

$$M\ddot{y} + C\dot{y} + Ky + \bar{C}(\dot{z} + \dot{y}) = F(t) \, , \tag{1.17}$$

in which the term $\overline{C}\dot{y}$ represents the passive control force generated
by the damper.

The control action $\dot{z}$ can be related to the available
information supplied by the sensors as

$$\dot{z}(t) = G_d y(t) \, , \qquad\qquad (1.18)$$

in which G_d is the gain operator. Equations (1.17) and (1.18) can
be represented by the block diagram of Figure 1.23 after Laplace
transformation.

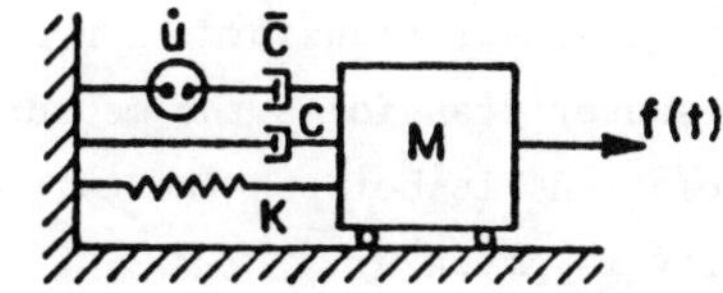

*Figure 1.22 - Active Control of ODOF System
through Auxiliary Damper*

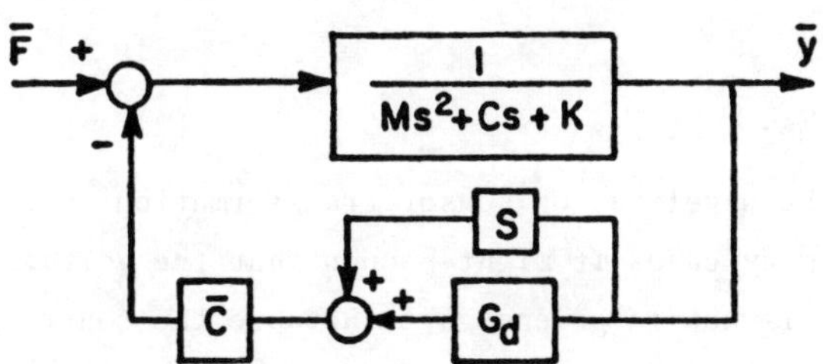

Figure 1.23 - Block Diagram

1.6 THE CONTROL DEVICES

It has been mentioned that in order to achieve feedback control,
the controller needs continuous information about the system's
response. Such information can be supplied by using measurement
devices called sensors. The sensor is an electrical transducer
which transforms the measured quantities into calibrated voltage.

This voltage provides the signal needed to actuate the controller.
Of the various types of sensors used in practice [1.12], a brief
description of only three of them shall be given here. These are,
respectively, the deflection sensor, the velocity sensor, and
the acceleration sensor (accelerometer).

The deflection sensor transforms the measured deflection
into a corresponding voltage. The relationship between the resulting voltage and the measured deflection can be expressed as

$$v(t) = T_d y(t) \; , \tag{1.19}$$

where T_d is the deflection sensor transformation operator.

The velocity sensor transforms the measured velocity
into a voltage. The relationship between the calibrated voltage
and the measured velocity is given by

$$v(t) = T_v \dot{y}(t) , \tag{1.20}$$

where T_v is the velocity sensor transformation operator.

Similarly, for the acceleration sensor, one has a voltage
given as

$$v(t) = T_a \ddot{y}(t) \; , \tag{1.21}$$

where T_a is the acceleration sensor transformation operator.

In many cases it might happen that the voltage produced
by the sensor is not large enough to actuate the controller. In
such cases, one may connect the sensor with an amplifier to magnify
the voltage generated. The magnified voltage is then fed through
an actuator which generates the control response u(t). The actuator
may take many forms, however, only three types are considered here.

1.6.1 *Electro-Hydraulic Servomechanism*

Servomechanism is a commonly used device for generating the control
action in automatic control fields. Its components are depicted
in Figure 1.24 and its mathematical model [1.6, 1.13] is shown in
Figure 1.25. In this device, the voltage supplied by the sensor

is magnified and fed through a servo-valve which transforms the
electrical power into hydraulic power. The hydraulic power works
on displacing a spool. This displacement regulates the flow of
the fluid into the actuator which produces the control action u(t).
The relation between the voltage supplied by the sensor and the
control u(t) is expressed [1.6] in the following closed-loop transfer function:

$$\overline{G}_c(s) = \frac{\overline{u}(s)}{\overline{v}(s)} = \frac{K_c}{K_f} \frac{1}{\tau s^2 + s + K_c} , \tag{1.22}$$

in which $\overline{G}_c(s)$ is the closed-loop transfer function; τ is a time
constant which can be neglected; $K_c = K_a K_v K_p K_f$ is a resultant gain;
K_f is the feedback transducer gain; and s is the Laplace variable.

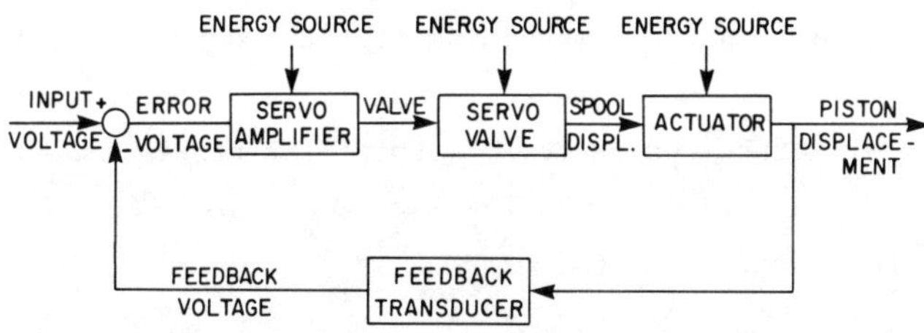

Figure 1.24 - Servomechanism's Components
Taken from Reference [1.6].

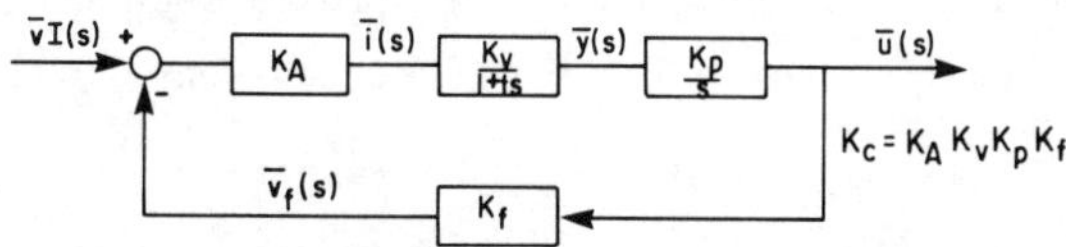

Figure 1.25 - Servomechanism's Parameters

The relationship between the control action u(t) and the
measured signal can easily be obtained from multiplying any of the
equations (1.19), (1.20) or (1.21) by equation (1.22) to yield the

closed-loop transfer functions of the control system. For example,
using the deflection sensor, the closed-loop transfer function of
the control system is

$$G_c(s) = \frac{\overline{u}(s)}{\overline{v}(s)} \frac{\overline{v}(s)}{\overline{y}(s)} = \frac{K_c}{K_f} \frac{T_d}{\tau s^2 + s + K_c} \ . \tag{1.23}$$

It can be noticed from equation (1.23) that the control
u(t) is not directly proportional to the quantities measured by
the sensor, but it is a combination of y, $\dot{y}$, $\ddot{y}$, etc., i.e., one
has

$$u(t) = f(y, \dot{y}, \ddot{y}, \dddot{y}, \ldots, c) \ , \tag{1.24}$$

in which the c are constants.

The above result can be obtained by solving for u(t) from
equation (1.23). The servomechanism may have detrimental effects
on the structure if the designer is only interested in controlling
one of the structural parameters, keeping the others fixed.

1.6.2 *Proportional Gain Controller*

This controller is similar to the servomechanism in its components
up to a few differences. These differences occur in the servo-
valve and actuator equations. The block diagram of this device
is shown in Figure 1.26, from which the closed loop transfer function of the control system using a deflection sensor, is obtained
as

$$G_c(s) = \frac{\overline{u}(s)}{\overline{v}(s)} \frac{\overline{v}(s)}{\overline{y}(s)} = KT_d \ . \tag{1.25}$$

In equation (1.25), K = $(K_A K_v K_p)/(1+K_c)$ is a constant.

Equation (1.25) indicates that the control response u(t)
will be proportional to the sensed response. For example, if the
measured response is y(t), then the control action will be
u(t) = $T_d K y(t)$. Therefore, this device is useful if the designer

is just interested in introducing active damping or in changing
only the stiffness of a one-degree-of-freedom or an infinite
degree of freedom uncoupled system.

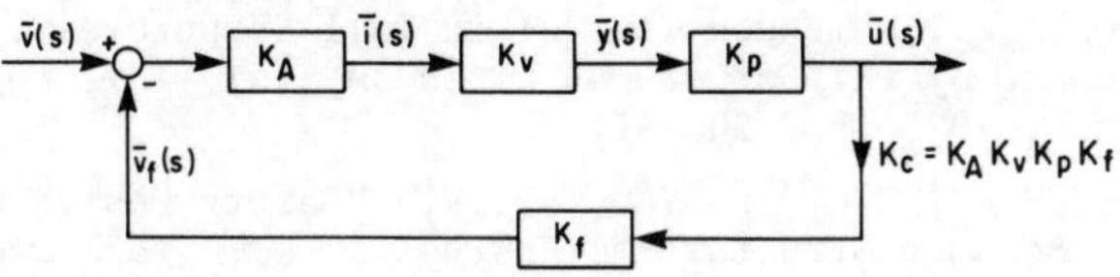

*Figure 1.26 - Proportional Gain Controller's
Parameters*

1.6.3 *Automatic Gain Controller*

The automatic gain controller is considered to be on the highest
level of controllers. It is similar to the proportional controller
except that the control is generated from measuring more than one
quantity (multivariable system). In mathematical terms, the control
gain is expressed by a matrix which could have constant or time-
varying elements. This controller is widely used in designing a control system by optimal control or the pole assignment method and will
frequently be applied in the next chapters.

Further details on control devices may be found in [1.12]
and [1.13].

1.7 REFERENCES

[1.1]   FARQUHARSON, F.B., "Aerodynamic Stability of Suspension
        Bridges", *Technical Report No. 16*, Part 1, 1949, Part 3,
        1952, Part 4, 1954, University of Washington.

[1.2]   YAO, J.T.P. and TANG, J-P., "Active Control of Civil
        Engineering Structures", *Technical Report No. CE-STR-73-1*,
        Department of Civil Engineering, Purdue University,
        Lafayette, Indiana, July, 1973.

[1.3]   ZUK, W. and CLARK, R.H., *Kinetic Architecture*, Van Nostrand
        Reinhold Co., New York, 1970.

[1.4]   YANG, J.N., "Application of Optimal Control Theory to Civil
        Engineering Structures", *ASCE, Journal of the Engineering
        Mechanics Division*, Vol. 101, No. EM6, December, 1975,
        pp. 819-838.

[1.5]   SCHORN, G., "Feedback Control of Structures", *Ph. D. Thesis*, University of Waterloo, Ontario, Canada, 1975.

[1.6]   ROORDA, J., "Active Damping in Structures", *Report Aero No. 8*, Cranfield Institute of Technology, England, July, 1971.

[1.7]   ROORDA, J., "Tendon Control in Tall Structures", *ASCE, Journal of the Structural Division*, Vol. 101, No. ST3, March, 1975, pp. 505-521.

[1.8]   WIRSCHING, P.H. and YAO, J.T.P., "Safety Design Concepts for Seismic Structures", *Journal of Computers and Structures*, Vol. 3, 1973, pp. 809-826.

[1.9]   YEH, H.Y. and YAO, J.T.P., "Response of Bilinear Structural Systems to Earthquake Loads", presented at the *ASME Vibrations Conference*, Philadelphia, Pa., U.S.A., March 30-April 2, 1969, ASME Preprint No. 69-TIBR-20.

[1.10] YAO, J.T.P., "Adaptive Systems for Seismic Structures", *Report on NSF-UCEER Earthquake Engineering Research Conference*, University of California, Berkeley, March 27-28, 1969, pp. 142-150.

[1.11] REED, W.H., "Hanging-Chain Impact Dampers: A Simple Method for Damping Tall Flexible Structures", *Wind Effects on Buildings and Structures*, University of Toronto Press, Canada, 1969, pp. 283-300.

[1.12] MANSFIELD, P.H., *Electrical Transducers for Industrial Measurement*, Butterworth and Comp. Ltd., Toronto, Canada, 1973.

[1.13] JAMES, H.M., NICHOLS, N.B. and PHILLIPS, R.S., *Theory of Servomechanism*, Dover Publications Inc., New York, 1965.

[1.14] BIGGS, J.M., *Introduction to Structural Dynamics*, McGraw-Hill Book Company, New York, New York, 1964.

Appendix A
Classical Control Theory

A.1 *Introduction*

Classical control theory deals primarily with linear, constant coefficient systems. The consideration of linear, time-invariant systems is greatly simplified by the use of Laplace transforms and frequency domain techniques. Thus, algebraic equations in the transformed variables are dealt with rather than considering

the system's differential equations. Manipulating these algebraic
equations is facilitated by the use of transfer functions and
block diagrams.

A.2 *Feedback Control Systems*

Most systems considered in classical control theory are feedback
control systems. A typical single input-single output feedback
system is shown in Figure A.1.

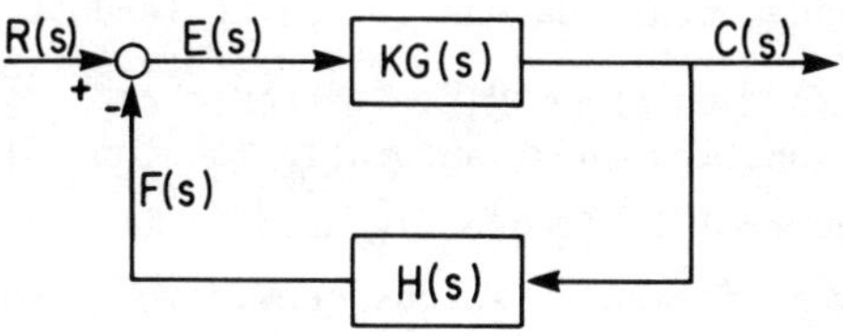

Figure A.1 - Feedback Control System

The notations shown in Figure A.1 are as follows: R is
the input or reference signal, the output or controlled signal is
C, the actuating or error signal is E, and the feedback signal is
F. The forward transfer function is KG(s), where K is an adjus-
table gain. The feedback transfer function is H(s). The open-
loop transfer function is KG(s)H(s). It represents the transfer
function around the loop, from E to F, when the feedback signal
is disconnected from the summing junction.

In the feedback system of Figure A.1, the actuating signal
is determined by comparing the feedback signal with the input signal.
When H(s) = 1, the system is called a unity feedfack, and the comparison is directly between the output and the input. In such
cases, the difference E(s) is an error signal.

In general, $G(s) = g_n(s)/g_d(s)$ and $H(s) = h_n(s)/h_d(s)$
are ratios of polynomials in s. The values of s for which the
numerator becomes zero are called zeros. Roots of the denominator

are called poles. The open-loop poles are roots of the denominator
of the open-loop transfer function, i.e., of KG(s)H(s). The closed-
loop transfer function is given by

$$P(s) = \frac{C(s)}{R(s)} = \frac{KG(s)}{1+KG(s)H(s)} = \frac{Kg_n(s)h_d(s)}{g_d(s)h_d(s)+Kg_n(s)h_n(s)} \ . \qquad (A.1)$$

The closed-loop zeros are the roots of $g_n(s)h_d(s) = 0$.
The closed-loop poles are roots of $1+KG(s)H(s) = 0$, or equivalently,
roots of $g_d(s)h_d(s)+Kg_n(s)h_n(s) = 0$.

The fundamental reasons for using feedback control are:

(1) The system output can be made to follow or track the
specified input function in an automatic fashion. That is why the
name automatic control is frequently used.

(2) System performance is less sensitive to variations of
parameter values.

(3) System performance is less sensitive to unwanted disturbances.

(4) The ease in controlling and adjusting the transient and
steady-state response.

However, the disadvantages of feedback control are:

(a) The possibility of instability. In fact feedback can
either stabilize or destabilize a system.

(b) The cost in general is considerable.

(c) Additional components of high precision are usually
required to provide the feedback signals.

A.3 *Design of a Control System*

The output C(s) of Figure A.1 is obtained from

$$C(s) = \frac{KG(s)R(s)}{1+KG(s)H(s)} \ . \qquad (A.2)$$

If the complete solutions for C(t) were available in
analytical form for every conceivable input R(s), the system response

could be assessed. However, the numerical solution of C(t) from
equation (A.2) is difficult to obtain, especially if one has to
check the effect of many inputs, and if the denominator of equation (A.2) is a high-degree polynomial.

Rather than seek complete analytical solutions, classical
control theory assesses only certain desirable features which C(s)
should possess in order to evaluate the system's performance.
Since these methods were developed before digital computers were
wide-spread, all techniques seek as much information as possible
about the behaviour of C(t) without actually solving for it. The
methods developed stress graphical techniques to ease their applications.

The problem of infinite variety for possible inputs is
dealt with by considering important aperiodic and periodic signals
as test inputs. Step function, ramp function, parabolic function,
and unit impulse functions are common examples. They are shown
in Figure A.2.

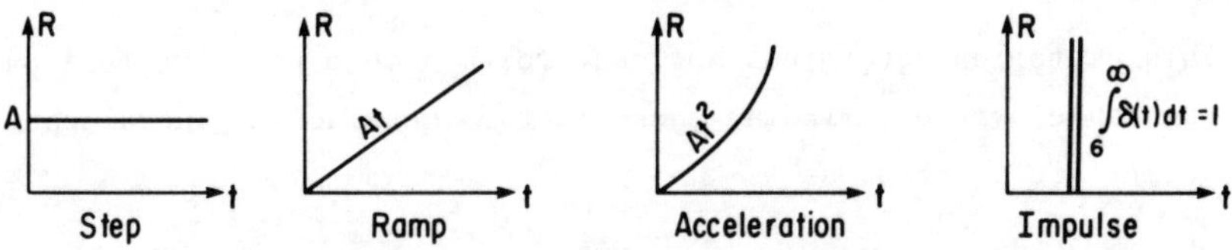

Figure A.2 - Standard Input Functions

The general characteristics which a well-designed control system should possess are: stability, steady-state accuracy,
satisfactory transient response, and/or satisfactory frequency
response. These criteria are summarized in the following.

A.3.1 Stability

Stability means that C(t) must not grow without bound due to a
bounded input. For linear time-invariant systems, stability depends

only on the locations of the roots of the closed-loop characteristic
equation. If any roots have positive real parts, the system is
unstable, as summarized in Figure A.3. Methods for investigating
the stability are given below

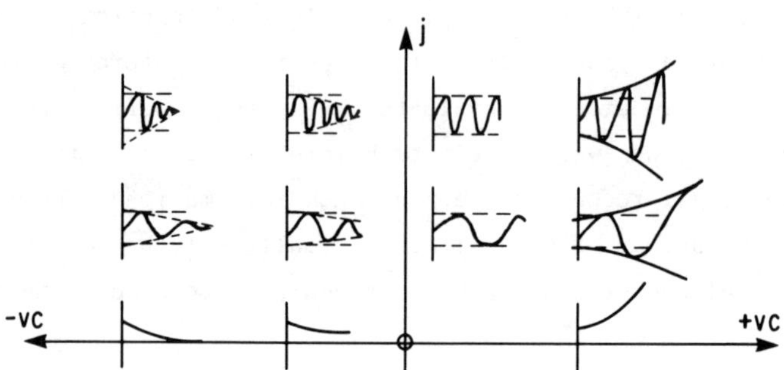

*Figure A.3 - Effect of Poles' Location
on Stability*

A.3.1.1 Routh's Criterion

This criterion determines how many roots have a positive real part
(unstable) without finding these roots. Consider a general characteristic equation as

$$s^n + a_{n-1}s^{n-1} + a_{n-2}s^{n-2} + \cdots + a_1 s + a_0 = 0 . \qquad (A.4)$$

Routh table consists of the following array of
quantities in which s^n etc. and a_{n-1} etc. are taken from (A.4) and

$$b_1 = \frac{a_{n-1}a_{n-2} - a_{n-3}}{a_{n-1}} \quad \text{etc.,} \qquad (A.5)$$

$$c_1 = \frac{b_1 a_{n-3} - a_{n-1} b_2}{b_1} \quad \text{etc.} \qquad (A.6)$$

The system is stable if the sign of the first column does not
change.

$$\text{Table A.1 - Routh Table}$$

s^n	1	a_{n-2}	$a_{n-4}\ldots\ldots\ldots a_2$	
s^{n-1}	a_{n-1}	a_{n-3}	$a_{n-5}\ldots\ldots\ldots a_1$	
s^{n-2}	b_1	b_2	$\ldots\ldots$	
$\vdots$	c_1	c_0	$\ldots\ldots$	
$\vdots$	$\vdots$			

A.3.1.2 Root Locus

It is a graphical way of factoring the characteristic equation.
The characteristic equation of equation (A.1) is written in the
form

$$KG(s)H(s) = -1 . \tag{A.7}$$

Since $G(s)H(s)$ is a complex number with a magnitude and
a phase angle, equation (A.7) imposes two conditions on the open-
loop transfer function. These conditions are

$$|KGH| = 1 , \tag{A.8}$$

$$\underline{/GH} = 180° . \tag{A.9}$$

Thus "root locus" determines the closed-loop roots by
working with the open-loop transfer function KGH.

A.3.1.3 Bode Plots

It is a graphical method which provides stability information for
open-loop systems. Magnitude and phase angle are considered

separately as in equations (A.8) and (A.9), but the only values of
s considered are s = jω. This corresponds to considering sinusoidal
input functions with frequencies ω. This method is greatly simplified by using decible units for magnitude and a logarithmic frequency scale for plotting. The critical point for stability, -1,
becomes the point of 0db, and -180° phase shift.

A.3.1.4 Polar Plots and Nyquist's Stability Criterion

Polar plots convey much the same information as Bode plots, but
the term KG(jω)H(jω) is plotted as a locus of phasors with ω as
the parameter. The critical point is -1, and Nyquist's stability
criterion which applies for stable as well as unstable open-loop
systems states that "the number of unstable closed-loop poles is
Z_r = P_r-N, where N is the number of encirclements of the critical
point -1 made by the locus of the phasors'. Counterclockwise encirclements are considered positive and P_r is the number of open-
loop poles in the right half of the s-plane.

A.3.1.5 Nichols' Charts

They contain the same information as Bode plots, but magnitude and
angle are combined on a single graph with ω as a parameter.

A.3.2 Steady-State Accuracy

Steady-state accuracy requires that the signal E(t), which is
often an error signal, approach a sufficiently small value for
large values of time. The final value theorem of Laplace transform facilitates analyzing this requirement without actually finding inverse transforms. By this theory one has

$$\lim_{t \to \infty}[E(t)] = \lim_{s \to 0}[sE(s)] \ . \tag{A.10}$$

By considering standard input functions, the error constants are developed. These constants are called position, velocity and acceleration for step, ramp and parabolic functions, respectively. They provide direct information on steady-state accuracy.

A.3.3 Satisfactory Transient Response

It means, there is no excessive overshoot, an acceptable level of oscillation in an acceptable frequency range, and satisfactory speed of response as well as settling time. These criteria are actually depending on the location of the closed-loop poles in the s-plane and their proximity to the stability boundary. Transient response of a feedback system is best studied using "root locus", since it is the only classical method which actually determines closed-loop pole locations. Bode, Nyquist and Nichols methods give indirectly information regarding the transient response. Gain margin, GM, is a measure of the additional gain which a system can tolerate with no change in phase and remaining stable. Phase margin, PM, is the additional phase shift that can be tolerated with no gain change and the system remaining stable. Experience has shown that acceptable transient response will require PM > 30°, and GM > 6db. A correlation between frequency domain and time domain specification is approximately given by:

$$\text{damping ratio} \simeq 0.01 \text{ PM} , \qquad\qquad (A.11)$$

$$\% \text{ overshoot} + \text{PM} = 75 , \qquad\qquad (A.12)$$

$$\begin{aligned}&\text{rise time x closed-loop bandwidth in}\\&\text{rad/sec} \simeq 0.45 \text{ x } 2\pi . \end{aligned} \qquad\qquad (A.13)$$

The previously presented terms are defined in Figures A.4, A.5 and A.6.

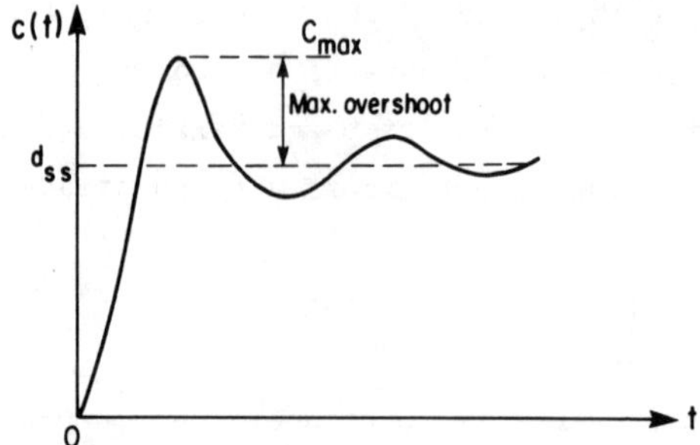

Figure A.4 - Overshoot in the Step Response

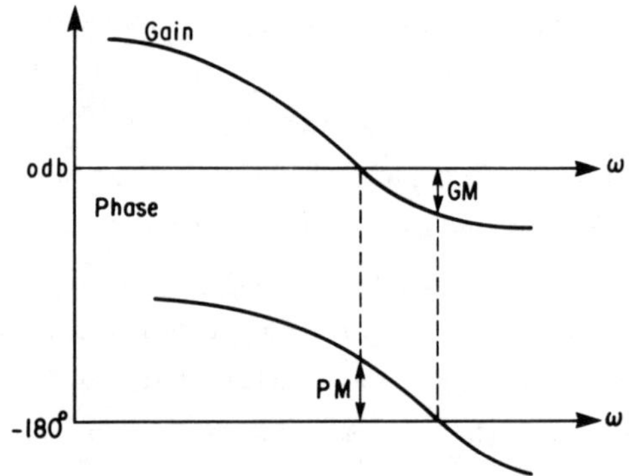

Figure A.5 - Bode Plot

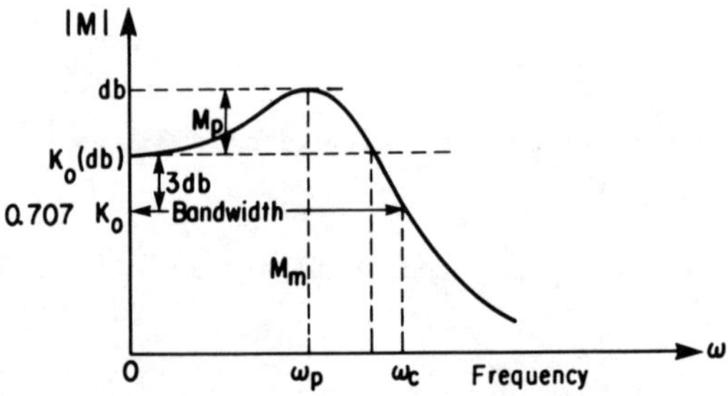

Figure A.6 - Gain Characteristic of a
Closed-Loop System

A.3.4 Satisfactory Frequency Response

It implies such things as satisfactory bandwidth, magnitude of
resonance peak, Mp, and its frequency, phase margin and gain margin.
Therefore, the methods used to assess these quantities are Nyquist,
Bode and Nichols charts.

A.4 *References*

[A.4.1]   BROGAN, W.L., *Modern Control Theory*, Quantum, QPI Series,
          New York, 1974.

[A.4.2]   DORF, R.C., *Modern Control Systems*, Addison-Wesley Publish-
          ing Co., Ontario, 1977.

[A.4.3]   DI STEFANO, J.J., STUBBERUD, A.R. and WILLIAMS, I.J.,
          Feedback Control Systems, Schaum's Outline Series,
          McGraw-Hill Co., New York, 1967.

Chapter II
Morphology of Structural Control

2.1   PRELIMINARY REMARKS

When considering structural control, a clear definition of that
type of control should be given first. As already stated in
Chapter 1, in the sense of this book, structural control is meant
to be control of the structural deformation resulting from the
effect of external loading to which the structure is being subjected.
There are of course broader interpretations of the term
"structural control": one may include automatic control of the
temperature and of the air condition in a building, control of the
building as a whole with respect to its respective position towards
a certain line of action of an external agent, like sunshine, wind,
etc., and in the broadest sense control of the lay-out of the structural components as to accommodate the structure to variations of
functional requirements.
However, here, in order to establish a clearly defined
task, only deformation control will be considered. Such restriction is quite acceptable, as this kind of control is indeed fundamental and essential.

There remains still the question why one would treat
structural control as a discipline distinct from automatic control,
which is well developed with respect to its foundations in physics
and its mathematical methodology. Is it not a more or less trivial
affair to apply the concepts and methods of classical control theory
to just one more of the many objects already subjected to control?
Not quite. From the theoretical and mathematical point of view,
a structure, which is to be controlled, is a rather complicated
and involved entity. It is for example, imperative that the person,
who is to deal with deformation control of a structure, is quite
familiar with the theory of structural dynamics. Alone this need
to include structural dynamics in the considerations of structural
control justifies that structural control is dealt with separately
in order to warrant that structural dynamics is broadly enough
taken into account. Yet, there is another, more important reason
why structural control must be seen as a more or less independent
branch of control theory: the occurrence of continuous structural
elements which require methods capable of functioning in the presence of "distributed parameters". In classical control theory,
it is mostly assumed that the system to be controlled has a few
degrees of freedom only. Even, if one would discretize a structure,
one would have to admit a large number of degrees of freedom, and
this would cause noticeable difficulties when applying the methods
of classical control theory. But, if one would ever decide to deal
with the structure as a system of continuous elements, what the
structure actually is, classical control theory would not provide
much help, as methods to deal with continuous objects have not
been developed broadly enough in control theory as it presently
stands. It is therefore quite reasonable to take up the question
of the control of continuous objects in the context of structural
control. In spite of some excellent work done for example by
J.L. Lions on the optimal control of systems governed by partial
differential equations [2.1], there remains the task to translate

findings of a theoretician into a formulation which allows application to practice. Also, when adjusting theory to practice, problems may come up, which may require extensions of the already
existing theory, the need of which the theoretician may not have
been able to perceive. Hence, there remains still much work to be
done that requires devoted and to a degree exclusive dedication to
structural control. In this context one must also realize that
structural control poses special problems with respect to the distribution of sensors, the choice of appropriate actuators and of
control devices that are specifically able to provide control
forces adequate to affect huge structures. Finally, in the presence
of large inertia forces, the possibility of time-gap occurrence in
the behaviour of the controlled structures may have to be taken into
account more urgently than when dealing with other kinds of control
systems.

2.2 ON THE BASICS AND THE ACTUAL NATURE OF STRUCTURAL CONTROL PROBLEMS

In order to be able to properly deal with upcoming problems of
structural control, one must have a clear notion of the actual
nature of such problems. In the following, an analysis of this
nature shall be performed. This analysis will be concerned with
a very fundamental and crisp formulation of these problems as to
enable one to see the essential and characteristic features of
structural control. Some familiarity with functional analysis
[2.2] is recommended.

Let the structure be described by a linear operator D.
This operator includes all the features of the structure which
come into play when the structure is exposed to an external input,
a perturbation, f. The response of the structure, i.e., its
deformation caused by the perturbation, shall be denoted by y_0.
Then, one has

$$Dy_0 = f \; , \qquad\qquad\qquad (2.1)$$

as the operator equation of the system. Actually, systems are limited in extension and are supported at their boundary B. Therefore, one has to add to (2.1) the boundary conditions

$$U[y_0]_B = 0 \; , \qquad\qquad\qquad (2.2)$$

in order to describe the behaviour of the structure fully. In (2.2) U is a "vector" of operators on y_0, and, for a description of the situation at the boundary, the quantities $U[y_0]$ have to be taken at the boundary B of the structure. For the sake of simplicity, *homogeneous* boundary conditions have been assumed in (2.2). This is fairly justified as such homogeneous conditions are very common in practice.

Concerning the operator D, it is assumed that it possesses an inverse D^{-1}. The inverse exists if

$$||Dy_0|| \geq m||y_0|| \; , \qquad m > 0 \; , \qquad\qquad (2.3)$$

and

$$||D^{-1}|| < \frac{1}{m} \; . \qquad\qquad\qquad (2.4)$$

Then, the deformation response of the structure caused by the perturbation is obtained explicitly as

$$y_0 = D^{-1}[f] \; . \qquad\qquad\qquad (2.5)$$

Such explicit representation of y is given very commonly in a form in which D^{-1} is an integral operator involving Green's function G. This Green function satisfies the boundary conditions in (2.2) so that also the solution y_0 in (2.5) satisfies automatically the prescribed boundary conditions.

Assume, the response y of the structure to the perturbation f were not satisfactory. Then, the idea of control may be

brought into play. Let u be the control. By virtue of u, (2.1),
(2.2) change into

$$Dy + u = f \; , \qquad U[y]_B = 0 \; . \qquad\qquad (2.6)$$

In (2.6), it has tacitly been assumed, that control is restricted
to the structure as such. However, in a more general case, there
could as well be a control action at the boundary. That possibility
has for example been pointed out in [2.1].

The control u is again considered to be an operator.
Therefore, there are many possibilities to choose a control. For
this reason, control cannot be seen to be a well-defined mathematical problem. Control is in general a *design* problem and as such
an "art". This means, one may in most of the cases not be able to
"calculate" a control. One may have to base the "design" of a
control on previously gained experience. Therefore, "trial and
error" approaches to control, as they are often proposed in
Chapters 3 and 4, are quite acceptable.

Firstly, let the perturbation f be predictable. Then,

$$u = Af \; , \qquad\qquad (2.7)$$

is a possible choice for the control, where A is an appropriate
operator. Using (2.7) in (2.6) yields

$$Dy + Af = f \; , \qquad U[y]_B = 0 \; ,$$

which can be rearranged to read

$$Dy = (E-A)f \; , \qquad U[y]_B = 0 \; . \qquad\qquad (2.8)$$

This is an open loop control system as pointed out in 1.5.1. In
(2.8), E is the unit operator. Also, it becomes obvious from
(2.8) why (2.6) and (2.8) have been written in terms of y and
not y_0. Referring to (2.8) it is clearly seen that the solution
cannot be y_0 any more but must be different from y_0, which is

indicated by writing y instead of y_0, since the right hand side
in (2.1) is different from that in (2.8).

The main goal of the control designer must be two-fold:
a solution to (2.8) should exist, and the response of the controlled structure, y, should satisfy certain requirements. That
means, the designer faces the *existence* and the *performance* requirements. Specifically, with respect to the performance requirements,
something must be said here. In classical control theory, certain
specific requirements have been established like steady-state
accuracy, satisfactory transient response, and stability of the
controlled response. No doubt, these are very reasonable requirements. However, they are also rather special ones. If investigating
the nature of control in principle, one should be much broader
minded and not follow dogmatic postulates justified in some narrow
sense only. The right attitude would be to prescribe to y, as it
is provided by (2.8), broad requirements which may follow from
some actual practical situations, not overlooking yet the fundamental
requirement for stability.

Since D^{-1} is supposed to exist, there is no problem to
determine y using (2.8). One obtains

$$y = D^{-1}[(E-A)f] \ . \qquad\qquad (2.9)$$

Hence, there is no *existence problem.*

Concerning the *performance problem,* let (2.9) be written
as

$$y = y_0 - D^{-1}[Af] \ , \qquad\qquad (2.10)$$

which is possible when using (2.5) in (2.9). Hence, one has

$$y_0 - y = D^{-1}[Af] \ . \qquad\qquad (2.11)$$

One may consider the expression $y_0 - y$ to be a *measure of performance.* By assigning to this measure of performance a specific mean-
ing, one creates a performance criterion which may be used for the

determination of the operator A. Although one may create in this way possibly a mathematically well-defined procedure for the calculation of A, the fixing of A continues to involve much arbitrariness, as the performance criterion is to some extent already arbitrary. No question though that it should contain certain features which at least guarantee that the mathematical problem, from which A is finally to follow, does have a solution.

Assume, A could be determined properly. Then, derive from (2.11) the relationship

$$||y-y_0|| = ||D^{-1}A[f]|| \leq ||D^{-1}||\ ||A||\ ||f|| \ . \qquad (2.12)$$

Correspondingly, (2.5) yields

$$||y_0|| \leq ||D^{-1}||\ ||f|| \ ,$$

which leads to

$$||y_0|| = p||D^{-1}||\ ||f|| \ , \qquad p < 1 \ , \qquad (2.13)$$

where p is an appropriate constant smaller than one. Using (2.13) in (2.12) yields

$$||y-y_0|| < p^{-1}||A||\ ||y_0|| \ . \qquad (2.14)$$

If, for example, one would require that

$$\frac{||y-y_0||}{||y_0||} < \varepsilon \qquad (2.15)$$

should hold, where ε were a prescribed number, one would have introduced a performance criterion, which when satisfied, would also guarantee stability of the controlled structure if the response y_0 of the uncontrolled structure were stable.

Inequality (2.15) together with (2.14) leads to a condition for operator A, if one would combine the two inequalities in the form

$$\frac{||y-y_0||}{||y_0||} < p^{-1}||A|| < \varepsilon \ . \tag{2.16}$$

As a consequence, one would have for the norm of A the inequality

$$||A|| < \varepsilon p \ . \tag{2.17}$$

This is of course a very vague condition for A, and it becomes obvious that control problems, involving an *unknown operator* to be determined, are usually *undeterminant* problems. This fact confirms what has been said earlier: control problems require an experienced designer rather than a sheer mathematician. The experience may have been gained by earlier having applied trial and error procedures with an appropriate amount of attention and understanding.

Secondly, feedback control shall be considered. In that case,

$$u = Ay \ , \tag{2.18}$$

holds for the control u, where A is again some operator to be appropriately determined. However, as before, restriction to linear operators shall be observed.

Using (2.18) in (2.6) yields

$$Dy = f - Ay \ , \qquad U[y]_B = 0 \ . \tag{2.19}$$

Since it had been postulated that D^{-1} exists, the operator equation in (2.19) can be rewritten to read

$$y = D^{-1}f - D^{-1}Ay \ ,$$

so that one finally has

$$(E + D^{-1}A)y = D^{-1}f = y_0 \ . \tag{2.20}$$

Let it be assumed that the inverse $(E+D^{-1}A)^{-1}$ exists.
Then, the response y of the structure becomes

$$y = (E + D^{-1}A)^{-1}y_0 \ . \tag{2.21}$$

This result emphasizes two facts: First, by virtue of the control
that had been chosen, the original nature of the structure specified by D has been noticeably changed by the introduction of the
operator A. Consequently, the actual nature of the controlled
structure is quite different than that of the uncontrolled structure. As a result of this change, the controlled structure may
mathematically have been transferred from one class (for example
self-adjoint) to a quite different one (for example nonselfadjoint).
Hence, facts which held firmly for the original structure, for
example: concerning the eigenvalue spectrum of the structural
operator, conservativeness of the structure, symmetry, may have
become invalid for the controlled structure. Thus, great care
has to be exerted when dealing with the controlled structure.
Increased attention has to be paid to the existence and stability
problems. Specifically, a modal approach to the solution, which
may have been quite adequate for the original problem (2.1), (2.2),
may be invalid for the control problem (2.19). The last mentioned
incidence of the modal approach becoming inapplicable to the controlled structure is something which can happen quite possibly to
continuous structures. Therefore, there is again a reason to study
structural control on its own merits, as for most of the control
problems encountered in mechanical and electrical engineering,
this phenomenon of inappropriateness of the modal approach seems
not to have had much importance.

One may interpret (2.21) as the problem of mapping y
into prescribed images y_0. In this way, one would look at the
performance question in the broadest sense. And one would also
have defined the design problem of the control in the most general
manner. The generality of such design problem again points out how

much, as before in the case of open-loop control, this problem is undeterminate. Hence, all that has been said before about the design of a control remains valid also here.

A minimum performance criterion, that could be derived from (2.21), might read

$$||y|| \leq ||(E+D^{-1}A)^{-1}||\ ||y_0|| \leq \frac{1}{m}||y_0|| \leq ||y_0||\ , \quad m > 1. \quad (2.22)$$

In that case, stability of the response of the original structure, y_0, would automatically imply stability of the response of the controlled structure, i.e., y. Moreover, the "magnitude" of the response y, i.e., $||y||$, would always be below the "magnitude" of the original structure's response, i.e., $||y_0||$. That is certainly a desirable and satisfactory situation.

From (2.22) follows

$$||(E + D^{-1}A)^{-1}|| \leq \frac{1}{m}\ , \tag{2.23}$$

which implies

$$||(E+D^{-1}A)y_0|| \geq m||y_0||\ . \tag{2.24}$$

The last condition guarantees existence of a solution to problem (2.20). Hence, one is facing the pleasant fact that the performance criterion is in complete agreement with the existence requirement.

Since

$$||E+D^{-1}A||\ ||y_0|| \geq ||(E+D^{-1}A)y_0||\ , \tag{2.25}$$

conditions (2.24) and (2.25) combined yield the inequality

$$1 + ||D^{-1}A|| = ||E+D^{-1}A|| \geq m > 1\ . \tag{2.26}$$

Consequently,

$$||D^{-1}A|| \geq m-1 > 0 \tag{2.27}$$

follows from (2.27), which by virtue of

$$||D^{-1}||\ ||A|| \geq ||D^{-1}A||\ ,$$ (2.28)

leads to

$$||A|| \geq \frac{m-1}{||D^{-1}||}\ .$$ (2.29)

Since operator D was supposed to have an inverse,

$$||D^{-1}|| \leqq \frac{1}{\mu}\ ,\qquad \mu > 0\ ,$$

must hold. Set

$$||D^{-1}|| = \frac{\varepsilon}{\mu}\ ,\qquad \varepsilon \leq 1\ .$$ (2.30)

Using (2.30) in (2.29) yields

$$||A|| \geq \frac{(m-1)\mu}{\varepsilon}\ ,$$ (2.31)

which can be viewed as a requirement on operator A. Again, one
has in that way obtained a design criterion for operator A and
thus for the control u. Yet, it is as before in general a vague
criterion only. Consequently, the design of the control remains
also now more an "art" than a well specified mathematical problem.

In summary, the morphology of the structural control
problem can be described as follows: Based on certain predetermined
requirements concerning the structure's performance, an operator
A has to be found which may be either an operator on the perturbation (open-loop system) or on the structure's response (feedback
system). In both cases, one is facing an undeterminate problem
in the first instance, as the prescribed requirements make up a
"vector" in mathematical terms, while the unknown quantity appears
as an operator, e.g., a matrix. In order to make the control
problem solvable, further conditions have to be introduced, for
example, so-called optimization criteria. Since choosing these

additional conditions depends on the skill and experience of the engineer concerned, control is not only a mathematical but to a great extent a design problem.

Depending on the nature of the chosen control operator A, the task to specify the control can become mathematically more or less difficult. The operator can for example be chosen to be simply a function involving a single undetermined parameter. In such a case, the problem of optimal control degenerates into an ordinary extremum problem of analysis leading to an optimal value of that parameter. Another possibility were to make the operator A indeed a true operator but of a certain predetermined structure, for example an integral operator with a Green function to be determined etc. In that way, a certain amount of indeterminacy involved in A can be removed a priori. However, one should not overlook that an undesirable consequence of adding conditions to the control problem in order to facilitate the calculation of A may be that the problem may become unsolvable. Therefore, an eminent task of the engineer is to pay attention to the problem of the *existence of a solution* to the control problem right at the beginning of the calculations.

2.3   EXAMPLES

(i)   Consider a string of length ℓ as shown in Figure 2.1. The string is elastically supported at $x = 0$ and has a simple support at $x = \ell$. The partial differential equation describing the transversal oscillations of the string reads

$$\ddot{v} = c^2 v'' \, , \quad \ddot{v} = \frac{\partial^2 v}{\partial t^2} \, , \quad v'' = \frac{\partial^2 v}{\partial x} \, . \tag{2.32}$$

At time $t = 0$, the string is assumed to be in its trivial equilibrium condition. Hence,

$$v(x,t) = 0 \quad \text{for} \quad t = 0 \tag{2.33}$$

must hold. In accordance with the support at the right end of
the string,

$$v(x,t) = 0 \qquad \text{for} \qquad x = \ell \qquad\qquad (2.34)$$

is also to be required as one of the boundary conditions. The
boundary condition at the left end of the string is to be chosen
so as to provide frequency control of the string's oscillation.
For that purpose, set

$$[Av]_{x=0} = 0 \ . \qquad\qquad (2.35)$$

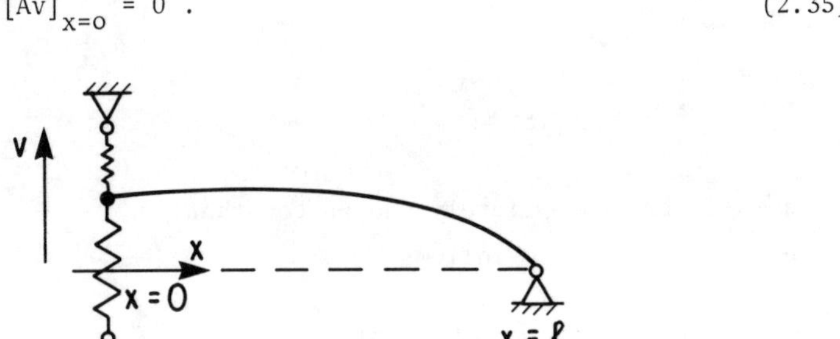

Figure 2.1

The control is determined by means of the operator A. In many
cases, as it will be done here, certain assumptions on the nature
of A will be made a priori so that the amount of effort needed to
specify operator A in a definite matter according to the performance
criteria imposed on the system is minimal. In doing so, one must
be careful not to set up conditions which would rule out a solution
to the problem at all.

Let the operator A be chosen as

$$A = \frac{\partial}{\partial x} - KE \ , \qquad\qquad (2.36)$$

where E is the unit operator and K the yet undetermined spring
constant of the elastic support at x = 0. With (2.36) in (2.35)
one has

$$v'(0,t) = Kv(0,t) \ , \tag{2.37}$$

as the second boundary condition.

Assume the solution to (2.32) to be of the form

$$v(x,t) = \phi(t)\psi(x) \ . \tag{2.38}$$

With (2.38) in (2.32) one obtains

$$\ddot{\phi}\psi = c^2\phi\psi'' \ ,$$

which yields

$$\frac{\ddot{\phi}}{\phi} = \frac{\psi''}{\psi} \ c^2 = -\omega^2 \ , \tag{2.39}$$

where ω is an appropriately chosen constant.

From (2.39) follows

$$\ddot{\phi} = -\omega^2\phi \ , \qquad \phi = A\sin\omega t \ , \tag{2.40}$$

and

$$\psi'' = -\left(\frac{\omega}{c}\right)^2\psi \ , \qquad \psi = \sin\frac{\omega}{c}(x-a) \ , \tag{2.41}$$

where A is an arbitrary amplitude and a is a yet undetermined constant.

By virtue of (2.38), (2.40) and (2.41), the solution to equation (2.32) becomes

$$v(x,t) = A\sin\omega t \ \sin\frac{\omega}{c}(x-a) \ . \tag{2.42}$$

Condition (2.33) is immediately satisfied by (2.42). In order to satisfy condition (2.34), which by virtue of (2.42) reads

$$A\sin\omega t \ \sin\frac{\omega}{c}(\ell-a) = 0 \ , \tag{2.43}$$

one has to require that

$$\frac{\omega}{c}(\ell-a) = n\pi \ , \qquad n = 1,2,3, \ \ldots \tag{2.44}$$

holds. In other words, the frequency ω of the string's oscillation must be

$$\omega = \frac{n\pi c}{\ell-a} \ . \tag{2.45}$$

Now, condition (2.37) shall be used in order to determine together with an appropriate performance requirement the control quantity K. When K is prescribed, also the spring at the elastic support of the string is prescribed, since K is, as already mentioned the "spring constant".

As the performance requirement choose the condition that the string oscillates with a certain given frequency ω^*. From (2.45) it follows that then the quantity a is equal to

$$a^* = \frac{\ell\omega^*-n\pi c}{\omega^*} \ . \tag{2.46}$$

By virtue of (2.46) and (2.45), (2.42) becomes

$$v(x,t) = A \sin \frac{n\pi c}{\ell-a^*} \ t\sin \frac{n\pi}{\ell-a^*} \ (x-a^*) \ . \tag{2.47}$$

Using (2.47) in (2.37) yields

$$\frac{n\pi}{\ell-a^*} \ A\sin \frac{n\pi c}{\ell-a^*} \ t\cos \frac{n\pi a^*}{\ell-a^*} = -KA\sin \frac{n\pi c}{\ell-a^*} \ t\sin \frac{n\pi a^*}{\ell-a^*} \ ,$$

and

$$K = - \frac{n\pi}{\ell-a^*} \ ctg \ \frac{n\pi a^*}{\ell-a^*} \ . \tag{2.48}$$

Yet, from (2.46) follows

$$\frac{n\pi}{\ell-a^*} = \frac{\omega^*}{c} \ , \qquad \frac{n\pi a^*}{\ell-a^*} = \frac{\ell\omega^*-n\pi c}{c} \ . \tag{2.49}$$

Hence,

$$K = - \frac{\omega^*}{c} \ ctg \ \frac{\ell\omega^*-n\pi c}{c} \tag{2.50}$$

is the value that must be assigned to the control parameter K in order to guarantee that the string oscillates with the desired frequency ω^*.

In this example it has been shown that sometimes one can formulate a control problem in such a way that for a prescribed performance requirement one has a simple and well determined condition by which the control can be specified. In subsequent examples, more sophisticated situations will be considered.

(ii) A beam of uniform cross section is subjected to a load $p(t,x)$, where t is the time and x the coordinate along the axis of the beam. If μ is the beam's mass density per unit length, α its flexural rigidity, and $w(x,t)$ the beam's lateral deflection, the transverse vibrations of the beam are described by

$$\mu\ddot{w} + \alpha w'''' = p(t,x) , \quad U[w]_B = 0 . \tag{2.51}$$

In (2.51),

$$\ddot{w} = \frac{\partial^2 w}{\partial t^2} , \qquad w'''' = \frac{\partial^4 w}{\partial x^4} ,$$

and $U[w]_B = 0$ are the conditions of the "boundary" of the beam, i.e., at $x = 0$ and $x = \ell$, where ℓ is the length of the beam. For a more compact presentation of the problem introduce the operator

$$D = \mu \frac{\partial^2}{\partial t^2} + \alpha \frac{\partial^4}{\partial x^4} , \tag{2.52}$$

so that (2.51) can be replaced by

$$Dw = p(t,x) , \quad U[w]_B = 0 . \tag{2.53}$$

Equations (2.53) govern the uncontrolled response w of the beam to the load (perturbation) p.

In order to adjust this response, let a control

$$u = Aw \tag{2.54}$$

be introduced, where A is an operator yet to be determined. Let the control of the beam be implemented by setting

$$Dw = p + Cu , \quad U[w]_B = 0 , \tag{2.55}$$

where C is another yet undetermined operator. Using (2.54) in (2.55), and rearranging, one has

$$(D-CA)w = p \ , \qquad U[w]_B = 0 \ . \tag{2.56}$$

Let the notation

$$D - CA = D^* \tag{2.57}$$

be proposed. Then, (2.56) changes into

$$D^*w = p \ , \qquad U[w]_B = 0 \ . \tag{2.58}$$

Comparing (2.58) with (2.53) shows clearly that controlling a system is essentially redesigning a *given* system characterized by an operator D into a *new* system with operator D* so that the response of (2.58) to a perturbation p may become different, hopefully more agreeable, than that of (2.53). Hence, it has again come out very clearly, that control is essentially a design problem and may in a first attempt not even be a "properly posed" one, mathematically speaking.

Let a few, very simple possibilities for the choice of the operators C and A be considered. First, choose $C \equiv E$. Then, let be

$$(i) \qquad Aw = Em\ddot{w} \ , \qquad m = m(x) \ , \tag{2.59}$$

so that

$$CAw \equiv EAw = Aw = m\ddot{w} \ . \tag{2.60}$$

With (2.60) in (2.56), and observing (2.52), one obtains

$$(\mu-m)\ddot{w} = \alpha w'''' = p \ , \tag{2.61}$$

i.e., the beam undergoes a change in inertia.

$$(ii) \qquad Aw = Ekw'''' \ , \qquad k = k(x) \ , \tag{2.62}$$

so that

$$CAw \equiv EAw = Aw = kw'''' \ . \tag{2.63}$$

With (2.63) in (2.56), and observing (2.52), one has

$$\mu\ddot{w} + (\alpha-k)w'''' = p \ ,$$
(2.64)

i.e., the beam suffers a change in stiffness.

(iii) $Aw = Ec\dot{w} \ , \quad c = const \ ,$
(2.65)

so that

$$CAw \equiv EAw = Aw = c\dot{w} \ .$$
(2.66)

Substituting (2.66) in (2.56), and observing (2.52), yields

$$\mu\ddot{w} + c\dot{w} + \alpha w'''' = p \ ,$$
(2.67)

i.e., the beam is now a damped one.

Obviously, all these controls affect the frequency of
the beam's vibration. If these frequencies were prescribed
a priori, one could interpret the control problem for this beam as
being the so-called "inverse" problem in the theory of vibrations.
This problem consists in designing a beam for a predetermined set
of frequencies.

In view of that fact one may now conclude that control
is in the broadest sense of this term an "inverse" problem: If
the "regular" problem consists in determining the output caused
by a given perturbation for a system characterized by a given
operator, then the "inverse" problem, i.e., the control problem,
consists in determining the operator of the system for a prescribed
output caused by a given perturbation.

The inverse problem is of course more difficult than the
regular problem as it is not well-posed a priori and requires
adding appropriate conditions in order to make the problem well-
posed posteriorily. It is the specific task of the control designer
to provide these additional conditions.

(iii) As before a beam may in an uncontrolled state be represented by

$$Dw_0 = p \ , \quad U[w_0]_B = 0 \ , \quad p = p(x) \ . \tag{2.68}$$

Let it be assumed that the operator D has the inverse

$$D^{-1} = \int_0^{\ell} G(x,\xi) \ \cdots \ d\xi \ , \tag{2.69}$$

where G is a Green function. Then, from (2.68) follows with (2.69)

$$w_0 = D^{-1}p = \int_0^{\ell} G(x,\xi)p(\xi)d\xi \ .$$

Introducing the passive control u = Aw, and changing for an implementation of the control equation (2.68) into

$$Dw = p + Cu \ , \quad Dw = p + CAw \ , \quad U[w]_B = 0 \ ,$$

one has finally for

$$CA = F \ ,$$

the control problem

$$Dw = p + Fw \ , \quad U[w]_B = 0 \ . \tag{2.70}$$

Making use of the inverse D^{-1} in (2.70), one arrives at

$$w = D^{-1}p + D^{-1}Fw \ . \tag{2.71}$$

Yet, by virtue of (2.68),

$$D^{-1}p = w_0 \ . \tag{2.72}$$

Hence, (2.71) can be rearranged to yield

$$w-w_0 = \Delta w = D^{-1}Fw \ . \tag{2.73}$$

With (2.69), one can rewrite (2.73) as

$$\Delta w = \int_0^{\ell} G(x,\xi)Fw(\xi)d\xi \ . \qquad (2.74)$$

This relationship is to be used to determine the control operator F for a prescribed performance measure Δw.

Without changing the morphology of the control problem, it may yet be simplified for the following considerations by setting

$$Fw = \lambda\phi(x)w \ , \qquad (2.75)$$

where λ is a parameter and $\phi(x)$ a yet undetermined function. Then, (2.73) yields

$$w = \lambda D^{-1}\phi w + D^{-1}p \ . \qquad (2.76)$$

With (2.69) and (2.72), equation (2.76) can be rewritten to read

$$w(x) - \lambda \int_0^{\ell} G(x,\xi)\phi(\xi)w(\xi)d\xi = w_0(x) \ . \qquad (2.77)$$

This is a relationship which is of the nature of an integral equation [2.4].

Assuming that p is given, w_0 represents a known function. Moreover, setting

$$G(x,\xi)\phi(\xi) = K(x,\xi) \ , \qquad (2.78)$$

equation (2.77) changes into

$$w(x) - \lambda \int_0^{\ell} K(x,\xi)w(\xi)d\xi = w_0(x) \ . \qquad (2.79)$$

If the kernel $K(x,\xi)$ were known and the response $w(x)$ of the beam were to be determined, equation (2.79) would represent a Fredholm integral equation of the second kind. However, in the case of control, the problem is different: w_0 is known, w is prescribed and the kernel K together with the parameter λ are to be determined.

For this purpose, apply to (2.79) one of the formulas
of approximate integration transforming (2.79) into the system of
equations

$$w(x_j) - \sum_{k=1}^{n} A_k \lambda K(x_j,\xi_k) w(\xi_k) = w_0(x_j) , \quad j = 1,2,\ldots,n . \quad (2.80)$$

In (2.80), the A_k are certain constants prescribed by the integration formula.

Setting

$$w(x_j) - w_0(x_j) = b_j , \qquad \lambda K(x_j,\xi_k) = \hat{K}(x_j,\xi_k) , \quad (2.81)$$

the system of equations changes into

$$\sum_{k=1}^{n} A_k \hat{K}(x_j,\xi_k) w(\xi_k) = b_j , \quad j = 1,2,\ldots,n . \quad (2.82)$$

Let now the (n×n) block diagonal matrix

$$\hat{\underline{A}} = \begin{bmatrix} A & & & & 0 \\ & A & & & \\ & & A & & \\ & & & \ddots & \\ 0 & & & & A \end{bmatrix} \quad (2.83)$$

be defined for which the A-blocks are the (1×n) matrices (row-
vector)

$$A = (A_1 w(\xi_1), A_2 w(\xi_2), A_3 w(\xi_3), \ldots, A_n w(\xi_n)) . \quad (2.84)$$

Moreover, let the $(n^2 \times 1)$ matrix (column-vector)

$$\hat{\underline{K}} = \begin{bmatrix} \hat{K}_{11} \\ \hat{K}_{12} \\ \vdots \\ \hat{K}_{1n} \\ \hat{K}_{21} \\ \hat{K}_{22} \\ \vdots \\ \hat{K}_{2n} \\ \vdots \\ \hat{K}_{n1} \\ \hat{K}_{n2} \\ \vdots \\ \hat{K}_{nn} \end{bmatrix} \quad , \quad \hat{K}_{jk} = \hat{K}(x_j, \xi_k) \quad , \quad j,k = 1,2,\ldots,n, \quad (2.85)$$

and the $(n \times 1)$ matrix (column-vector)

$$\underline{b} = \begin{bmatrix} b_1 \\ b_2 \\ \vdots \\ b_n \end{bmatrix} \tag{2.86}$$

be introduced. Then, the system (2.82) assumes the form

$$\underset{(n \times n^2)}{\underline{A}} \quad \underset{(n^2 \times 1)}{\hat{\underline{K}}} = \underset{(n \times 1)}{\underline{b}} \quad . \tag{2.87}$$

It clearly represents a linear, inhomogeneous system of n equations for the n^2 unknowns $\hat{K}_{ij}$. Hence, it becomes obvious that one is confronted with the "ill-posed" problem of an *undetermined* system of equations [5].

Let the "augmented" $[n \times (n^2+1)]$ matrix

$$\underline{W} = [\hat{\underline{A}} \vdots \underline{b}] \tag{2.88}$$

be introduced. Let $r_{\hat{A}}$ be the rank of $\hat{\underline{A}}$ and r_W that of $\underline{W}$. If

$$r_W = r_{\hat{A}} \tag{2.89}$$

the system (2.87) has at least one solution. But since $r_A < n^2$,
there will even be an infinite set of solutions for (2.87). Thus,
there exists one condition for the prescribed output $w(x)$, which
for the design of the control must be taken into account, and that
is (2.89).

In order to make the control problem well-posed so that
there exists a unique solution for the unknowns $\hat{K}_{jk}$, one must add
further conditions to (2.87), for example the *linear programming*
conditions:

$$\text{maximize:} \quad Z = \sum_{j=1}^{n} \sum_{k=1}^{n} p_{jk} \hat{K}_{jk}$$

and satisfy the requirement:

$$\hat{K}_{jk} \geq 0 , \quad j,k = 1,2,\ldots,n .$$

$$\tag{2.90}$$

Of course, other conditions than (2.90) may be chosen. It is at
this point, where skill and experience of the designer has bearing
on the appropriateness of the control problem formulation.

Here, again, the complexity of the problem can be reduced
by making additional assumptions about the operator that is to be
determined. Assume for example that there exists an appropriate
complete, orthonormal system $\phi_i(x)$ of coordinate functions which
may be true by virtue of the nature of the uncontrolled system.
Then,

$$w(x) - w_0(x) = \sum_i c_i \phi_i(x) \tag{2.91}$$

may hold, where the c_i are prescribed quantities if the controlled
response $w(x)$ is being specified to the designer.

Again, by virtue of the nature of the uncontrolled system,
the kernel K may have the form

$$K(x,\xi) = \sum_i \frac{\phi_i(x)\phi_i(\xi)}{\kappa_i} , \tag{2.92}$$

where the κ_i are undetermined quantities on which the structure of
the system depends.

With (2.91) and (2.92) in (2.79), one has

$$\sum_i c_i \phi_i(x) = \lambda \int_o^\ell \sum_i \frac{\phi_i(x)\phi_i(\xi)}{\kappa_i} w(\xi)d\xi \ . \tag{2.93}$$

Yet, according to (2.91),

$$\int_o^\ell w(\xi)\phi_j(\xi)d\xi = c_j + \int_o^\ell w_0(\xi)\phi_j(\xi)d\xi \ . \tag{2.94}$$

Let

$$\int_o^\ell w_0(\xi)\phi_j(\xi)d\xi = e_j \ . \tag{2.95}$$

Then, (2.93) changes into

$$\sum_i c_i \phi_i(x) = \lambda \sum_i \frac{c_i + e_i}{\kappa_i} \phi_i(x) \ , \tag{2.96}$$

$$\sum_i \left[c_i - \lambda \frac{c_i + e_i}{\kappa_i} \right] \phi_i(x) = 0 \ . \tag{2.97}$$

The last relationship can only hold true if

$$\kappa_i = \frac{\lambda(e_i + c_i)}{c_i} \tag{2.98}$$

is satisfied. Condition (2.98) specifies the control: for given
λ, c_i, and e_i, the κ_i are to be determined using (2.98). Once
the quantities κ_i have been found, the system's parameter are prescribed, and the control problem is solved.

2.4   THE MODAL APPROACH

Consider the control problem

$$\mu\ddot{w} + D_0 w = p(t,x) + Fw \ , \qquad U[w]_B = 0 \ , \tag{2.99}$$

which is slightly more general than (2.70) insofar as it involves two variables, t and x. In (2.99), the operators D_0, F, and U are assumed to act on x only.

Let a complete set of orthonormal coordinate functions $y_i(x)$ be generated by means of the auxiliary problem

$$D_0 y_i = \lambda_i y_i \ , \qquad U[y_i]_B = 0 \ . \tag{2.100}$$

Assume, the problem (2.99) is such that the expansion

$$w(x,t) = \sum_i f_i(t) y_i(x) \tag{2.101}$$

holds, in which the $f_i(t)$ are modal amplitudes and the $y_i(x)$ are the modes [2.6].

Substituting in (2.99) w by (2.101) yields

$$\mu \sum_i \ddot{f}_i y_i + \sum_i f_i D_0 y_i = p + \sum_i F(f_i y_i) \ . \tag{2.102}$$

Let Galerkin's method be applied to (2.102). Then, one has

$$\int_\Omega \left[\mu \sum_i \ddot{f}_i y_i + \sum_i f_i \lambda_i y_i \right] y_j d\Omega = \int_\Omega p y_j d\Omega + \int_\Omega \sum_i F(f_i y_i) y_j d\Omega \ , \tag{2.103}$$

$$j = 1,2,\ldots,n \ .$$

The perturbation p is supposed to have the Fourier series

$$p = \sum_k P_k(t) y_k(x) \ , \qquad P_k(t) = \int_\Omega p y_k(x) dx \ . \tag{2.104}$$

Moreover, set

$$\Phi_{ij} = \int_\Omega F(f_i y_i) y_j d\Omega \ . \tag{2.105}$$

Under those circumstances, and observing the orthonormality of the coordinate functions, the system of equations (2.103) can be brought into the form

$$\mu \ddot{f}_j + \lambda_j f_j = P_j(t) + \sum_{i=1}^{n} \phi_{ij}(f_i) \ , \qquad j = 1,2,\ldots,n \ . \tag{2.106}$$

With that, one has obtained a system of ordinary, coupled differential equations for f(t). At the same time, one can say that one has discretized the originally continuous structure, so that one is facing the control problem of a system with n degrees of freedom. Having arrived at this conclusion, one may assume that the further procedure is a simple one: all that is left is to apply classical control theory methods to (2.106). But, at this point, one realizes immediately that complications do arise, as predicted, if n is a large number: For example, if optimal control theory is applied to (2.106), one is led to a matrix Riccati differential equation of high order for the determination of the control, which can cause insurmountable difficulties.

The stability problem connected with (2.99) is related to

$$\mu\ddot{w} + (D_0+F)w = 0 , \qquad U[w]_B = 0 . \tag{2.107}$$

Assume that the operators D_0 and F are such that the solution to (2.107) has the form

$$w(x,t) = e^{i\omega t}y(x) , \tag{2.108}$$

where ω is a frequency. With (2.108) in (2.107), one can derive

$$-\mu\omega^2 y + (D_0+F)y = 0 , \qquad U[y]_B = 0 . \tag{2.109}$$

Using the coordinate functions y_i generated by (2.100), one can set

$$y = \sum_i a_i y_i . \tag{2.110}$$

With (2.110), and applying Galerkin's method, one obtains from (2.109) the system of algebraic equations

$$-\mu\omega^2 a_j + \sum_i a_i \int_\Omega [(D_0+F)y_i]y_j d\Omega = 0 . \tag{2.111}$$

Setting

$$\int_{\Omega} [(D_0+F)y_i]y_j \, d\Omega = \rho_{kj} \; , \tag{2.112}$$

one can rewrite (2.111) to obtain

$$\sum_{i=1}^{n} (-\mu\omega^2\delta_{ij}+\rho_{ij})a_i = 0 \; , \quad j = 1,2,\ldots,n \; . \tag{2.113}$$

Since there must be a nontrivial solution for the a_i, it follows
from (2.113) that

$$\det(-\mu\omega^2\delta_{ij}+\rho_{ij}) = 0 \; ; \quad i,j = 1,2,\ldots,n \; , \tag{2.114}$$

must be satisfied. Equation (2.114) is an algebraic equation of
the n-th order for the quantity ω^2. It follows from (2.108) that
for stability, the ω_i^2, i = 1,2,...,n, must all be positive, real.
Now, one realizes another difficulty caused by the modal
approach: not only may it be cumbersome to solve (2.114) for ω^2
in the case of large n, but also, one may have to face the fact
that by virtue of operator F, the determinant in (2.114) may not
be symmetric. Then, complex solutions for ω^2 cannot be excluded.
As a result, instability by flutter can occur, [2.7], [2.8]. In
that case the flutter phenomenon may turn up for a frequency of
a certain very high order, while all the frequencies of an order
smaller than that one correspond to (stable) vibrations. Let the
discretization of the actually continuous system as shown have led
to n degrees of freedom, and, therefore, to the algebraic equation
(2.114) which is then of the n-th order. Let all the quantities
ω_i^2, i = 1,2,...,n, have been obtained as positive, real quantities,
but let the determinant in (2.114) not be symmetric. Then, one
cannot safely claim stability: the possibility is given that
ω_j, j > n, leads to flutter. Thus, the modal approach is deficient
for unsymmetric, i.e., nonselfadjoint problems, as it does not
allow one to detect whether higher modes, which have not been

considered in the calculations, might after all cause instability
by "spillover".

The conclusion must be, that one should avoid the modal
approach in the case of continuous systems, if, as a consequence
of the control, the system has become nonselfadjoint, or if one
is not able to safely exclude nonselfadjointness. One should
instead strive for closed form solutions and for global statements
on stability. Under such circumstances, new approaches to the
control problem have to be developed. How that can be done, will
be reported in Chapter 5.

In spite of these cautioning statements, the modal
approach will be applied throughout in Chapters 3 and 4. This can
be justified by the fact that the controlled systems may still be
selfadjoint, in which case flutter is ruled out and the modal
approach remains admissible. Also, one may claim that dangerous
modes may be of such a high order that they will possibly not be
excited, thus causing no difficulties. Then, results provided by
the modal approach are fairly reliable. Finally, the modal approach may reveal important features of the problem at least
approximately, and certainly easily. One has therefore good reasons not to exclude an application of the modal approach completely.
In fact, in the following Chapters 3 and 4, modal approach will
be used extensively.

2.5 REFERENCES

[2.1] LIONS, J.L., *Optimal Control of Systems Governed by Partial
       Differential Equations*, Springer-Verlag, Berlin-Heidelberg-
       New York, 1971.

[2.2] ODEN, J.T., *Applied Functional Analysis*, Prentice-Hall Inc.,
       Englewood Cliffs, New Jersey 07632, 1979.

[2.3] TRICOMI, F.G., *Integral Equations*, Interscience Publishers
       Inc., New York, Fourth Edition, 1967.

[2.4] HOCHSTADT, H., *Integral Equations*, John Wiley and Sons,
       New York-Chichester-Brisbane-Toronto-Sydney, 1973.

[2.5]   PAYNE, L.E., *Improperly Posed Problems in Partial Differential Equations*, SIAM, Philadelphia, Pennsylvania, 19103, 1975.

[2.6]   MEIROVITCH, L., *Computational Methods in Structural Dynamics*, Sijthoff and Noordhoff, Aalphen aan den Rijn, The Netherlands, 1980.

[2.7]   LEIPHOLZ, H., *Direct Variational Methods and Eigenvalue Problems in Engineering*, Noordhoff International Publishers, Leyden, The Netherlands, 1977.

[2.8]   LEIPHOLZ, H., *Stability of Elastic Systems*, Sijthoff and Noordhoff, Alphen aan den Rijn, The Netherlands, 1980.

Chapter III

Automatic Active Control of Simple Span Bridges

3.1 MATHEMATICAL MODELS OF SIMPLE SPAN BRIDGES

A bridge is a very complex system. Its analysis is rather involved
for the following reasons:

(1) It is difficult to analyse the combined action of stringers
and road slab in the presence of dynamic loading.

(2) The vehicle, which is passing on the bridge inducing the
dynamic loading, is also a very complex system with multi-degrees
of freedom, thus complicating the analysis.

(3) The bridge has infinite degrees of freedom, and as many
of the vibrational modes as possible should be considered.

(4) Many unknown variables are introduced into the analysis
by the moving vehicle such as the vehicles' position with respect
to the centreline of the bridge, its speed, its axle spacing and
the unevenness of the road slab.

Many investigations, experimentally and theoretically,
have been carried out in order to identify the bridge behaviour
and to correlate the theoretical results with experimental data
[3.1-3.4]. These investigations have shown the bridge to be representable in terms of a mathematical model which allows some

simplifications, thus easing the theoretical analysis. Among such admissible simplifications [3.5] are the following.

(1) The bridge is regarded as a single beam with a rigidity equivalent to that of the combined system of floor, stringers and main girders.

(2) Only the fundamental mode of vibration is considered, and all higher modes are neglected.

(3) The vehicle is represented by a system with a few degrees of freedom only, e.g., as a constant force moving with constant speed, or as spring and unsprung masses moving with constant or accelerated speed, etc.

(4) The weight of the vehicle is applied at the vehicle's mass centre rather than at the wheels' point of contact with the bridge floor.

The first assumption produces a small error only if the bridge is relatively narrow and the vehicle is positioned on the centreline of the bridge. The second assumption is permissible for most purposes since the higher modes contribute little to the deflection of mid-span. However, the effect of higher order modes is mainly responsible for human discomfort, and by neglecting it, the public confidence in the structure may be jeopardized. The third assumption may cause an essential error since the vehicle has actually many degrees of freedom, and it is uncertain which model may represent the vehicle properly. The fourth assumption is valid as long as the ratio of the bridge's span to the vehicle-axle spacing is greater than 5 [3.5].

In the past, vibration and safety problems have been overcome by using structural material abundantly in order to keep the bridge heavy in weight. However, in view of the modern trend towards the building of long-span bridges, those traditional techniques to deal with vibration and safety problems are unsuitable and uneconomical. The purpose of this chapter is to illustrate how automatic active control techniques can be used to suppress

the vibration and to guarantee the safety of simple span bridges
even in the case of light-weight structures.

Based on the results of Section 1.5 in Chapter 1, some
control mechanisms for the control of simple span bridges are
proposed. One of these mechanisms is explicitly analysed in this
chapter. Active control of one mode and then of many modes of
vibration using classical control and modern control design methods
is discussed. The effect of the modes' interaction on structural
stability and the stabilization of the structure is shown in
numerical examples.

3.2　ACTIVE CONTROL MECHANISMS IN BRIDGES

It has been shown in Section 1.5 that control of a structure can
be implemented by generating a system of control forces. These
forces are provided by auxiliary structural elements such as
masses, tendons (springs), or dampers. Using each of these elements,
it is shown in this section how a control mechanism can be designed.

3.2.1　*Control through Auxiliary Masses*

A system of mass blocks can be supported by springs between the
main girders of the bridge, as shown in Figure 3.1. Control devices
at each mass, which are supported by the cross girders, generate
two opposite forces. One of these forces is absorbed by the auxiliary mass, and the other force is applied to the cross girder.
The forces applied to the cross girders are the ones which bring
about the control of the structural response. The springs which
support the auxiliary masses should be designed in such a way that
some energy is dissipated. For this purpose, one may connect these
springs with dampers. The control devices must be designed such
as to supply the appropriate control forces.

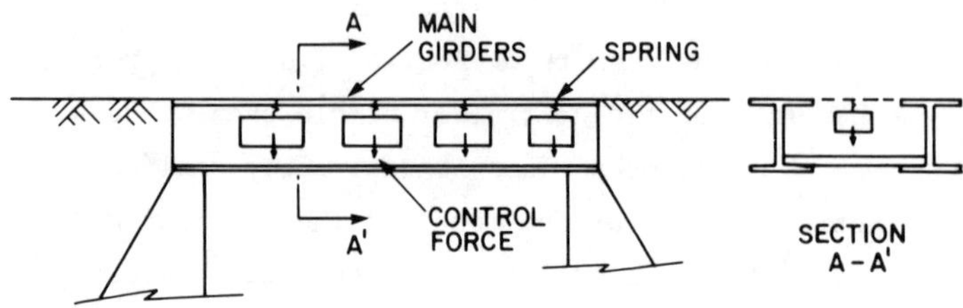

*Figure 3.1 - Simple Span Bridge Controlled by Auxiliary Mass
        Absorbers*

3.2.2 *Control through Auxiliary Tendons*

Tendons provide suitable means for the control of most structures
[3.10]. For a simple span bridge, the control mechanisms shown
in Figure 3.2 are suggested. The bridge shown in Figure 3.2(a) is
controlled by a central control which results from tensioning the
tendons. The bridge shown in Figure 3.2(b) is controlled by two
symmetrical control forces, whereas the one shown in Figure 3.2(c)
is controlled by two symmetrical control moments. One should note
that tendons provide the control forces by either being in tension
or in compression.

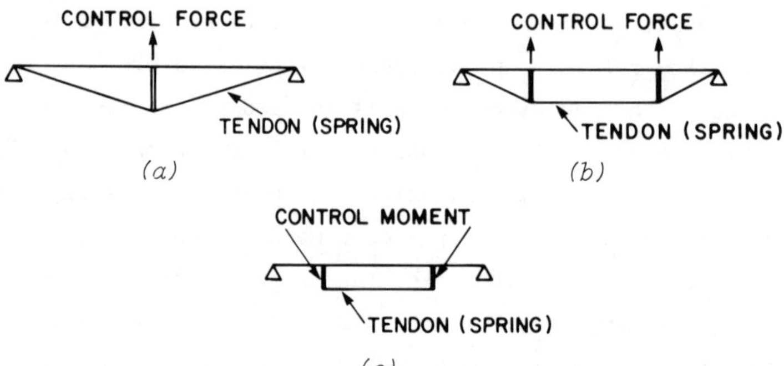

Figure 3.2 - Simple Span Bridge Controlled by Auxiliary Tendons

3.2.3 *Control through Auxiliary Dampers*

A system of dampers (dashpots) may be installed in some hangers of
the suspension bridge. Control devices will generate control forces
by the dampers through velocity response. An illustration of the
mechanism is given in Figure 3.3.

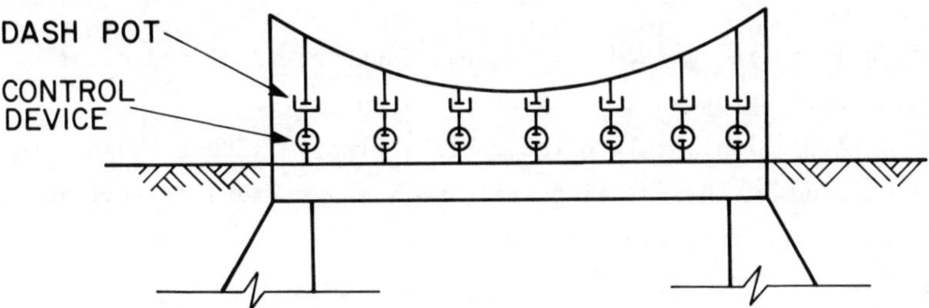

Figure 3.3 - Simple Span Bridge Controlled by Auxiliary Dampers

3.2.4 *Control using Aerodynamic Appendages*

Also, a system of appendages can be installed along the mid-portion
of the bridge. The appendages can be deployed or folded automatically in order to control the bridge against torsional instability.
This mechanism is shown in Figure 3.4.

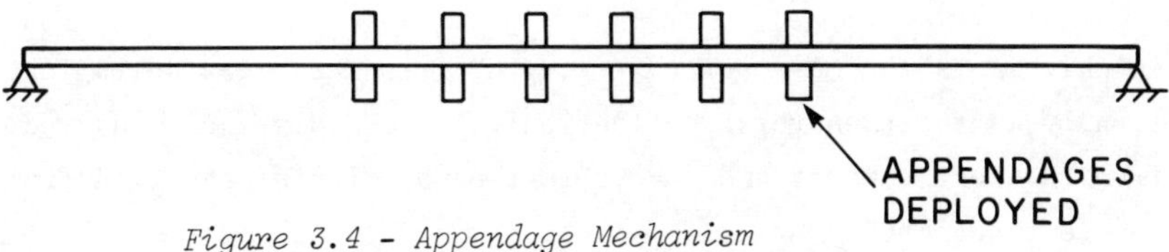

Figure 3.4 - Appendage Mechanism

In order to keep the presentation in the beginning as
simple as possible, only tendon control shall be considered in
this chapter. In fact, control mechanisms involving tendons can

be implemented in practice. Moreover, tendons have already frequently been used in the static control of structures. In general,
one may use any of the various types of control mechanisms, as shall
be shown in Chapter 4, or a combination of them in order to achieve
a desirable control.

3.3 ACTIVE CONTROL BY CLASSICAL CONTROL METHODS

Classical control methods are mainly based on trial and error procedures using graphical or theoretical analysis. In order to provide a simple insight into the control problem, first these methods
will be used in this section while more modern methods will be considered in the following sections.

3.3.1 *Equation of Motion of the System*

Consider a simple span bridge with constant flexural rigidity EI
and span L under a concentrated load P of constant magnitude,
which moves with a constant speed v. Secondary effects coming
from the inertia of the moving mass, from the control force, and
from the uneveness of the road slab shall be investigated later.
The equation of motion of the structure shown in Figure 3.5, neglecting internal damping, is given by the equation

$$EI \frac{\partial^4 y}{\partial x^4} + m \frac{\partial^2 y}{\partial t^2} = P\delta(x-\bar{x}) \, , \qquad (3.1)$$

in which m is the mass per unit length; δ is the Dirac delta function; y is the transverse deflection; x is the distance measured
from the left support and $\bar{x} = vt$ is the position of the load from
the left support.

The mechanism shown in Figure 3.6 allows only a control
of the sag. Therefore, the bridge is not being dampened efficiently.
In order to improve the control effect, the arrangement shown in
Figure 3.7 is proposed. In this mechanism, the central cable is

controlling the sagging deflection whereas the outward cables control the hogging deflection. Using spring control instead of tendon control will eliminate the above problems, and therefore the mechanism shown in Figure 3.8 is used throughout this chapter.

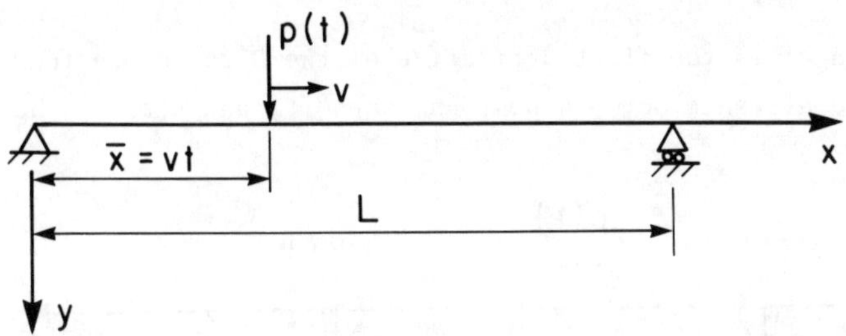

Figure 3.5 - Simple Span Bridge under Concentrated Moving Load

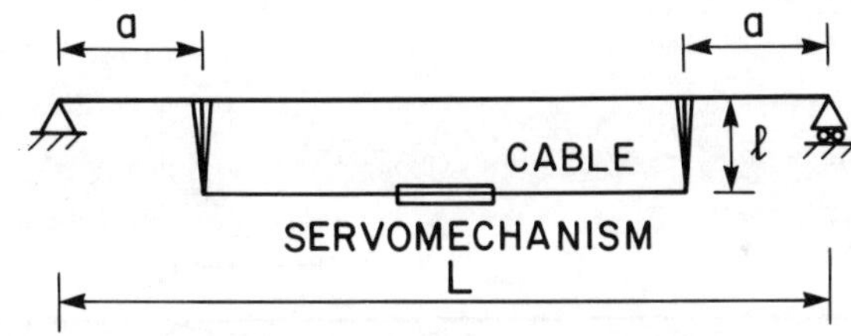

Figure 3.6 - Moment Control by Tendons

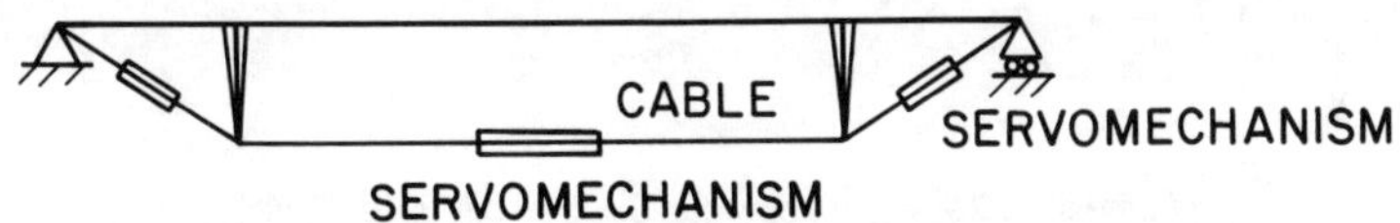

Figure 3.7 - Force Control by Tendons

Figure 3.8 - Moment Control by Spring

Using the free body diagrams of Figures 3.9 and 3.10,
the following equation of motion is obtained:

$$EI \frac{\partial^4 y}{\partial x^4} + m \frac{\partial^2 y}{\partial t^2} = P\delta(x-vt) + M(t)\delta'(x-a) - M(t)\delta'(x-L+a) , \qquad (3.2)$$

in which δ' is the first derivative of the Dirac delta function,
a is the distance between post and support, and M(t) is the control
moment.

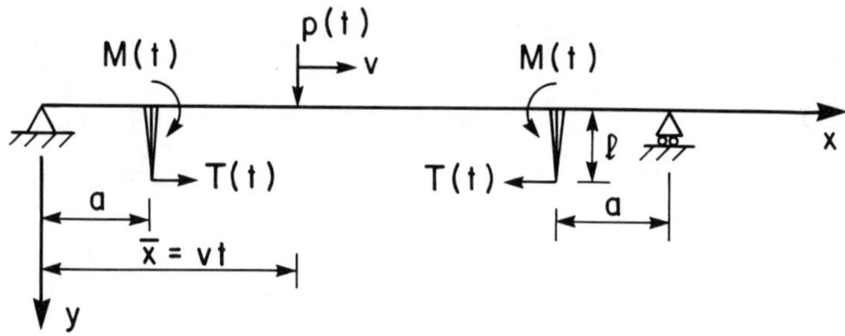

Figure 3.9 - Free Body Diagram

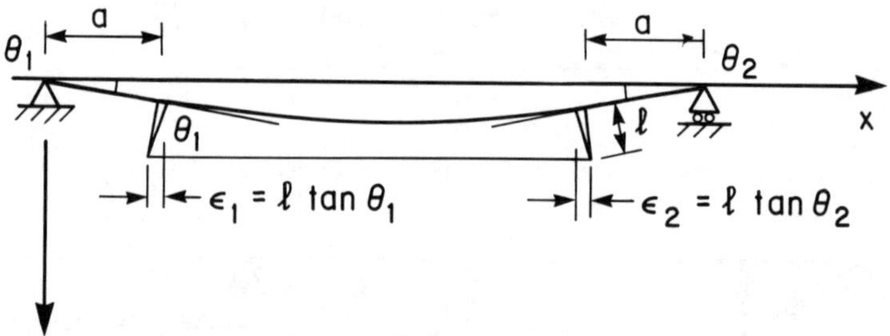

Figure 3.10 - Deformed Shape of the Bridge

Assuming very rigid posts, as shown in Figure 3.10, the
control moment is obtained as

$$M(t) = S\ell\Delta(t) = S\ell[u(t)+\ell y'(a,t)-\ell y'(L-a,t)] , \qquad (3.3)$$

in which u(t) is the displacement of the spring caused by active
control; S is the stiffness of the spring; $\Delta(t)$ is the active and
passive displacement of the spring; and ℓ is the post's length.

The first term between brackets on the right hand side of equation (3.3) represents the active elongation or shortening of the spring, and the other two terms represent the passive elongation or shortening due to the beam's deformation.

The solution of equation (3.2) is assumed to be given by

$$y(x,t) = \sum_{j=1}^{\infty} \phi_j(x)A_j(t) , \qquad (3.4)$$

in which $A_j(t)$ is the generalized coordinate of mode j; and $\phi_j(x)$ is the characteristic function of mode j.

The functions $\phi_j(x)$ are to be chosen such that they satisfy the boundary conditions. A possible choice for $\phi_j(x)$ is for example

$$\phi_j(x) = \sin \frac{j\pi x}{L} . \qquad (3.5)$$

Substituting equation (3.4) into equation (3.2) and applying an integral transformation, one arrives at the following mode equations:

$$\ddot{A}_1(t)+\omega_1^2 A_1(t) = \frac{2P}{mL} \sin \frac{\pi vt}{L} - \frac{4S\ell}{m} \frac{\pi}{L} \cos \frac{\pi a}{L} [u(t)+\ell y'(a,t)-\ell y'(L-a,t)], \qquad (3.6)$$

$$\ddot{A}_2(t)+\omega_2^2 A_2(t) = \frac{2P}{mL} \sin \frac{2\pi v}{L} t , \qquad (3.7)$$

$$\ddot{A}_3(t)+\omega_3^2 A_3(t) = \frac{2P}{mL} \sin \frac{3\pi v}{L} t -$$

$$\frac{4S\ell}{mL} \frac{3\pi}{L} \cos \frac{3\pi a}{L} [u(t)+\ell y'(a,t)-\ell y'(L-a,t)] , \qquad (3.8)$$

$$\vdots$$

in which

$$\omega_K^2 = \frac{\pi^4 K^4}{L^4} \frac{EI}{m} , \qquad K = 1,2,\ldots,n . \qquad (3.9)$$

Equations (3.5) - (3.8) indicate that by applying symmetrical control moments with respect to the centreline of the beam,

the second mode, and in general each even mode is uncontrolled.
In order to control the even modes as well, a careful study is
required for determining the best position for the application of
the control moments. This shall be done at the end of this chapter.

The control device to be used in this section is the
servomechanism which was presented in Section 1.6.1. The equations of motion of a servomechanism augmented by a deflection
sensor, velocity sensor and acceleration sensor, respectively, and
neglecting the time constant τ, are, respectively, given by

$$\dot{u}(t) + K_c u(t) = \alpha_d K_c y(x_c,t) \; , \tag{3.10}$$

$$\dot{u}(t) + K_c u(t) = \alpha_v K_c \dot{y}(x_c,t) \; , \tag{3.11}$$

$$\dot{u}(t) + K_c u(t) = \alpha_a K_c \ddot{y}(x_c,t) \; , \tag{3.12}$$

in which x_c is the position of the sensor on the bridge; $\alpha_d = T_d/K_f$,
$\alpha_v = T_v/K_f$, and $\alpha_a = T_a/K_f$.

3.3.2 *Control of the Fundamental Mode of Vibration*

Let, for the present, the contributions of the second and of higher
order modes be neglected. The design of the control devices will
be based on considering the first mode only. The results will
enable one to check the validity of the assumption that high order
modes can be neglected and that only the fundamental mode of vibration must be considered.

For one mode, equation (3.4) becomes

$$y(x,t) = \sin \frac{\pi x}{L} A_1(t). \tag{3.13}$$

Substituting equation (3.13) into equation (3.6) one
obtains

$$\ddot{A}_1(t) + \omega_1^2 A_1(t) = \frac{2P}{mL} \sin \frac{\pi v}{L} t - \frac{4S\ell}{mL} \frac{\pi}{L} \cos\left[\frac{\pi a}{L} u(t) + 2\ell \frac{\pi}{L} \cos \frac{\pi a}{L} A_1(t) \right].$$
$$\tag{3.14}$$

Investigating equation (3.14), one arrives at the conclusion [3.6] that, if u(t) is a function of $\dot{A}_1(t)$, damping is induced to the bridge. Mass and stiffness can also be modified if u(t) is chosen as a function of $\ddot{A}_1(t)$ and $A_1(t)$, respectively. Let now Laplace transformation be applied to equation (3.14). Considering zero initial conditions one has

$$\overline{A}_1(s)[s^2+\omega_1^2] = \frac{2P}{mL}\frac{\pi v/L}{s^2+(\pi v/L)^2} - \frac{4S\ell}{mL}\frac{\pi}{L}\cos\frac{\pi a}{L}\left[\overline{u}(s)+2\ell\frac{\pi}{L}\cos\frac{\pi a}{L}\overline{A}_1(s)\right].$$

$$(3.15)$$

For the sake of brevity, the following constants are defined for the odd modes

$$B_j = \frac{4S\ell}{mL}\frac{j\pi}{L}\cos\frac{j\pi a}{L}, \qquad (3.16)$$

$$c_j = 2\ell\frac{j\pi}{L}\cos\frac{j\pi a}{L}. \qquad (3.17)$$

The active control $\overline{u}(s)$ can be introduced using any of the equations (3.10) - (3.12). Equation (3.15) can now be represented by the block diagram shown in Figure 3.11.

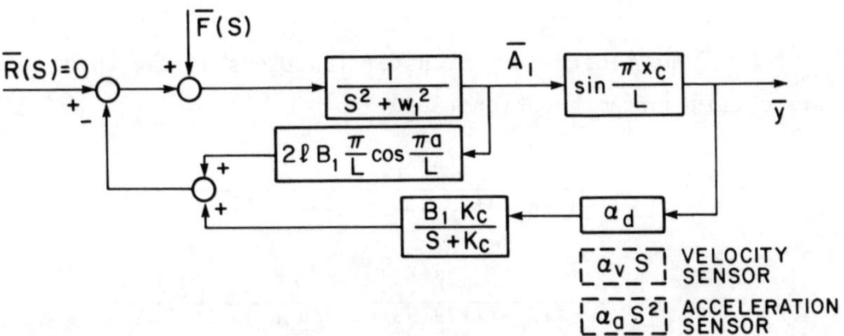

Figure 3.11 - Block Diagram Considering One Mode

The design is carried out using the classical transfer function method. The simplicity of this method is well known from

the design of one-degree-of-freedom systems [3.7]. By definition
[3.8], the forward, feedback, and closed-loop transfer functions
are obtained, respectively, from Figure 3.11 as

$$\overline{G}(s) = \frac{1}{s^2 + \omega_1^2} \, , \tag{3.18}$$

$$\overline{H}(s) = B_1 c_1 + \frac{\alpha K_c B_1}{s + K_c} \sin \frac{\pi X_c}{L} \, , \tag{3.19}$$

$$\overline{P}(s) = \frac{\overline{A}_1(s)}{\overline{F}(s)} = \frac{\overline{G}(s)}{1 + \overline{G}(s)\overline{H}(s)} \, . \tag{3.20}$$

Consider for the sake of simplicity the following data:
$L = 100$ ft, $a = 10$ ft, $\ell = 3$ ft, $P = 20$ kips, $EI = 12 \times 10^{10}$ lb.in^2,
$m = 0.3$ lb.sec^2/in^2, $v = 60$ ft/sec, $S = 62.5$ kips/in, and $X_c = L/2$.
With these data one obtains the following results:

$$\text{maximum static deflection at mid-span} = 6 \text{ in.} \tag{3.21}$$

$$\text{first mode}$$
$$\text{natural frequency}, \omega_1 = 4.334 \text{ rad/sec.} \, , \tag{3.22}$$

$$B_1 = \frac{4S\ell}{mL} \frac{\pi}{L} \cos \frac{\pi a}{L} = 62.246 \text{ sec}^{-2} \, . \tag{3.23}$$

The closed-loop transfer functions of the three systems
are obtained from equation (3.20) as

$$\overline{P}_d(s) = \frac{(s + K_c)}{s^3 + K_c s^2 + 29.95s + K_c(29.95 + 62.24\alpha_d)} \, , \tag{3.24}$$

$$\overline{P}_v(s) = \frac{(s + K_c)}{s^3 + K_c s^2 + (29.95 + 62.24\alpha_v K_c)s + 29.95 K_c} \, , \tag{3.25}$$

$$\overline{P}_a(s) = \frac{(s + K_c)}{s^3 + (K_c + 62.24 K_c \alpha_a)s^2 + 29.95s + 29.95 K_c} \, , \tag{3.26}$$

in which the subscripts $\overline{P}(s)$ indicate the type of sensor; $\overline{P}_d$ stands
for deflection sensor, $\overline{P}_v$ for velocity sensor, and $\overline{P}_a$ for acceleration sensor.

The characteristic equation [3.8] of each system is given by the denominator of its closed-loop transfer function. For deflection, velocity, and acceleration sensor systems, the characteristic equations are, respectively, given by

$$s^3 + K_c s^2 + 29.95s + K_c(29.95 + 62.24\alpha_d) = 0 \; , \tag{3.27}$$

$$s^3 + K_c s^2 + (29.95 + 62.24 K_c \alpha_v)s + 29.95 K_c = 0 \; , \tag{3.28}$$

$$s^3 + (K_c + 62.24 K_c \alpha_a)s^2 + 29.95s + 29.95 K_c = 0 \; . \tag{3.29}$$

The design is now carried out with the intent of satisfying certain criteria. These criteria involve in the first place the requirement for dynamic stability. Other design requirements are an appropriate damping ratio, minimum steady state error, minimum sensitivity, etc. [3.7,3.8].

3.3.2.1 Stability Requirement

The stability of a one-degree-of-freedom time-invariant system can be checked by using a variety of methods (see Appendix A). Herein, the stability is checked by using the Routh criterion [3.8]. According to it, a stable system exists if the sign of the first column in the Routh-table is not changing. The Routh-tables for the three systems with the conditions of stability are shown in Tables 3.1, 3.2 and 3.3, respectively.

3.3.2.2 Damping Requirement

Introducing damping to the structure is an important objective. The damping ratio should be as large as possible in order to suppress the vibration of the structure quickly. Therefore, a parametric study is to be carried out in order to find the best combination of the controller's parameters, and to obtain maximum damping. In this context, the root-locus method [3.7,3.8] is applied. In

root-locus form (see Appendix A), the characteristic equations
(3.27)-(3.29), respectively, become

$$1 + \frac{K_c[s^2+(29.95+62.24\alpha_d)]}{s^3+29.95s} = 0 \ , \qquad (3.30)$$

$$1 + \frac{K_c[s^2+62.24\alpha_v s+29.95]}{s^3+29.95s} = 0 \ , \qquad (3.31)$$

$$1 + \frac{K_c[(1+62.24\alpha_a)s^2+29.95]}{s^3+29.85s} = 0 \ . \qquad (3.32)$$

Table 3.1 - Routh-Table of the System with Deflection Sensor

s^3	1.0	29.95
s^2	K_c	$K_c(29.95 + 62.24\alpha_d)$
s^1	$-62.24\alpha_d$	0
s^0	$K_c(29.95 + 62.24\alpha_d)$	
The system is stable if $K_c > 0$, and $-0.481 < \alpha_d < 0$		

Table 3.2 - Routh-Table of the System with Velocity Sensor

s^3	1.0	$(29.95 + 62.24\alpha_v)$
s^2	K_c	$29.95K_c$
s^1	$62.24\alpha_v$	0
s^0	$29.95K_c$	
The system is stable if $K_c > 0$, and $\alpha_v > 0$		

Table 3.3 - Routh-Table of the System with Acceleration Sensor

s^3	1.0	29.95
s^2	$(K_c + 62.24\alpha_a K_c)$	$29.95K_c$
s^1	$(29.95 + 62.24\alpha_a)/(1 + 62.24\alpha_a)$	0
s^0	$29.95K_c$	
The system is stable if $K_c > 0$, and $\alpha_a > 0$		

The best values of the variable α are found by trial and error, and the results are

$$\alpha_d = -0.35 \ , \tag{3.33}$$

$$\alpha_v = 0.104 \ , \tag{3.34}$$

$$\alpha_a = 0.044 \ . \tag{3.35}$$

The root-loci for the three systems were computed [3.9] and are plotted in Figures 3.12, 3.13 and 3.14, respectively. The design parameter K_c for the three systems is chosen to be 6.53, 12.67 and 1.736, respectively. These are the values necessary to obtain a constant damping ratio of 0.43 for the three systems so that a comparison of their behaviour is possible.

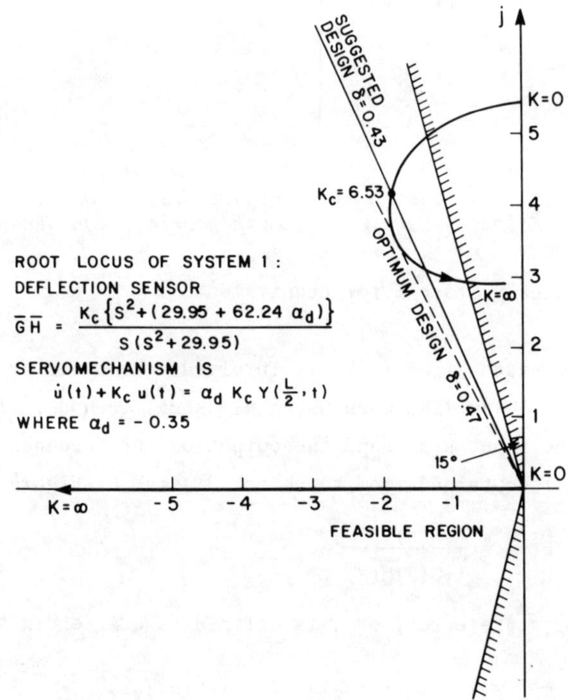

Figure 3.12 - Root Locus of System with Deflection Sensor

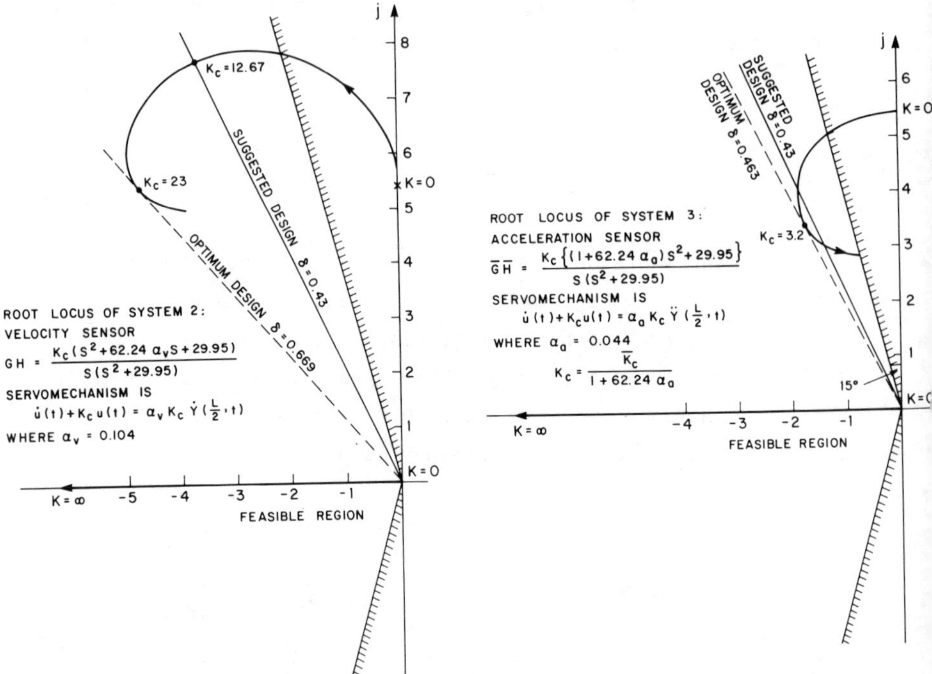

Figure 3.13 - Root System of Locus with Velocity Sensor

Figure 3.14 - Root Locus of System with Acceleration Sensor

3.3.2.3 Steady State Error Requirement

The steady state error [3.7] is involved in another criterion which
should also be applied when using classical methods. The error
between the input $\overline{R}(s)$, and the output of the feedback loop,
$\overline{H}(s)\overline{Y}(s)$, is obtained from the block diagram of Figure 3.15 as

$$\overline{E}(s) = \frac{\overline{R}(s)}{1+\overline{G}(s)\overline{H}(s)} \quad . \tag{3.36}$$

The steady state error, e_{ss}, is defined (see Appendix A) by

$$e_{ss} = \lim_{t \to \infty} L^{-1}[\overline{E}(s)] = \lim_{s \to \infty} [s\overline{E}(s)] \quad . \tag{3.37}$$

For the present case, in which $\overline{R}(s) = \overline{F}(s)$, $\overline{R}(s)$ is given by

$$\overline{R}(s) = \frac{2P}{mL} \frac{\pi v/L}{s^2 + (\pi v/L)^2} \; .$$

(3.38)

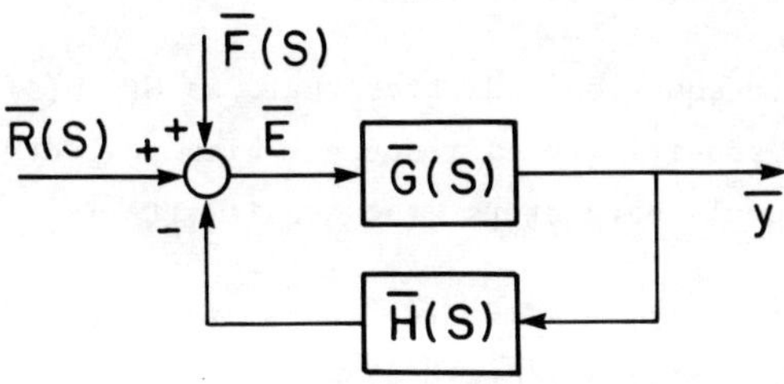

Figure 3.15 - Feedback Control System

One finds that the steady state error is zero. However, one has to check the effect of standard inputs such as the unit step input function, the unit ramp input function, etc., [3.8,3.9]. Studying the effect of the unit step input function, one finds that the velocity and acceleration sensor systems will yield a better performance than the deflection sensor system, since the steady state errors are, respectively, given by

$$e_{ss_d} = \frac{1}{1 + \frac{B_1}{\omega_1^2}(c_1 + \alpha_d)} = 2.3 \; ,$$

(3.39)

$$e_{ss_v} = \frac{1}{1 + \frac{B_1}{\omega_1^2} c_1} = 0.627 \; ,$$

(3.40)

$$e_{ss_a} = \frac{1}{1 + \frac{B_1}{\omega_1^2} c_1} = 0.627 \; .$$

3.3.2.4 Sensitivity Requirement

Sensitivity is defined as the effect of small changes in the system's parameters on the system's performance. These changes could happen due to aging, change of the environment, or any other

natural influence. The sensitivity of the system due to plant's
parameters variations is given [3.8] as

$$S_G^P = \frac{\partial \overline{P}(s)}{\partial \overline{G}(s)} \frac{\overline{G}(s)}{\overline{P}(s)} = \frac{1}{1+\overline{G}(s)\overline{H}(s)} \; . \tag{3.42}$$

The above equation indicates that, as $\overline{G}(s)\overline{H}(s)$ increases
in magnitude, the sensitivity is reduced. With respect to changes
in the feedback loop's parameters, the sensitivity is

$$S_H^P = \frac{\partial \overline{P}(s)}{\partial \overline{H}(s)} \frac{\overline{H}(s)}{\overline{P}(s)} = \frac{-\overline{G}(s)\overline{H}(s)}{1+\overline{G}(s)\overline{H}(s)} \; . \tag{3.43}$$

Equation (3.43) indicates that when $\overline{G}(s)\overline{H}(s)$ is large,
the sensitivity approaches unity, and any changes in $\overline{H}(s)$ will
directly affect the system's response. Therefore, the system is
sensitive to any changes in the feedback parameters, and it is
recommended that the feedback components be maintained continuously.

3.3.2.5 Controlled System's Response

The controlled response of the three systems in the time domain
can be obtained by taking the inverse Laplace transform of equations
(3.24-3.26), i.e.,

$$A_1(t) = L^{-1}[\overline{P}(s)\overline{F}(s)] \; , \tag{3.44}$$

in which $\overline{F}(s)$ is the applied disturbance given by equation (3.38).
The time domain response of the deflection at mid-span
of the three systems is plotted in Figure 3.16. Investigating
the response of the three systems, one arrives at the conclusion
that the vibration decays very fast as a result of introducing
active damping to the structure. The transient response of the
velocity and acceleration sensor systems is much smaller than the
uncontrolled response, and a reduction of more than 50 percent
can be observed. On the contrary, the transient response of the

deflection sensor system is larger than that of the uncontrolled
response. Therefore, it is not recommended to use a deflection
sensor in this case.

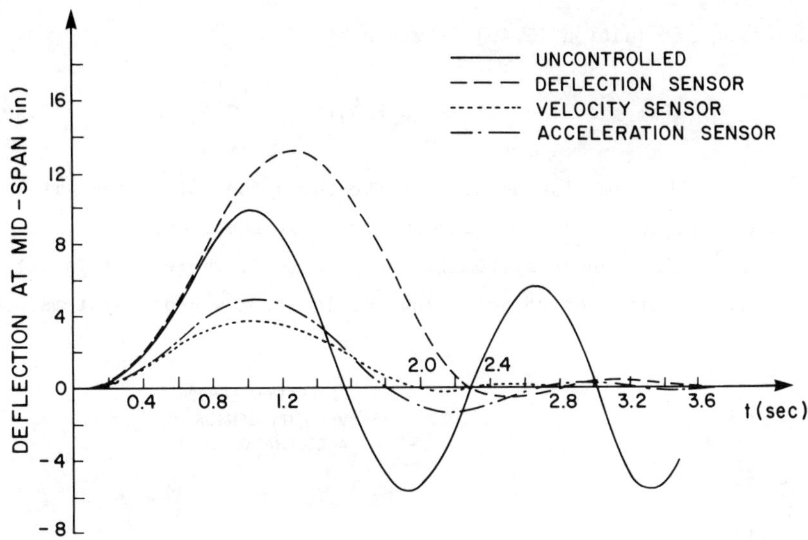

Figure 3.16 - Deflection at Mid-Span for Three Systems

The improved response resulting from the use of velocity
and acceleration sensors instead of a deflection sensor (whereas
the three systems have the same damping ratio) is credited to the
increase in stiffness of the system resulting from introducing
active stiffness. This conclusion is drawn from an investigation
of the new poles' locations comparing them with their original
locations in Figures 3.12 to 3.14. The above results indicate
that active damping is not the only important factor in the control
of the structure's response, but active stiffness and active mass
effects are such factors as well.

The ram displacement of the servomechanism, $u(t)$, is
obtained by solving for $u(t)$ from equations (3.10) to (3.12). For

example, the velocity sensor control system's equation is obtained
from equation (3.11) as

$$\dot{u}(t) + K_c u(t) = \alpha_v K_c \dot{A}_1(t) \; . \qquad (3.45)$$

Solution of equation (3.45) is given by

$$\dot{u}(t) = e^{-K_c t} u(t_0) + \int_o^t \alpha_v K_c \dot{A}_1(t-\tau) e^{-K_c(t-\tau)} \, d\tau \; . \qquad (3.46)$$

The ram displacement of the three control systems is
shown in Figure 3.17. One can observe that the ram response of
the deflection sensor system is not practical, whereas it is within
practical limits for velocity and acceleration sensors systems.

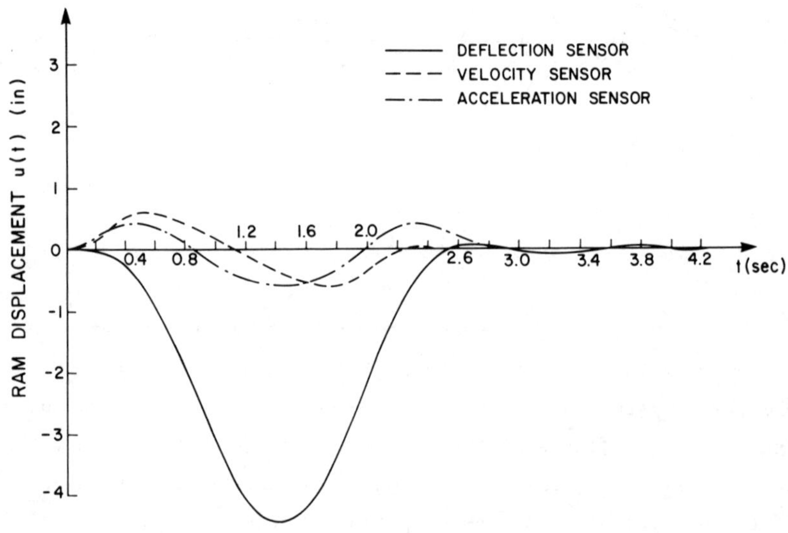

Figure 3.17 - Ram's Displacement Response

Finally, the control moment is evaluated from equation
(3.3). Part of this moment is generated by the active control,
and the other part is produced by the passive deformation. The
total moment is plotted in Figure 3.18, from which one can conclude
that it is not economical to use a deflection sensor.

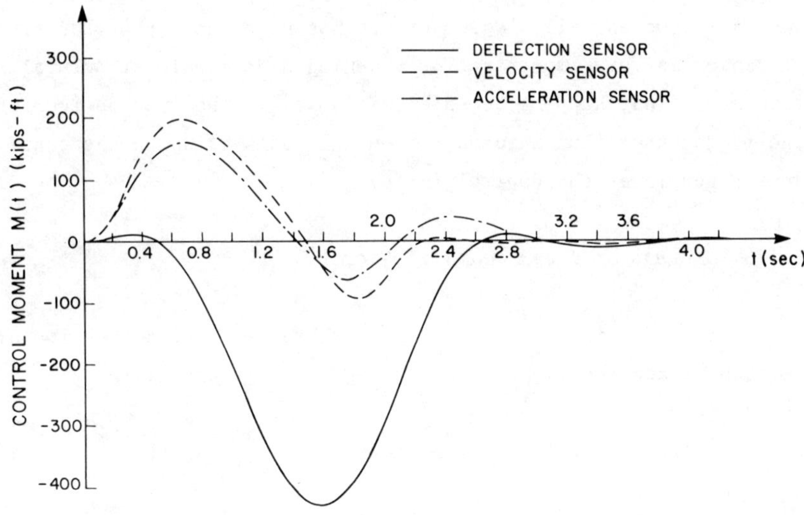

Figure 3.18 - Control Moment Response

3.3.2.6 Summary and Conclusion

The numerical examples presented in Section 3.3.2 have shown the
effectiveness of the feedback control in suppressing the vibration
of the simple span bridge. Although only one mode of vibration has
been considered in the analysis, the approach presented so far has
given a first insight into the treatment of the control problem.
The controller design was carried out by one of the classical
control theory design methods (root-locus method). This method
allows information about how much damping and stiffness is being
introduced to the structure by the control.

        The results obtained are interesting from the practical
point of view. They have shown that, in general, by means of
active feedback control, damping and stiffness of the structure
can significantly be increased. It has also been shown that
active damping should not be considered as the only important factor

for improving the structure response but that active stiffness is such a factor as well. Moreover, it has been shown that the type of sensor is very important for reaching a desirable structural response. That happens because the sensor is the main device for the feedback of that information on the basis of which the control device generates the control force.

3.3.3 *Control of Three Modes of Vibration*

Considering the first three modes of vibration, the deflection of section x, and the slope at $x = a$ and $x = L-a$ are given by

$$y(x,t) = \sin\frac{\pi x}{L} A_1(t) + \sin\frac{2\pi x}{L} A_2(t) + \sin\frac{3\pi x}{L} A_3(t) , \qquad (3.47)$$

$$y'(a,t) = \frac{\pi}{L} \cos\frac{\pi a}{L} A_1(t) + \frac{2\pi}{L} \cos\frac{2\pi a}{L} A_2(t) + \frac{3\pi}{L} \cos\frac{3\pi a}{L} A_3(t) , \qquad (3.48)$$

$$y'(L-a,t) = \frac{-\pi}{L} \cos\frac{\pi a}{L} A_1(t) + \frac{2\pi}{L} \cos\frac{2\pi a}{L} A_2(t) - \frac{3\pi}{L} \cos\frac{3\pi a}{L} A_3(t) , \qquad (3.49)$$

respectively.

Substituting equations (3.48) and (3.49) into equations (3.6) and (3.8), one arrives at

$$\ddot{A}_1(t)+\omega_1^2 A_1(t) = \frac{2P}{mL} \sin\frac{\pi v}{L} t - B_1[u(t)+c_1 A_1(t)+c_3 A_3(t)] , \qquad (3.50)$$

$$\ddot{A}_3(t)+\omega_3^2 A_3(t) = \frac{2P}{mL} \sin\frac{3\pi v}{L} t - B_3[u(t)+c_1 A_1(t)+c_3 A_3(t)] . \qquad (3.51)$$

The second mode equation is given by equation (3.7), and it will not be written here since it is not controlled. In order to substitute for u(t) in equations (3.50) or (3.51), equation (3.50), for example, is multiplied by K_c, and the result is added to the differentiated equation (3.50). This differentiation is with respect to time. For the deflection sensor system one has

$$\ddot{\ddot{A}}_1(t)+K_c\ddot{A}_1(t)+\omega_1^2\dot{A}_1(t)+K_c\omega_1^2A_1(t) = \frac{2P}{mL}\left[K_c \sin \frac{\pi vt}{L} + \frac{\pi v}{L} \cos \frac{\pi vt}{L}\right]$$

$$- B_1[\alpha_d K_c(A_1(t)-A_3(t))+c_1K_cA_1(t)+c_3K_cA_3(t)+c_1\dot{A}_1(t)+c_3\dot{A}_3(t)] \ .$$

$$(3.52)$$

Similarly, equation (3.51) becomes

$$\ddot{\ddot{A}}_3(t)+K_c\ddot{A}_3(t)+\omega_3^2\dot{A}_3(t)+K_c\omega_3^2A_3(t) = \frac{2P}{mL}\left[K_c \sin \frac{3\pi v}{L} t + \frac{3\pi v}{L} \cos \frac{3\pi v}{L} t\right]$$

$$- B_3[\alpha_d K_c(A_1(t)-A_3(t))+c_1K_cA_1(t)+c_3K_cA_3(t)+c_1\dot{A}_1(t)+c_3\dot{A}_3(t)] \ ,$$

$$(3.53)$$

whereas x_c = L/2 has been substituted in equations (3.47) and (3.10).

Introducing the following state variables: $x_1(t) = A_1(t)$, $x_2(t) = \dot{A}_1(t)$, $x_3(t) = \ddot{A}_1(t)$, $x_4(t) = A_3(t)$, $x_5(t) = \dot{A}_3(t)$, and $x_6(t) = \ddot{A}_3(t)$, and using $\Omega_j = j\pi v/L$, one may express equations (3.52) and (3.53) in matrix form as

$$\dot{\underline{X}}(t) = \underline{A}_d\underline{X}(t) + \underline{F} \ , \qquad (3.54)$$

in which $\underline{X}(t)$ is a 6×1 state vector. The matrix $\underline{A}_d$ and the vector $\underline{F}$ are, respectively, given by

$$\underline{A}_d = \begin{bmatrix} 0 & 1 & 0 & 0 & 0 & 0 \\ 0 & 0 & 1 & 0 & 0 & 0 \\ -K_c\omega_1^2-B_1(\alpha_dK_c+K_cc_1) & -\omega_1^2-B_1c_1 & -K_c & -B_1(-\alpha_dK_c+K_cc_3) & -B_1c_3 & 0 \\ 0 & 0 & 0 & 0 & 1 & 0 \\ 0 & 0 & 0 & 0 & 0 & 1 \\ -B_3(\alpha_dK_c+K_cc_1) & -B_3c_1 & 0 & -K_c\omega_3^2-B_3K_c(-\alpha_d+c_3) & -\omega_3^2-B_3c_3 & -K_c \end{bmatrix},$$

$$(3.55)$$

$$\underline{F} = \begin{bmatrix} 0 \\ 0 \\ \dfrac{2P}{mL} [K_c \sin \Omega_1 t + \Omega_1 \cos \Omega_1 t] \\ 0 \\ 0 \\ \dfrac{2P}{mL} [K_c \sin \Omega_3 t + \Omega_3 \cos \Omega_3 t] \end{bmatrix} . \tag{3.56}$$

Similarly, the matrix $\underline{A}_v$ of the velocity sensor system and the matrix $\underline{A}_a$ of the acceleration sensor system are, respectively,

$$\underline{A}_v = \begin{bmatrix} 0 & 1 & 0 & 0 & 0 & 0 \\ 0 & 0 & 1 & 0 & 0 & 0 \\ -K_c(\omega_1^2+B_1c_1) & -\omega_1^2-B_1(\alpha_v K_c+c_1) & -K_c & -B_1K_c c_3 & -B_1(-\alpha_v K_c+c_3) & 0 \\ 0 & 0 & 0 & 0 & 1 & 0 \\ 0 & 0 & 0 & 0 & 0 & 1 \\ -B_3K_c c_1 & -B_3(\alpha_v K_c+c_1) & 0 & -K_c(\omega_3^2+B_3c_3) & -\omega_3^2-B_3(-\alpha_v K_c+c_3) & -K_c \end{bmatrix} \tag{3.57}$$

$$\underline{A}_a = \begin{bmatrix} 0 & 1 & 0 & 0 & 0 & 0 \\ 0 & 0 & 1 & 0 & 0 & 0 \\ -K_c(\omega_1^2+B_1c_1) & -\omega_1^2-B_1c_1 & -K_c-B_1\alpha_a K_c & -B_1K_c c_3 & -B_1c_3 & B_1\alpha_a K_c \\ 0 & 0 & 0 & 0 & 1 & 0 \\ 0 & 0 & 0 & 0 & 0 & 1 \\ -B_3K_c c_1 & -B_3c_1 & -B_3\alpha_a K_c & -K_c(\omega_3^2+B_3c_3) & -\omega_3^2-B_3c_3 & -K_c+B_3\alpha_a K_c \end{bmatrix} . \tag{3.58}$$

The solution of equation (3.54) is given [3.12] by

$$\underline{X}(t) = \underline{\Phi}(t-t_0)\underline{X}(t_0) + \int_{t_0}^{t} \underline{\Phi}(t-\tau)\underline{F}(\tau)d\tau \ , \qquad (3.59)$$

in which $\underline{\Phi}(t)$ is the transition matrix. There are a number of
methods [3.12] available to evaluate this transition matrix.
However, a solution for X(t) has been obtained by numerical
integration, using a Runge-Kutta routine.

3.3.3.1 Controlled Systems' Responses

The three systems have been analysed for the same design parameters
determined previously. The result was that each system is unstable
and cannot be used to control the bridge vibration. To check the
stability of the system, one has to check the eigenvalues of the
closed loop matrix. The closed loop matrices for the three systems
dealt with herein are, respectively, $\underline{A}_d$, $\underline{A}_v$, and $\underline{A}_a$. The responses
of the three systems and their eigenvalues are displayed in
Figures 3.19, 3.20 and 3.21, respectively.

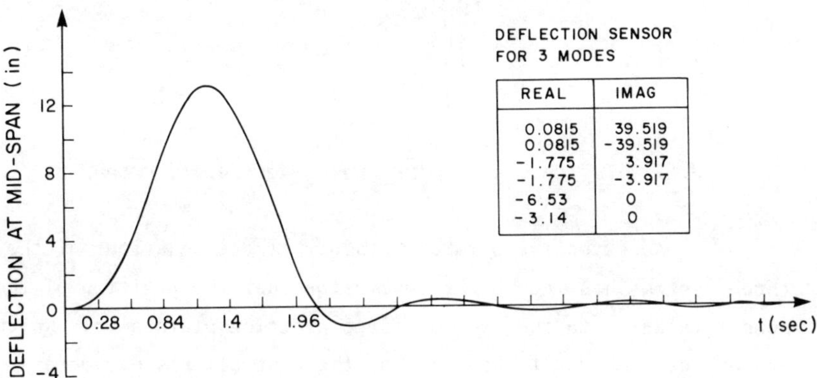

REAL	IMAG
0.0815	39.519
0.0815	-39.519
-1.775	3.917
-1.775	-3.917
-6.53	0
-3.14	0

Figure 3.19 - Deflection Response During Deflection Sensor

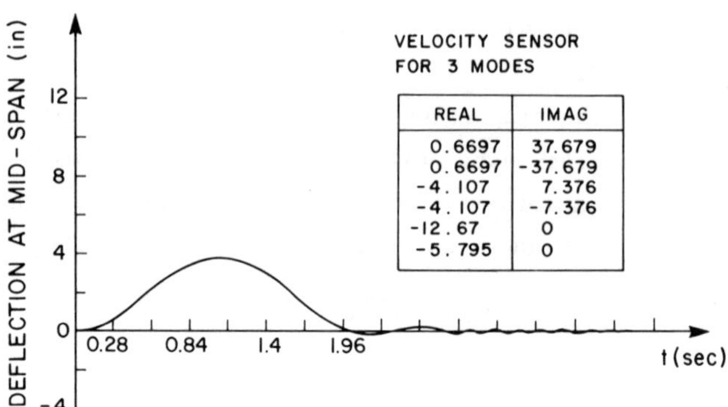

Figure 3.20 - Deflection Response Using Velocity Sensor

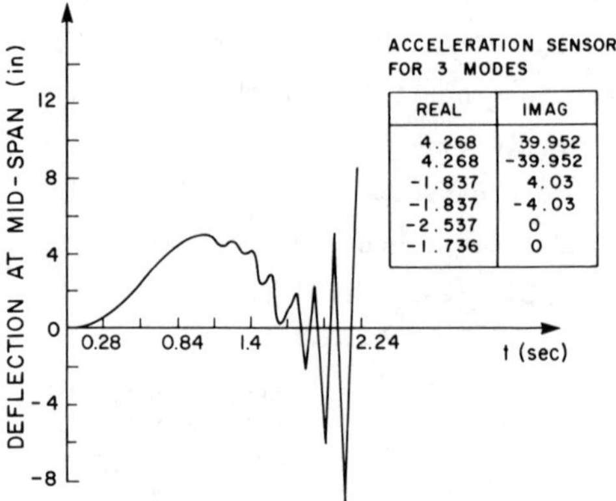

Figure 3.21 - Deflection Response using Acceleration Sensor

An extensive parametric study of the behaviour of the
three systems has led to the conclusion that the position of the
sensor relative to the control force position plays a role which
is not less important than that of the controller's parameters
in establishing a stable system. In fact, by inserting the sensor
at the control force location, and measuring the rate of rotation
at this location, the systems become stable.

3.3.3.2 Design of the Control System using a Servomechanism

A new design was carried out to determine the servomechanism parameters. One recalls that a servomechanism enables one to change
the stiffness and damping in the structure simultaneously. The
sensors are now located at the control moment positions. They
measure the difference in the rate of rotation of these positions.
In this case, the servomechanism's equation is given by

$$\dot{u}(t)+K_c u(t) = K_c \alpha_v [\dot{y}'(a,t)-\dot{y}'(L-a,t)]/\ell$$

or

$$\dot{u}(t)+K_c u(t)=K_c \alpha_v [c_1\dot{A}_1(t)+c_3\dot{A}_3(t)]/\ell \; , \qquad (3.60)$$

in which c_j is given by equation (3.17).

Performing the same calculations as in Section 3.3.3.1,
the closed-loop matrix A_v becomes

$$\underline{A}_{-v} = \begin{bmatrix} 0 & 1 & 0 & 0 & 0 & 0 \\ 0 & 0 & 1 & 0 & 0 & 0 \\ -K_c(\omega_1^2+B_1c_1) & -\omega_1^2-B_1(c_1+c_1\frac{\alpha K_c}{\ell}) & -K_c & -B_1K_c c_3 & -B_1(c_3+c_3\frac{\alpha K_c}{\ell}) & 0 \\ 0 & 0 & 0 & 0 & 1 & 0 \\ 0 & 0 & 0 & 0 & 0 & 1 \\ -B_3K_c c_2 & -B_3(c_1+c_1\frac{\alpha K_c}{\ell}) & 0 & -K_c\omega_3^2-B_3K_c c_3 & -\omega_3^2-B_3(c_3+c_3\frac{\alpha K_c}{\ell}) & -K_c \end{bmatrix} .$$

$$(3.61)$$

In order to illustrate high damping to the structure,
the parameters a = 11 ft, K_c = 12.67 and α_v = 22.5 were chosen after
some trials. The response of the controlled bridge considering the
three modes as well as the bridge's response when considering only
one mode are plotted in Figure 3.22. The passive control response
(no active control) and the eigenvalues of the closed-loop matrix
$\underline{A}_{-v}$ are also displayed in Figure 3.22. One can observe that not much
difference exists between the controlled behaviour of the structure
considering one mode or three modes of vibration.

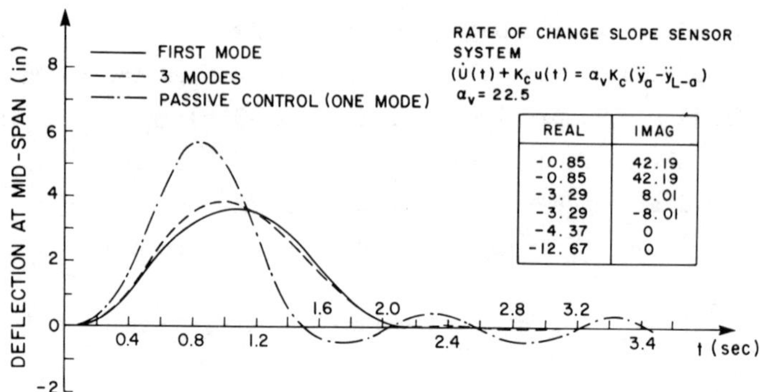

Figure 3.22 - Deflection Response using Servomechanism

It has been noticed that increasing the magnitude of the variable α_V increases the active stiffness of the structure. For example, considering α_V = 80 improves the stiffness of the structure but it reduces, to a certain extent, the active damping. The response of the structure in this case, as well as the eigenvalues of the closed-loop matrix $\underline{A}_V$, are displayed in Figure 3.23.

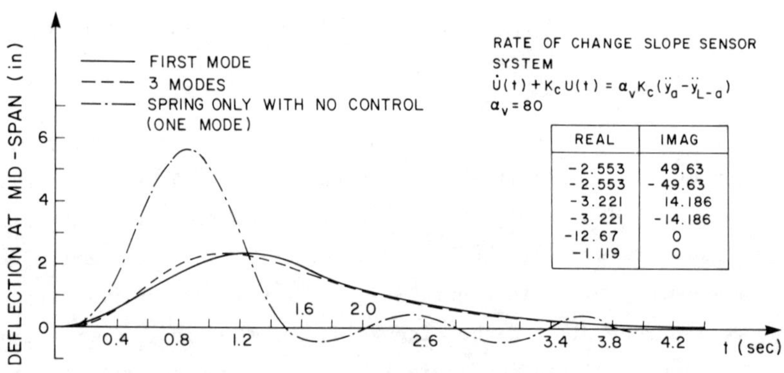

Figure 3.23 - Deflection Response using Servomechanism

3.3.3.3 Design of the Control System using a Proportional Controller

This controller, as outlined in Section 1.6.2, amplifies the measured response. The active control in this case, using the same type of measurement as in the previous section, is given by

$$u(t) = K_c[\dot{y}'(a,t)-\dot{y}'(L-a,t)]/\ell \ . \tag{3.62}$$

It is noticed that the active control provided by this controller introduces only damping to the structure if the structure is a one-degree-of-freedom system. Considering now seven modes of vibration, equation (3.62) becomes

$$u(t) = \frac{K_c}{\ell}\ [c_1\dot{A}_1(t)+c_3\dot{A}_3(t)+c_5\dot{A}_5(t)+c_7\dot{A}_7(t)] \ . \tag{3.63}$$

Substituting into equations (3.6) to (3.8) yields

$$\ddot{A}_1(t)+\omega_1^2 A_1(t) \ = \ \frac{2P}{mL}\ \sin\Omega_1 t \ - \ \frac{B_1}{\ell}\ [K_c c_1\dot{A}_1(t)+K_c c_3\dot{A}_3(t)+K_c c_5\dot{A}_5(t)$$

$$+ \ K_c c_7\dot{A}_7(t)+\ell c_1 A_1(t)+\ell c_3 A_3(t)+\ell c_5 A_5(t)+\ell c_7 A_7(t)] \ , \tag{3.64}$$

$$\ddot{A}_3(t)+\omega_3^2 A_3(t) \ = \ \frac{2P}{mL}\ \sin\Omega_3 t \ - \ \frac{B_3}{\ell}\ [K_c c_1\dot{A}_1(t)+K_c c_3\dot{A}_3(t)+K_c c_5\dot{A}_5(t)$$

$$+ \ K_c c_7\dot{A}_7(t)+\ell c_1 A_1(t)+\ell c_3 A_3(t)+\ell c_5 A_5(t)+\ell c_7 A_7(t)] \ , \tag{3.65}$$

$$\ddot{A}_5(t)+\omega_5^2 A_5(t) \ = \ \frac{2P}{mL}\ \sin\Omega_5 t \ - \ \frac{B_5}{\ell}\ [K_c c_1\dot{A}_1(t)+K_c c_3\dot{A}_3(t)+K_c c_5\dot{A}_5(t)$$

$$+ \ K_c c_7\dot{A}_7(t)+\ell c_1 A_1(t)+\ell c_3 A_3(t)+\ell c_5 A_5(t)+\ell c_7 A_7(t)] \ , \tag{3.66}$$

$$\ddot{A}_7(t)+\omega_7^2 A_7(t) \ = \ \frac{2P}{mL}\ \sin\Omega_7 t \ - \ \frac{B_7}{\ell}\ [K_c c_1\dot{A}_1(t)+K_c c_3\dot{A}_3(t)+K_c c_5\dot{A}_5(t)$$

$$+ \ K_c c_7\dot{A}_7(t)+\ell c_1 A_1(t)+\ell c_3 A_3(t)+\ell c_5 A_5(t)+\ell c_7 A_7(t)] \ . \tag{3.67}$$

Using the following state variables: $x_1(t) = A_1(t)$, $x_2(t) = \dot{A}_1(t)$, $x_3(t) = A_3(t)$, $x_4(t) = \dot{A}_3(t)$, $x_5(t) = A_5(t)$, $x_6(t) = \dot{A}_5(t)$, $x_7(t) = A_7(t)$, and $x_8(t) = \dot{A}_7(t)$, one may express equations (3.64) to (3.67) in matrix form as

$$\dot{\underline{X}}(t) = \underline{A}_v \underline{X}(t) + \underline{R}(t) , \tag{3.68}$$

in which $\underline{X}(t)$ is the state vector of dimension 8×1. The matrix $\underline{A}_v$ and vector $\underline{R}(t)$ are, respectively given by

$$\underline{A}_v = \begin{bmatrix}
0 & 1 & 0 & 0 & 0 & 0 & 0 & 0 \\
-\omega_1^2-B_1c_1 & -B_1\dfrac{K_c c_1}{\ell} & -B_1c_3 & -B_1\dfrac{K_c c_3}{\ell} & -B_1c_5 & -B_1\dfrac{K_c c_5}{\ell} & -B_1c_7 & -B_1\dfrac{K_c c_7}{\ell} \\
0 & 0 & 0 & 0 & 0 & 0 & 0 & 0 \\
-B_3c_1 & -B_3\dfrac{K_c c_1}{\ell} & -\omega_3^2-B_3c_3 & -B_3\dfrac{K_c c_3}{\ell} & -B_3c_5 & -B_3\dfrac{K_c c_5}{\ell} & -B_3c_7 & -B_3\dfrac{K_c c_7}{\ell} \\
0 & 0 & 0 & 0 & 0 & 1 & 0 & 0 \\
-B_5c_1 & -B_5\dfrac{K_c c_1}{\ell} & -B_5c_3 & -B_5\dfrac{K_c c_3}{\ell} & -\omega_5^2-B_5c_5 & -B_5\dfrac{K_c c_5}{\ell} & -B_5c_7 & -B_5\dfrac{K_c c_7}{\ell} \\
0 & 0 & 0 & 0 & 0 & 0 & 0 & 1 \\
-B_7c_1 & -B_7\dfrac{K_c c_1}{\ell} & -B_7c_3 & -B_7\dfrac{K_c c_3}{\ell} & -B_7c_5 & -B_7\dfrac{K_c c_5}{\ell} & -\omega_7^2-B_7c_7 & -B_7\dfrac{K_c c_7}{\ell}
\end{bmatrix} ,$$

$$R(t) = \frac{2P}{mL}\begin{bmatrix} 0 & \sin\Omega_1 t & 0 & \sin\Omega_3 t & 0 & \sin\Omega_5 t & 0 & \sin\Omega_7 t \end{bmatrix}^T .$$

Solution of Equation (3.68) is obtained by numerical integration, or by using equation (3.59). The deflection at mid-span is obtained from

$$y(L/2,t) = A_1(t)-A_3(t)+A_5(t)-A_7(t) . \tag{3.69}$$

The design was chosen as to provide high damping to the structure. The parameter K_c is found, after trials, to be $K_c = 30$. The deflection response as well as the eigenvalues of the closed loop matrix $\underline{A}_v$ are displayed in Figure 3.24 where also the effect of damping on the structure's response is compared with the passive controlled response. Investigating the eigenvalues of the

system, one arrives at the conclusion that the interaction between
the modes of vibration is affecting the stiffness of each mode,
even in the presence of a proportional controller. That again
leads to the conclusion that active damping is not the only factor
which improves the structure's response, but that active stiffness
is also such a factor. One should note that the interaction of
vibrational modes increases the stiffness of some modes and decreases the stiffness of others. Therefore, one has to include
as many modes as possible for a representation of the structure,
in order to be sure of the stability of the system.

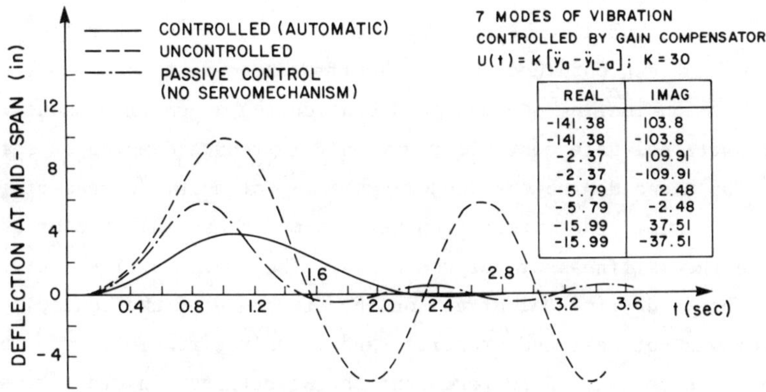

$$U(t) = K\left[\ddot{y}_a - \ddot{y}_{L-a}\right]; \quad K = 30$$

REAL	IMAG
-141.38	103.8
-141.38	-103.8
-2.37	109.91
-2.37	-109.91
-5.79	2.48
-5.79	-2.48
-15.99	37.51
-15.99	-37.51

Figure 3.24 - Deflection Response using Proportional Controller

3.3.3.4 Summary and Conclusions

It has been shown that when designing a control system, one should
include as many vibrational modes as possible. It may turn out
that the design of the control system for a system approximated as
a one-degree-of-freedom system is unsafe. Such failure of the
control system can be attributed to the unavoidable interaction of
the various modes.

Factors affecting the stability of the controlled system
are the controller's parameters, the type of measurement, and the
relative position of the sensor with respect to the control force
position.

A conclusion from the previous examples is that locating
the sensor at the control force position and a proper measurement
matching the type of control force, as well as a proper design of
the controller's parameters, ensures the stability of the controlled
system. This conclusion is logical because the feedback signal
generated by the controller intends only to control the sensor's
measurement, since the controller does not have any information
about the response at other locations.

It has also been shown that once a stable controlled
system has been obtained, not much difference exists between the
response considering one mode and considering a number of modes.
A proportional gain controller may only introduce damping or stiffness for a one-degree-of-freedom system. For multi-degrees-of-
freedom system, the interaction between modes may increase or decrease the stiffness of various modes. Therefore, active control
may cause a detrimental effect on the structure if the control
system was not designed properly, and if only a few modes of vibrations were considered to represent the structure. In order to show
how to design the control system properly, modern control design
techniques are illustrated and used in the next sections.

Modern design methods consist of optimal control methods
[3.13,3.14], the modal control method [3.15], and the pole assignment
method [3.12]. An application of optimal control and pole assignment methods to the bridge treated in the previous section is
shown in the following, using numerical examples.

3.4   ACTIVE STRUCTURAL CONTROL BY POLE ASSIGNMENT METHOD

It has been demonstrated in the previous section that the classical
control design methods consist mainly in forcing the closed-loop

poles to be suitably located in the s-plane. These locations are
obtained by specifying certain items like damping ratio, stability,
sensitivity, etc. In designing single-input, single-output systems,
the classical control design methods provide a fast and appropriate
design. However, in designing multivariable systems, a trial and
error process has been used in order to obtain an acceptable controlled response.

The pole assignment method provides the design of the
feedback components of a multivariable, linear, time-invariant
system in a systematic way such that the designed closed-loop
poles' locations, and hence a satisfactory response, is obtained.
In order to ease the understanding of the examples, a brief summary [3.12] of the method and the related concepts is given now
rather than in an appendix. In this summary, some unnecessary
details which are not appropriate for structural systems have been
modified or omitted. The work presented in this section is dependent on results obtained from the previous section. One of these
results is the fact that a sensor should be located at the control
force's position in order to warrant a stable system. Another
result is that the rate of rotation at the control moment's position should be measured in order to introduce damping to the
structure, as well as to match the type of the control forces
which are to ensure the stability of the structure.

3.4.1 *The Pole Assignment Method*

The pole assignment method is mainly depending on the complete
controllability and/or observability of the open-loop system.
An open-loop system is defined, in general, as

$$\dot{\underline{X}} = \underline{A}\,\underline{X} + \underline{B}\,\underline{U}\ ,\qquad\qquad (3.70)$$

$$\underline{Y} = \underline{C}\,\underline{X}\ ,\qquad\qquad (3.71)$$

in which $\underline{X}$ is the state vector of dimension $n{\times}1$; $\underline{U}$ is the input
vector of dimension $r{\times}1$; $\underline{Y}$ is the output vector of dimension $m{\times}1$;
and $\underline{A}$, $\underline{B}$ and $\underline{C}$ are time-invariant matrices of appropriate dimensions.

3.4.1.1 Controllability

A linear system is said to be controlled at t_0 if it is possible
to find some control signals which will transfer the initial state
$\underline{X}(t_0)$ to the origin at some finite time $t_1 > t_0$. If this is true
for all initial times t_0 and all initial states $\underline{X}(t_0)$, the system
is said to be completely controllable. If the system is not completely controllable, then for some initial states no control signals exist which can drive the system to the zero state.

    The easiest and more general approach to check the con-
trollability of a time-invariant linear system is to check the
matrix

$$\underline{P} = [\underline{B} \mid \underline{A}\,\underline{B} \mid \underline{A}^2\underline{B} \mid ----- \mid \underline{A}^{n-1}\underline{B}] \; . \tag{3.72}$$

If it has rank n, then the system is completely controllable.

3.4.1.2 Observability

A linear system is said to be observable at t_0 if $\underline{X}(t_0)$ can be
determined from the output function $\underline{Y}(t_0,t_1)$, where $t_1 > t_0$. If
this is true for all t_0 and all $\underline{X}(t_0)$, then the system is said to
be completely observable. If a system is not completely observable,
then the initial states $\underline{X}(t_0)$ cannot be determined from the available output, no matter how long the output is being observed.

    A linear time-invariant system is completely observable
if the matrix

$$Q = \left[\overline{\underline{C}}^T \mid \overline{\underline{A}}^T\overline{\underline{C}}^T \mid \overline{\underline{A}}^{2^T}\overline{\underline{C}}^T \mid ----- \mid \overline{\underline{A}}^{n-1^T}\overline{\underline{C}}^T\right], \tag{3.73}$$

has rank n, where $\overline{\underline{C}}$ and $\overline{\underline{A}}$ are, respectively, the conjugate complex
of $\underline{C}$ and $\underline{A}$.

3.4.1.3 Design Procedure

If the linear time-invariant system is completely controllable and
observable, the eigenvalues of the matrix $\underline{A}$ are always the poles
of the system. The eigenvalues may or may not represent the poles
of a system which is not completely controllable or observable,
depending on the pole-zero cancellation in the transfer function.
The pole assignment method has been developed for completely controllable and observable systems, where the specified poles are
set to be the eigenvalues of the closed-loop system. It has been
proved [3.17-3.19] that if and only if the open-loop system ($\underline{A}$ and
$\underline{B}$) is completely controllable, then any set of desired closed-loop
poles (eigenvalues) can be implemented by using a constant state
feedback gain matrix.

At this state, one has to differentiate between state
feedback and output feedback. Generally, a state feedback is an
ideal system, where all the state variables are measured and used
in the feedback signals. An output feedback is the usual practical
system, where the feedback signal is generated from the output
which is a linear combination of the state variables, i.e.,
$\underline{Y} = \underline{C}\,\underline{X}$. However, it is advantageous to study the design of state
feedback since an estimator (observer) [3.20] could be designed to
estimate the state variables from the available output, and hence
a state feedback can be used. Simulation diagrams for state
feedback and output feedback are shown in Figures 3.25 and 3.26
respectively.

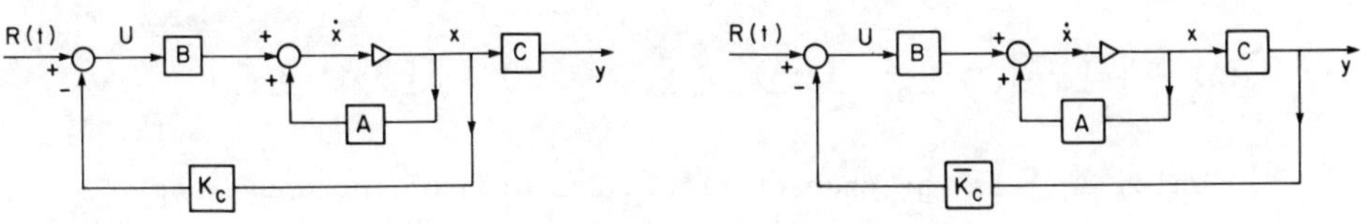

Figure 3.25 - State Feedback   *Figure 3.26 - Output Feedback*

3.4.1.4   Design of State Feedback

From Figure 3.25, the control vector $\underline{U}$ is obtained as

$$\underline{U} = \underline{R} - \underline{K}_c \underline{X} \, , \tag{3.74}$$

in which $\underline{R}$ is the input vector of dimension r×1; $\underline{K}_c$ is the state feedback gain matrix of dimension r×n; and $\underline{U} = \underline{R}$ in the case of zero feedback.

Substituting equation (3.74) into equation (3.70), the closed-loop system is obtained as

$$\underline{\dot{X}} = (\underline{A} - \underline{B}\,\underline{K}_c)\underline{X} + \underline{B}\,\underline{R} \, . \tag{3.75}$$

The eigenvalues of the closed-loop system are determined from

$$\overline{\Delta}(\lambda) = \left| \lambda\underline{I}_n - \underline{A} + \underline{B}\,\underline{K}_c \right| = 0 \, , \tag{3.76}$$

in which λ are the eigenvalues; $\underline{I}_n$ is the identity matrix of dimension n×n; and $\overline{\Delta}(\lambda)$ is the characteristic equation of the closed-loop system.

Solution of equation (3.76) gives the eigenvalues λ_i, i = 1,2,3,...,n, of the closed-loop system. For a completely controllable system, these eigenvalues represent the poles of the system. In the following, it is shown how to determine $\underline{K}_c$ such that the poles can be obtained in a specified way.

Equation (3.76) can be written as

$$\overline{\Delta}(\lambda) = \left| \lambda\underline{I}_n - \underline{A} \right| \cdot \left| \underline{I}_n + (\lambda\underline{I}_n - \underline{A})^{-1}\underline{B}\,\underline{K}_c \right| = \Delta(\lambda) \cdot \left| \underline{I}_n + \underline{\Phi}(\lambda)\underline{B}\,\underline{K}_c \right| = 0 \, , \tag{3.77}$$

in which $\Delta(\lambda)$ is the characteristic equation of the open-loop system and $\underline{\Phi}(\lambda) = (\lambda\underline{I}_n - \underline{A})^{-1}$ is the transition matrix in the s-plane, except that λ replaces the Laplace variable s.

For a specified pole λ_i, equation (3.77) becomes zero if a row or a column of $[\underline{I}_n + \underline{\Phi}(\lambda_i)\underline{B}\ \underline{K}_c]$ is set to be zero. This is the way to determine $\underline{K}_c$. For each specified pole, a column or a row from $[\underline{I}_n + \underline{\Phi}(\lambda)\underline{B}\ \underline{K}_c]$ is made equal to zero. Thus, a system of equations is obtained. The complete controllability of the closed-loop system ensures that it is always possible to find a set of independent equations to determine $\underline{K}_c$. In order to show the computation scheme, a column is now chosen to be zero.

Equation (3.77) can be written as

$$\overline{\Delta}(\lambda) = \Delta(\lambda) \cdot \left|\underline{I}_n + \underline{\Phi}(\lambda)\underline{B}\ \underline{K}_c\right| = \Delta(\lambda) \cdot \left|\underline{I}_r + \underline{K}_c\underline{\Phi}(\lambda)\underline{B}\right| = 0 , \qquad (3.78)$$

in which $\underline{I}_r$ is the identity matrix of dimension r×r.

Equation (3.78) was obtained using the determinant identity [3.12]. For the ith pole λ_i, the jth column of equation (3.78) becomes

$$\underline{e}_j + \underline{K}_c\underline{\psi}_j(\lambda_i) = 0 , \qquad (3.79)$$

in which $\underline{e}_j$ is the jth column of $\underline{I}_r$; and $\underline{\psi}_j = [\underline{\Phi}(\lambda)\underline{B}]_j$.

Repeating equation (3.79) for each pole λ_i, $i = 1,2,3,\ldots,n$, and assuming that these poles are distinct, the following relationship is obtained

$$[\underline{e}_{j1}\!\mid\!\underline{e}_{j2}\!\mid\!----\!\mid\ \underline{e}_{jn}] + \underline{K}_c[\underline{\psi}_{j1}(\lambda_1)\!\mid\!\underline{\psi}_{j2}(\lambda_2)\!\mid\!----\!\mid\ \underline{\psi}_{jn}(\lambda_n)] = 0 .$$
$$(3.80)$$

The matrix $\underline{K}_c$ is now obtained as

$$\underline{K}_c = -[\underline{e}_{j1}\!\mid\!\underline{e}_{j2}\!\mid\!----\!\mid\ \underline{e}_{jn}] \cdot [\underline{\psi}_{j1}(\lambda_1)\!\mid\!\underline{\psi}_{j2}(\lambda_2)\!\mid\!----\!\mid\ \underline{\psi}_{jn}(\lambda_n)]^{-1} .$$
$$(3.81)$$

It is worthwhile to note that the matrix $\underline{K}_c$ is not unique, because it depends on the choice of the columns in equation (3.78). However, the matrix ensures the existence of the required poles. Note also that the state feedback gain matrix $\underline{K}_c$ does not alter the controllability of the open-loop system [3.12], if the system is as described in Figure 3.25.

3.4.1.5 Design of Output Feedback

The control vector $\underline{U}$ is obtained from Figure 3.26 as

$$\underline{U} = \underline{R} - \underline{K}'_c \, \underline{C} \, \underline{X} \, , \qquad (3.82)$$

in which $\underline{K}'_c$ is the output feedback gain matrix, and $\underline{C}$ is a constant matrix of dimension m×n.

Substituting equation (3.82) into equation (3.70) yields

$$\dot{\underline{X}} = (\underline{A} - \underline{B}\,\underline{K}'_c - \underline{C})\underline{X} + \underline{B}\,\underline{R} \, . \qquad (3.83)$$

The eigenvalues of the closed-loop system are obtained from

$$\overline{\Delta}(\lambda) = \left| \lambda \underline{I}_n - \underline{A} + \underline{B}\,\underline{K}'_c\,\underline{C} \right| = 0 \, . \qquad (3.84)$$

Using a similar approach as in Section 3.4.1.4, equation (3.84) can be written as

$$\overline{\Delta}(\lambda) = \Delta(\lambda) \cdot \left| \underline{I}_n + \underline{\Phi}(\lambda)\underline{B}\,\underline{K}'_c\,\underline{C} \right| = \Delta(\lambda) \cdot \left| \underline{I}_m + \underline{C}\,\underline{\Phi}(\lambda)\underline{B}\,\underline{K}'_c \right|$$

$$= \Delta(\lambda) \cdot \left| \underline{I}_r + \underline{K}'_c\,\underline{C}\,\underline{\Phi}(\lambda)\underline{B} \right| = 0 \, . \qquad (3.85)$$

The number of linearly independent columns which can be obtained from the matrices $[\underline{C}\,\underline{\Phi}(\lambda_i)\underline{B}]$, $i = 1,2,3,\ldots,n$, will never exceed the rank of $\underline{C}$ [3.22]. In fact, it has been shown in [3.21] that if the open-loop system $(\underline{A},\underline{B})$ is completely controllable, and if $\underline{C}$ has rank m, where $m \leq n$, then only m out of the n eigenvalues of the closed-loop system can be arbitrarily specified such that the nonsingular m×m matrix $[\underline{C}\,\underline{\Phi}(\lambda_i)\underline{B}]$, $i = 1,2,3,\ldots,m$, can be formed. For the pole λ_i, the jth column of $[\underline{I}_r + \underline{K}'_c\,\underline{C}\,\underline{\Phi}(\lambda)\underline{B}]$ is

$$\underline{e}_j + \underline{K}'_c\,\underline{\psi}'_{ji}(\lambda_i) = 0 \, , \qquad (3.86)$$

in which $\underline{\psi}'_{ji} = [\underline{C}\,\underline{\Phi}(\lambda_i)\underline{B}]_j \, .$

Repeating equation (3.87) for each of the m poles, one finally has

$$[\underline{e}_{j1} \mid \underline{e}_{j2} \mid \text{------} \mid \underline{e}_{jm}] + \underline{K}'_c [\psi'_{j1}(\lambda_1) \mid \psi'_{j2}(\lambda_2) \mid \text{-----} \mid \psi'_{jm}(\lambda_m)] = 0 \ .$$

(3.87)

The output feedback gain matrix $\underline{K}'_c$ is now obtained from

$$\underline{K}'_c = -[\underline{e}_{j1} \mid \underline{e}_{j2} \mid \text{------} \mid \underline{e}_{jm}] \cdot [\underline{\psi}'_{j1}(\lambda_1) \mid \underline{\psi}'_{j2}(\lambda_2) \mid \text{-----} \mid \underline{\psi}'_{jm}(\lambda_m)]^{-1} \ .$$

(3.88)

The matrix $\underline{K}'_c$ ensures the existence of m poles out of the n poles. Therefore, there is a possibility that the remaining (n-m) poles become located in the right half of s-plane, making the system unstable. That depends entirely on the choice of the columns from $\underline{\psi}'_{ji}(\lambda_i)$ and the specified m poles. However, for a completely observable and controllable system, it is always possible to transform the output feedback into a state feedback by using another dynamic system called the observer [3.20]. It is worthwhile to note that the output feedback gain matric $\underline{K}'_c$ does not alter the controllability or observability of the open-loop system [3.12], if the system is described as shown in Figure 3.26.

3.4.1.6 Design of the Observer

The observer is nothing more than an estimator to estimate the state variables from the available output. Hence, a state feedback gain matrix can be designed as described in Section 3.4.1.4. The input to the observer depends on $\underline{Y}$ and $\underline{U}$, and its output should be a good approximation to the states $\underline{X}(t)$. The dynamic equation of an observer is given by

$$\dot{\hat{\underline{X}}} = \underline{A}_c \hat{\underline{X}} + \underline{B}_c \underline{Y} + \underline{Z} \ ,$$

(3.89)

in which $\hat{\underline{X}}$ is an approximation of the state vector $\underline{X}$; $\underline{A}_c$ is a constant matrix of dimension n×n; $\underline{B}_c$ is a constant matrix of dimension n×m; and $\underline{Z}$ is a vector of dimension n×1.

The error, $\underline{e}(t)$, between the exact state $\underline{X}(t)$ and the approximated state $\underline{\hat{X}}(t)$, is given by

$$\underline{e}(t) = \underline{X}(t) - \underline{\hat{X}}(t) \ . \tag{3.90}$$

Differentiating equation (3.90) and substituting from equations (3.70), (3.71) and (3.89), one has

$$\underline{\dot{e}}(t) = \underline{\dot{X}}(t) - \underline{\dot{\hat{X}}}(t) = \underline{A}\,\underline{X} + \underline{B}\,\underline{U} - \underline{A}_c\underline{\hat{X}} - \underline{B}_c\underline{C}\,\underline{X} - \underline{Z}$$

$$= (\underline{A} - \underline{B}_c\underline{C})\underline{X} - \underline{A}_c\underline{\hat{X}} + \underline{B}\,\underline{U} - \underline{Z} \ . \tag{3.91}$$

Letting $\underline{Z} = \underline{B}\,\underline{U}$, and $\underline{A}_c = \underline{A} - \underline{B}_c\underline{C}$, equation (3.91) becomes

$$\underline{e}(t) = \underline{A}_c\underline{e}(t) \ . \tag{3.92}$$

If all the eigenvalues of $\underline{A}_c$ have negative real parts, then equation (3.92) is asymptotically stable, and the error $\underline{e}(t) \rightarrow 0$ as $t \rightarrow \infty$. Therefore, for the design of an observer, the following conditions must be satisfied: $\underline{Z} = \underline{B}\,\underline{U}$; $\underline{A}_c = \underline{A} - \underline{B}_c\underline{C}$; and $\underline{A}_c$ must be asymptotically stable.

If the open-loop system is completely observable, then, it is always possible to find a $\underline{B}_c$ which will yield any set of desired eigenvalues for $\underline{A}_c$. To specify the eigenvalues of $\underline{A}_c$, one has to try to minimize the time delay for $\underline{\hat{X}} = \underline{X}$. Experience indicates that a good design is obtained if the observer's poles are selected to the left of the desired closed-loop state feedback poles as shown for example in Table 3.4 and reference [3.12]. The eigenvalues of the observer are given by

$$\overline{\Delta}(\overline{\lambda}) = |\overline{\lambda}\underline{I}_n - \underline{A}_c| = |\overline{\lambda}\underline{I}_n - \underline{A} + \underline{B}_c\underline{C}| = 0 \ ,$$

$$\overline{\Delta}(\overline{\lambda}) = \Delta(\overline{\lambda}) \cdot |\underline{I}_n + \underline{\Phi}(\overline{\lambda})\underline{B}_c\underline{C}| \ ,$$

$$\overline{\Delta}(\overline{\lambda}) = \Delta(\overline{\lambda}) \cdot |\underline{I}_m + \underline{C}\,\underline{\Phi}(\overline{\lambda})\underline{B}_c| = 0 \ , \tag{3.93}$$

in which $\overline{\lambda}$ is an eigenvalue of the observer.

Table 3.4 - Eigenvalues of Observer and Structure

Eigenvalues of controlled structure (λ)		Eigenvalues of observer ($\bar{\lambda}$)	
Real	Imaginary	Real	Imaginary
-3.5	3.0	-5.0	3.0
-3.5	-3.0	-5.0	-3.0
0.0	17.339	-6.0	17.339
0.0	-17.339	-6.0	-17.339
-4.0	36.0	-8.0	36.0
-4.0	-36.0	-8.0	-36.0

Equation (3.93) holds if any row or column of $[\underline{I}_m + \underline{C}\ \Phi(\lambda)\underline{B}_c]$ is set to zero. The complete observability of the system guarantees that n linearly independent rows can be selected from the rows of $\underline{C}\ \Phi(\bar{\lambda})$. For the ith row and the kth pole one has

$$\underline{e}_i + \underline{\psi}_i(\bar{\lambda}_k)\underline{B}_c = 0\ , \tag{3.94}$$

in which $\underline{\psi}_i(\bar{\lambda}_k) = [\underline{C}\ \Phi(\bar{\lambda}_k)]_i$.

For the eigenvalues $\bar{\lambda}_k$, $k = 1,2,\ldots,n$, equation (3.94) becomes

$$[\underline{e}_{i1} \mid \underline{e}_{i2} \mid ----- \mid \underline{e}_{in}]^T + [\underline{\psi}_{i1}(\bar{\lambda}_1) \mid \underline{\psi}_{i2}(\bar{\lambda}_2) \mid ----- \mid \underline{\psi}_{in}(\bar{\lambda}_n)]^T \underline{B}_c = 0\ . \tag{3.95}$$

The matrix $\underline{B}_c$ is obtained from

$$\underline{B}_c = -[\underline{\psi}_{i1}(\bar{\lambda}_1)^T \mid \underline{\psi}_{i2}(\lambda_2)^T \mid ----- \mid \underline{\psi}_{in}(\bar{\lambda}_n)T]^{-1}[\underline{e}_{i1} \mid \underline{e}_{i2} \mid ----- \mid \underline{e}_{in}]^T\ . \tag{3.96}$$

A simulation diagram for the overall system is shown in Figure 3.27. The overall system becomes of order 2n, and the design

procedure consists in determining $\underline{K}_c$ and $\underline{B}_c$ so that the required
system performance is being ensured. According to the separation
principle [3.12], the other n eigenvalues can be specified so that
matrix $\underline{K}_c$ is determined. Such a procedure is justified since the
eigenvalues of the overall system are the sum of the eigenvalues
of the observer and the eigenvalues of the closed-loop system.

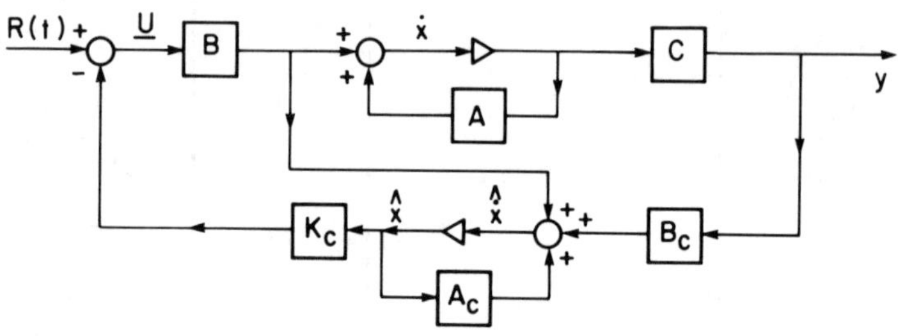

Figure 3.27 - Simulation Diagram of an Observer

3.4.2 *Application of the Pole Assignment Method*

Structural response is considered to be coming from vibrational
modes. Each vibration mode can be described by an equation of the
form $m\ddot{x} + Kx = r(t)$, in which $r(t)$ is the input force for that
mode. Fortunately, it is always possible to find an input force
(control force) which will drive both, the deflection and the
velocity of the mode to zero in a finite amount of time. That
means each one-degree-of-freedom system is completely controllable.
To show this, using the pole assignment concepts, the above equation is written in state form as

$$\dot{\underline{X}} = \underline{A}\,\underline{X} + \underline{B}\,\underline{U} , \tag{3.97}$$

in which

$$\underline{X} = [x \ \dot{x}]^T; \quad \underline{U} = r(t); \quad \underline{B} = \left[0 \ \frac{1}{m}\right]^T; \quad \text{and } A = \begin{bmatrix} 0 & 1 \\ \dfrac{-k}{m} & 0 \end{bmatrix}.$$

The controllability matrix $\underline{P}$ is given by

$$\underline{P} = [\underline{B} \vdots \underline{A} \ \underline{B}] = \begin{bmatrix} 0 & \dfrac{1}{m} \\ \dfrac{1}{m} & 0 \end{bmatrix} \qquad (3.98)$$

which has rank 2. Therefore the system is completely controllable.

It can also be shown that the above system is completely observable. Suppose, for example, the deflection x(t) is being measured. Then the output equation is

$$\underline{Y} = \underline{C} \ \underline{X} = [1 \quad 0]\underline{X} \ . \qquad (3.99)$$

Hence, the observability matrix is

$$\underline{Q} = [\underline{C}^T \vdots \overline{A}^T \underline{C}^T] = \begin{bmatrix} 1 & 0 \\ 0 & 1 \end{bmatrix}, \qquad (3.100)$$

which has rank 2. Therefore the system is completely observable.

However, the above results may or may not hold for multi-degree-of-freedom systems due to the interaction between vibrational modes and if the feedback components are different from those presented in Figures 3.25 and 3.26 in such a way that they affect the controllability of the closed-loop system.

3.4.2.1 Example 1

The open-loop system of the bridge considered in Section 3.3.1 is obtained by assuming a zero control moment M(t). Hence, one has

$$\ddot{A}_1(t) + \omega_1^2 A_1(t) = \frac{2P}{mL} \sin\Omega_1 t \; ,$$

$$\ddot{A}_2(t) + \omega_2^2 A_2(t) = \frac{2P}{mL} \sin\Omega_2 t \; ,$$

$$\ddot{A}_3(t) + \omega_3^2 A_3(t) = \frac{2P}{mL} \sin\Omega_3 t \; , \qquad\qquad (3.101)$$

$$\vdots$$

$$\ddot{A}_n(t) = \omega_n^2 A_n(t) = \frac{2P}{mL} \sin\Omega_n t \; ,$$

in which $\Omega_j = j\pi v/L$, has been used.

Using the state variables $x_1 = A_1$, $x_2 = \dot{A}_1$, $x_3 = A_2$, $x_4 = \dot{A}_2$, $x_5 = A_3$, $x_6 = \dot{A}_3$, etc., equation (3.101) can be expressed in a matrix form like equation (3.97), where for three modes

$$\underline{A} = \begin{bmatrix} 0 & 1 & 0 & 0 & 0 & 0 \\ -\omega_1^2 & 0 & 0 & 0 & 0 & 0 \\ 0 & 0 & 0 & 1 & 0 & 0 \\ 0 & 0 & -\omega_2^2 & 0 & 0 & 0 \\ 0 & 0 & 0 & 0 & 0 & 1 \\ 0 & 0 & 0 & 0 & -\omega_3^2 & 0 \end{bmatrix}, \quad \underline{B} = \begin{bmatrix} 0 & 0 & 0 \\ 1 & 0 & 0 \\ 0 & 0 & 0 \\ 0 & 1 & 0 \\ 0 & 0 & 0 \\ 0 & 0 & 1 \end{bmatrix}, \quad \underline{U} = \begin{bmatrix} \frac{2P}{mL} \sin \Omega_1 t \\ \frac{2P}{mL} \sin \Omega_2 t \\ \frac{2P}{mL} \sin \Omega_3 t \end{bmatrix}.$$

The controllability of the open-loop system is obtained, using equation (3.72) in the form

$$\underline{P} = \begin{bmatrix} 0 & 0 & 0 & 1 & 0 & 0 & 0 & 0 & 0 & -\omega_1^2 & 0 & 0 & 0 & 0 & 0 & -\omega_1^4 & 0 & 0 \\ 1 & 0 & 0 & 0 & 0 & 0 & -\omega_1^2 & 0 & 0 & 0 & 0 & 0 & \omega_1^4 & 0 & 0 & 0 & 0 & 0 \\ 0 & 0 & 0 & 0 & 1 & 0 & 0 & 0 & 0 & 0 & -\omega_2^2 & 0 & 0 & 0 & 0 & 0 & -\omega_2^4 & 0 \\ 0 & 1 & 0 & 0 & 0 & 0 & 0 & -\omega_2^2 & 0 & 0 & 0 & 0 & 0 & \omega_2^4 & 0 & 0 & 0 & 0 \\ 0 & 0 & 0 & 0 & 0 & 1 & 0 & 0 & 0 & 0 & 0 & -\omega_3^2 & 0 & 0 & 0 & 0 & 0 & -\omega_3^4 \\ 0 & 0 & 1 & 0 & 0 & 0 & 0 & 0 & -\omega_3^2 & 0 & 0 & 0 & 0 & 0 & \omega_3^4 & 0 & 0 & 0 \end{bmatrix}$$

Since $\underline{P}$ has rank 6, the open-loop system is completely controllable. Let the measurement be taken at $x = a$ when measuring the rate of rotation at this position. The output in this case is given by

$$\underline{Y} = \dot{y}'(a,t) = \dot{A}_1(t) \frac{\pi}{L} \cos \frac{\pi a}{L} + \dot{A}^2(t) \frac{2\pi}{L} \cos \frac{2\pi a}{L} + \dot{A}^3(t) \frac{3\pi}{L} \cos \frac{3\pi a}{L} ,$$

$$\underline{Y} = \begin{bmatrix} 0 & \frac{c_1'}{\ell} & 0 & \frac{c_2'}{\ell} & 0 & \frac{c_3'}{\ell} \end{bmatrix} \underline{X} = \underline{C}\,\underline{X} , \tag{3.102}$$

in which $c_j' = (\ell j\pi/1)\cos(j\pi a/L)$, and ℓ is the post's length.

From Figure 3.28, the output will actuate the automatic gain controller $\underline{K}_c'$. It is required to find K_c' such that some assumed pole locations are actually determined. Since only one output signal is available, and therefore the rank of $\underline{C}$ is one, one is allowed to specify only one pole. A simulation diagram of the overall system is shown in Figure 3.28.

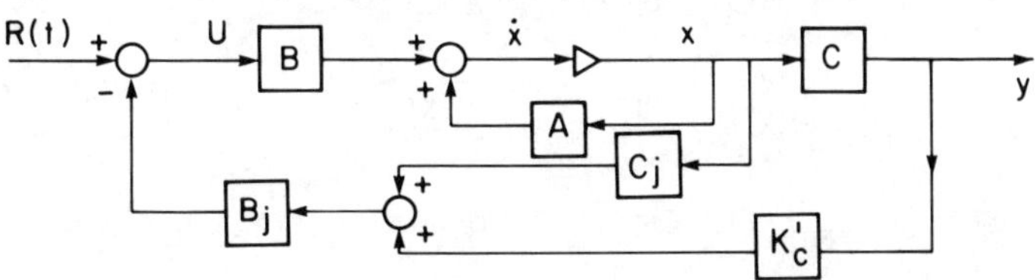

Figure 3.28 - Simulation Diagram of the System

The vector $\underline{U}(t)$ for the closed-loop system is obtained from Figure 3.28 as

$$\underline{U}(t) = \underline{R}(t) - \underline{B}_j(\underline{K}_c'\,\underline{C}\,\underline{X} + \underline{C}_j\underline{X}) , \tag{3.103}$$

in which $\underline{U}(t) = \underline{R}(t)$ for the open-loop system; and $\underline{B}_j$ and $\underline{C}_j$ are, respectively, given by

$$\underline{B}_j = [B_1 \quad 0 \quad B_3]^T \, , \tag{3.104}$$

$$\underline{C}_j = [c_1 \quad 0 \quad 0 \quad 0 \quad c_3 \quad 0] \, . \tag{3.105}$$

Investigating the simulation diagram of Figure 3.28 and comparing it with the output feedback of Figure 3.26, one finds that the controllability of the closed-loop system of Figure 3.28 may be lost as a result of $\underline{B}_j$. That is true because, as shown in equation (3.104), the second mode is always uncontrolled, $B_2 = 0$, and it should be ignored during the design such that it does not impair the calculations. However, since in this example only one measurement is being obtained (this case is similar to that of reference [3.16]), only one independent equation results and the calculations are without any problem.

The eigenvalues of the closed-loop system are obtained as

$$\overline{\Delta}(\lambda) = \left| \lambda\underline{I}_n - [\underline{A}+\underline{B}(-\underline{B}_j\underline{K}'_c\underline{C}-\underline{B}_j\underline{C}_j)] \right| = 0 \, ,$$

$$\overline{\Delta}(\lambda) = \Delta(\lambda) \cdot \left| \underline{I}_n + \underline{\Phi}(\lambda)\underline{B} \ \underline{B}_j\underline{K}'_c\underline{C} + \underline{\Phi}(\lambda)\underline{B} \ \underline{B}_j\underline{C}_j \right| = 0 \, , \tag{3.106}$$

$$\overline{\Delta}(\lambda) = \Delta(\lambda) \cdot \left| \underline{I}_1 + \underline{C} \ \underline{\Phi}(\lambda)\underline{B} \ \underline{B}_j\underline{K}'_c + \underline{C}_j\underline{\Phi}(\lambda)\underline{B} \ \underline{B}_j \right| = 0 \, .$$

The transition matrix $\underline{\Phi}(\lambda)$ is obtained as

$$\underline{\Phi}(t) = e^{\underline{A}t} = \underline{I}_n + \underline{A}t + \underline{A}^2 \frac{t^2}{2!} + \underline{A}^3 \frac{t^3}{3!} + \cdots \, , \tag{3.107}$$

in which $\underline{A}^2$, $\underline{A}^3$ are, respectively, given by

$$\underline{A}^2 = \begin{bmatrix} -\omega_1^2 & 0 & 0 & 0 & 0 & 0 \\ 0 & -\omega_1^2 & 0 & 0 & 0 & 0 \\ 0 & 0 & -\omega_2^2 & 0 & 0 & 0 \\ 0 & 0 & 0 & -\omega_2^2 & 0 & 0 \\ 0 & 0 & 0 & 0 & -\omega_3^2 & 0 \\ 0 & 0 & 0 & 0 & 0 & -\omega_3^2 \end{bmatrix} \quad \text{and} \quad \underline{A}^3 = \begin{bmatrix} 0 & -\omega_1^2 & 0 & 0 & 0 & 0 \\ \omega_1^4 & 0 & 0 & 0 & 0 & 0 \\ 0 & 0 & 0 & -\omega_2^2 & 0 & 0 \\ 0 & 0 & \omega_2^4 & 0 & 0 & 0 \\ 0 & 0 & 0 & 0 & 0 & -\omega_3^2 \\ 0 & 0 & 0 & 0 & \omega_3^4 & 0 \end{bmatrix} .$$

Hence, the transition matrix is given by

$$\underline{\Phi}(t) = \begin{bmatrix} (1-\omega_1^2\frac{t^2}{2!}+.)(t-\omega_1^2\frac{t^3}{3!}+.) & 0 & 0 & 0 & 0 \\ (-\omega_1^2 t+\omega_1^4\frac{t^3}{3!}.)(1-\omega_1^2\frac{t^2}{2!}+.) & 0 & 0 & 0 & 0 \\ 0 & 0 & (1-\omega_2^2\frac{t^2}{2!}+.)(t-\omega_2^2\frac{t^3}{3!}) & 0 & 0 \\ 0 & 0 & (-\omega_2^2 t+\omega_2^4\frac{t^3}{3!}.)(1-\omega_2^2\frac{t^2}{2!}+.) & 0 & 0 \\ 0 & 0 & 0 & 0 & (1-\omega_3^2\frac{t^2}{2!}+.)(t-\omega_3^2\frac{t^3}{3!}+..) \\ 0 & 0 & 0 & 0 & (-\omega_3^2+\omega_5^4\frac{t^3}{3!}.)(1-\omega_3^2\frac{t^2}{2!}+..) \end{bmatrix}, \quad (3.108)$$

This matrix can easily be written in a closed form as

$$\underline{\Phi}(t) = \begin{bmatrix} \cos\omega_1 t & \frac{\sin\omega_1 t}{\omega_1} & 0 & 0 & 0 & 0 \\ -\omega_1\sin\omega_1 t & \cos\omega_1 t & 0 & 0 & 0 & 0 \\ 0 & 0 & \cos\omega_2 t & \frac{\sin\omega_2 t}{\omega_2} & 0 & 0 \\ 0 & 0 & -\omega_2\sin\omega_2 t & \cos\omega_2 t & 0 & 0 \\ 0 & 0 & 0 & 0 & \cos\omega_3 t & \frac{\sin\omega_3 t}{\omega_3} \\ 0 & 0 & 0 & 0 & -\omega_3\sin\omega_3 t & \cos\omega_3 t \end{bmatrix} . \quad (3.109)$$

Taking Laplace transform of equation (3.109) and inserting λ instead of s, one has

$$\underline{\Phi}(\lambda) = \begin{bmatrix} \dfrac{\lambda}{\lambda^2+\omega_1^2} & \dfrac{1}{\lambda^2+\omega_1^2} & 0 & 0 & 0 & 0 \\[2ex] \dfrac{-\omega_1^2}{\lambda^2+\omega_1^2} & \dfrac{\lambda}{\lambda^2+\omega_1^2} & 0 & 0 & 0 & 0 \\[2ex] 0 & 0 & \dfrac{\lambda}{\lambda^2+\omega_2^2} & \dfrac{1}{\lambda^2+\omega_2^2} & 0 & 0 \\[2ex] 0 & 0 & \dfrac{-\omega_2^2}{\lambda^2+\omega_2^2} & \dfrac{\lambda}{\lambda^2+\omega_2^2} & 0 & 0 \\[2ex] 0 & 0 & 0 & 0 & \dfrac{\lambda}{\lambda^2+\omega_3^2} & \dfrac{1}{\lambda^2+\omega_3^2} \\[2ex] 0 & 0 & 0 & 0 & \dfrac{-\omega_3^2}{\lambda^2+\omega_3^2} & \dfrac{\lambda}{\lambda^2+\omega_3^2} \end{bmatrix} \cdot \tag{3.110}$$

Substituting into equation (3.106), the following terms are obtained

$$\underline{C}\,\underline{\Phi}(\lambda)\underline{B}\,\underline{B}_j = \frac{c_1'B_1}{\ell(\lambda^2+\omega_1^2)} + \frac{c_3'B_3}{\ell(\lambda^2+\omega_3^2)} \,, \tag{3.111}$$

$$\underline{C}_j\underline{\Phi}(\lambda)\underline{BB}_j = \frac{c_1B_1}{\lambda^2+\omega_1^2} + \frac{c_3B_3}{\lambda^2+\omega_3^2} \,. \tag{3.112}$$

Using the data of Section 3.3.2, one has $\omega_1^2 = 18.79$, $\omega_3^2 = 1522.02$, $B_1 = 63.154$, $B_3 = 102.51$, $c_1 = 0.1173$, $c_3 = 0.2878$, $c_1' = c_1/2$, and $c_3' = c_3/2$. Substituting these data in equation (3.106) and assuming that the required pole is supposedly located at -5.0, i.e., $\lambda_1 = -5.0$, one arrives at the condition

$$1.0 + (-0.01908\ K_c') + 0.27493 = 0 \,. \tag{3.113}$$

The gain K'_c is found to be 66.8. The response of the
system as well as the closed-loop eigenvalues are displayed in
Figure 3.29. It can be seen that -5.0 is one of the eigenvalues,
moreover, the system is asymptotically stable. However, these
results may not always be obtained, and there is a possibility that
the system becomes unstable if the remaining eigenvalues are in
the right half of s-plane. In order to overcome this dilemma,
an observer may be used to estimate the state variables, as will
be shown in Section 3.4.2.3. In the next section, the previous
example will be solved by considering seven modes of vibration

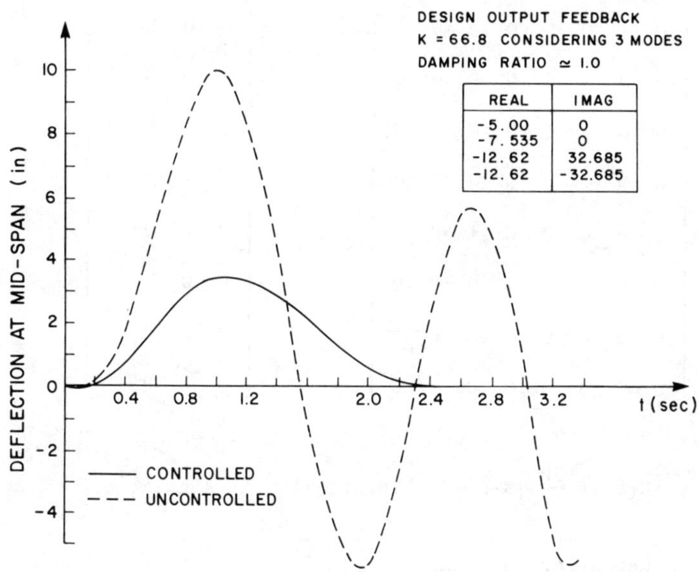

Figure 3.29 - Deflection Response for Three Modes

3.4.2.2 Example 2

Let the measurement be taken at $x = a$ and $x = L-a$, such that the
difference in the rate of rotation at these positions is considered.
The output in this case, considering seven modes of vibration, but
disregarding the even modes is

$$\underline{Y} = \dot{y}'(a,t) - \dot{y}'(L-a,t) = [0 \;\; \frac{c_1}{\ell} \;\; 0 \;\; \frac{c_3}{\ell} \;\; 0 \;\; \frac{c_5}{\ell} \;\; 0 \;\; \frac{c_7}{\ell}] \; \underline{X} \; ,$$

$$\underline{Y} = \underline{C} \; \underline{X} \; . \tag{3.114}$$

From Figure 3.28, one has

$$\underline{U}(t) = \underline{R}(t) - \underline{B}_j(\underline{K}'\underline{C} \; \underline{X} + \underline{C}_j\underline{X}) \; . \tag{3.103}$$

In the present example, matrix $\underline{A}$, matrix $\underline{B}$, and vector $\underline{R}(t)$ are respectively, given by

$$\underline{A} = \begin{bmatrix} 0 & 1 & 0 & 0 & 0 & 0 & 0 & 0 \\ -\omega_1^2 & 0 & 0 & 0 & 0 & 0 & 0 & 0 \\ 0 & 0 & 0 & 1 & 0 & 0 & 0 & 0 \\ 0 & 0 & -\omega_3^2 & 0 & 0 & 0 & 0 & 0 \\ 0 & 0 & 0 & 0 & 0 & 1 & 0 & 0 \\ 0 & 0 & 0 & 0 & -\omega_5^2 & 0 & 0 & 0 \\ 0 & 0 & 0 & 0 & 0 & 0 & 0 & 1 \\ 0 & 0 & 0 & 0 & 0 & 0 & -\omega_7^2 & 0 \end{bmatrix}, \quad \underline{B} = \begin{bmatrix} 0 & 0 & 0 & 0 \\ 1 & 0 & 0 & 0 \\ 0 & 0 & 0 & 0 \\ 0 & 1 & 0 & 0 \\ 0 & 0 & 0 & 0 \\ 0 & 0 & 1 & 0 \\ 0 & 0 & 0 & 0 \\ 0 & 0 & 0 & 1 \end{bmatrix},$$

$$\underline{R}(t) = \frac{2P}{mL} \; [\sin\Omega_1 t \;\; \sin\Omega_3 t \;\; \sin\Omega_5 t \;\; \sin\Omega_7 t]^T \; .$$

Repeating the same calculations of the previous section, the terms in equation (3.106) become

$$\underline{C} \; \Phi(\lambda)\underline{B} \; \underline{B}_j = \frac{c_1 B_1 \lambda}{\ell(\lambda^2+\omega_1^2)} + \frac{c_3 B_3 \lambda}{\ell(\lambda^2+\omega_3^2)} + \frac{c_5 B_5 \lambda}{\ell(\lambda^2+\omega_5^2)} + \frac{c_7 B_7 \lambda}{\ell(\lambda^2+\omega_7^2)} \; , \tag{3.115}$$

$$\underline{C}_j\Phi(\lambda)\underline{B} \; \underline{B}_j = \frac{c_1 B_1}{\lambda^2+\omega_1^2} + \frac{c_3 B_3}{\lambda^2+\omega_3^2} + \frac{c_5 B_5}{\lambda^2+\omega_5^2} + \frac{c_7 B_7}{\lambda^2+\omega_7^2} \; . \tag{3.116}$$

Assuming again that the required pole location is -5.0, substitution into equation (3.106) leads to

$$1 + (-0.03933\ K'_c) + 0.2831 = 0\ . \tag{3.117}$$

The gain K'_c is found to be 32.64. This gain is approximately one half of the gain obtained in the previous example. That difference in gain is caused by the doubling of the sensor's gain in this example as compared to the one in the previous example. Moreover, beyond doubling this gain as compared to the gain in the previous example, 7 modes of vibration instead of the three modes are considered. The system response, as well as the eigenvalues are displayed in Figure 3.29. Again one finds that -5.0 is one of the eigenvalues, and the system is asymptotically stable.

3.4.2.3 Example 3

The output feedback systems designed in Sections 3.4.2.1 and 3.4.2.2 were asymptotically stable. However, these results might not always be obtained, since only a number of poles less than the order of the system is specified. In order to overcome this problem, one may use an observer to estimate the state variables from the available output, and hence use the state feedback. The design of an observer and state feedback, as well as the controlled response are shown in this section.

Assuming that the output is again given by equation (3.102) the observability matrix, equation (3.73), is

$$\underline{Q} = \begin{bmatrix}
0 & \dfrac{-c_1'\omega_1^2}{\ell} & 0 & \dfrac{c_1'\omega_1^4}{\ell} & 0 & \dfrac{c_1'\omega_1^6}{\ell} \\[2ex]
\dfrac{c_1'}{\ell} & 0 & \dfrac{-\omega_1^2 c_1'}{\ell} & 0 & \dfrac{c_1'\omega_1^4}{\ell} & 0 \\[2ex]
0 & \dfrac{-c_2'\omega_2^2}{\ell} & 0 & \dfrac{c_2'\omega_2^4}{\ell} & 0 & \dfrac{c_2'\omega_2^6}{\ell} \\[2ex]
\dfrac{c_2'}{\ell} & 0 & \dfrac{-\omega_2^2 c_2'}{\ell} & 0 & \dfrac{c_2'\omega_2^4}{\ell} & 0 \\[2ex]
0 & \dfrac{-c_3'\omega_3^2}{\ell} & 0 & \dfrac{c_3'\omega_3^4}{\ell} & 0 & \dfrac{c_1'\omega_3^6}{\ell} \\[2ex]
\dfrac{c_3'}{\ell} & 0 & \dfrac{-\omega_3^2 c_3'}{\ell} & 0 & \dfrac{c_3'\omega_3^4}{\ell} & 0
\end{bmatrix} \qquad (3.118)$$

which indicates that the system is completely observable.

The simulation diagram, using an observer, is shown in Figure 3.31. The design process consists in determining the matrices $\underline{B}_c$ and $\underline{K}_c$ such that the desired eigenvalues of the overall system are obtained. The eigenvalues of the observer are located a little farther to the left than the eigenvalues of the controlled structure. These eigenvalues are assumed as shown in Table 3.4 (see p.   ). The eigenvalues $(0,17.339)$ and $(0,-17.339)$ were chosen in this specific form because it is known a priori that the second mode is uncontrolled. The remaining eigenvalues of the controlled structure were assumed with the intent of obtaining a damped structure, but with no interest in increasing the stiffness of the system (note the reduction in stiffness of the first and third modes as compared with the corresponding modes of the original system).

The eigenvalues of the observer are obtained from equation (3.93) which can be written as

$$\overline{\Delta}(\overline{\lambda}) = \Delta(\overline{\lambda}) \cdot \left| \underline{I}_n + \underline{\Phi}(\overline{\lambda})\underline{B}_c\underline{C} \right| = \Delta(\overline{\lambda}) \cdot \left| \underline{I}_1 + \underline{C}\underline{\Phi}(\overline{\lambda})\underline{B}_c \right| = 0 \ . \qquad (3.119)$$

From equations (3.102) and (3.110), the term $\underline{C}\,\underline{\Phi}(\overline{\lambda})$ is given by

$$\underline{C}\,\underline{\Phi}(\overline{\lambda}) = \left[\frac{-c_1'\omega_1^2}{\ell(\overline{\lambda}^2+\omega_1^2)} \quad \frac{c_1'\overline{\lambda}}{\ell(\overline{\lambda}^2+\omega_1^2)} \quad \frac{-c_2'\omega_2^2}{\ell(\overline{\lambda}^2+\omega_2^2)} \quad \frac{c_2'\overline{\lambda}}{\ell(\overline{\lambda}^2+\omega_2^2)} \quad \frac{-c_3'\omega_3^2}{\ell(\overline{\lambda}^2+\omega_3^2)} \quad \frac{c_3'\overline{\lambda}}{\ell(\overline{\lambda}^2+\omega_3^2)} \right].$$

$$(3.120)$$

Substituting $\overline{\lambda}$ in equation (3.120) by each of the $\overline{\lambda}_i$, $i = 1,2,3,\ldots,6$ of Table 3.4, the matrix $\underline{C}\,\underline{\Phi}(\overline{\lambda}_i)$ is formed from the rows given by equation (3.120). The matrix $\underline{B}_c$ is found from

$$\begin{bmatrix} 1 \\ 1 \\ 1 \\ 1 \\ 1 \\ 1 \end{bmatrix} + \underline{C}\underline{\Phi}(\lambda_i)\underline{B}_c = 0, \qquad \underline{B}_c = \begin{bmatrix} -249.76 \\ 8716.88 \\ -166.28 \\ 5384.96 \\ -83.60 \\ 8204.56 \end{bmatrix}. \qquad (3.121)$$

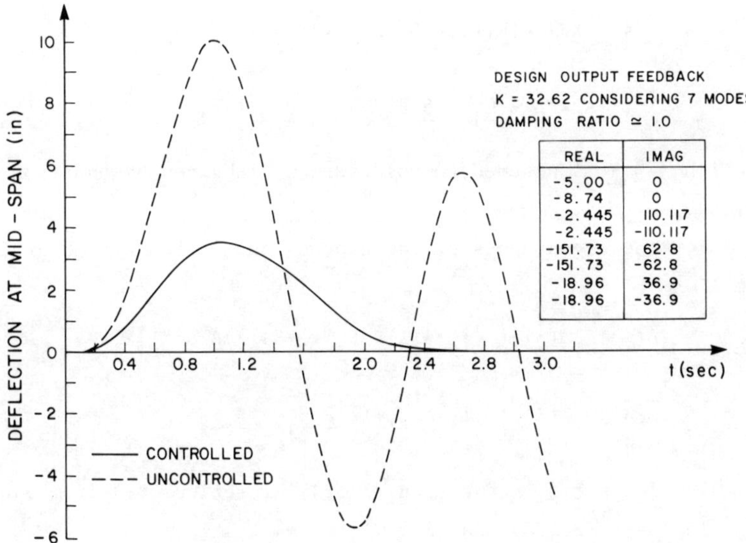

Figure 3.30 - Deflection Response for Seven Modes

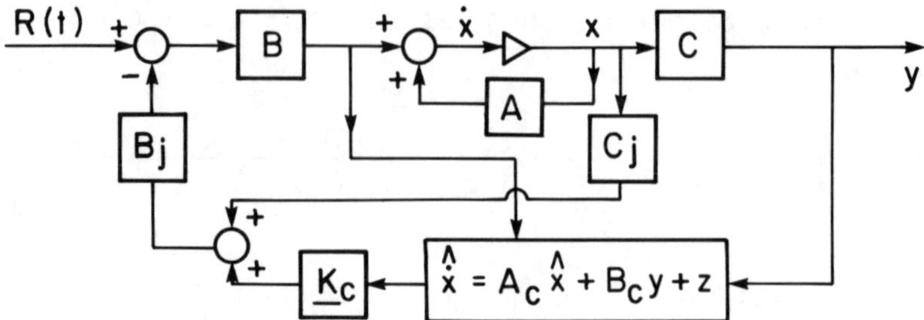

Figure 3.31 - Simulation Diagram of the Structure and the Observer

To design the state feedback gain matrix $\underline{K}_c$, the vector $\underline{U}$ is obtained from Figure 3.31 as

$$\underline{U}(t) = \underline{R}(t) - \underline{B}_j [\underline{C}_j \underline{X} + \underline{K}_c \hat{\underline{X}}] , \qquad (3.122)$$

in which $\hat{\underline{X}}$ is equal to the approximated state vector coming from the observer.

Substituting equation (3.122) into equation (3.97), the closed-loop system is defined by

$$\dot{\underline{X}} = \underline{A}\,\underline{X} + \underline{B}[\underline{R}(t) - \underline{B}_j (\underline{C}_j \underline{X} + \underline{K}_c \hat{\underline{X}})] ,$$

$$\dot{\underline{X}} = (\underline{A} - \underline{B}\,\underline{B}_j\underline{C}_j - \underline{B}\,\underline{B}_j\underline{K}_c)\underline{X} + \underline{B}\,\underline{R}(t) , \qquad (3.123)$$

in which $\underline{X} \cong \hat{\underline{X}}$ is assumed to hold; and $\underline{C}_j$ is given by equation (3.105).

The eigenvalues of the closed-loop system are obtained from

$$\overline{\Delta}(\lambda) = \Delta(\lambda) \cdot \left| \underline{I}_n + \underline{\Phi}(\lambda)\underline{B}\,\underline{B}_j\underline{C}_j + \underline{\Phi}(\lambda)\underline{B}\,\underline{B}_j\underline{K}_c \right| ,$$

$$\overline{\Delta}(\lambda) = \Delta(\lambda) \cdot \left| I_1 + \underline{C}_j\underline{\Phi}(\lambda)\underline{B}\,\underline{B}_j + \underline{K}_c\underline{\Phi}(\lambda)\underline{B}\,\underline{B}_j \right| = 0 . \qquad (3.124)$$

The terms in equation (3.124) are calculated from equations (3.104), (3.105) and (3.110) to be

$$\underline{C}_j \underline{\Phi}(\lambda) \underline{B}\ \underline{B}_j = \frac{c_1 B_1}{\lambda^2 + \omega_1^2} + \frac{c_3 B_3}{\lambda^2 + \omega_3^2} , \tag{3.125}$$

$$\underline{\Phi}(\lambda) \underline{B}\ \underline{B}_j = \left[\frac{B_1}{\lambda^2 + \omega_1^2} \quad \frac{\lambda B_1}{\lambda^2 + \omega_1^2} \quad 0 \quad 0 \quad \frac{B_3}{\lambda^2 + \omega_3^2} \quad \frac{\lambda B_3}{\lambda^2 + \omega_3^2} \right]^T . \tag{3.126}$$

Substituting for each of λ_i, $i = 1,2,\ldots,6$ from Table 3.4 the matrix $\underline{\psi} \triangleq \underline{\Phi}(\lambda) \underline{B}\ \underline{B}_j$, $j = 1,\ldots,2$ can be formed from the columns of equation (3.126). However, that matrix is singular due to the third and fourth zero rows. That happened because the closed-loop system is not completely controllable as a result of including the second mode, $B_2 = 0$, in the system's equations. This emphasizes the fact that the controllability of the closed-loop system should be checked if the system is in a different configuration than that of Figures 3.25 or 3.26 [3.26].

The design is continued while omitting the second mode equations in the previous process. In this case, the matrix $\underline{\psi} \triangleq \underline{\Phi}(\lambda) B\ B_j$, $j = 1,\ldots,4$ becomes

$$\psi = \begin{bmatrix}
\dfrac{B_1}{\lambda_1^2 + \omega_1^2} & \dfrac{B_1}{\lambda_2^2 + \omega_1^2} & \dfrac{B_1}{\lambda_5^2 + \omega_1^2} & \dfrac{B_1}{\lambda_6^2 + \omega_1^2} \\[2mm]
\dfrac{B_1 \lambda_1}{\lambda_1^2 + \omega_1^2} & \dfrac{B_1 \lambda_2}{\lambda_2^2 + \omega_1^2} & \dfrac{B_1 \lambda_5}{\lambda_5^2 + \omega_1^2} & \dfrac{B_1 \lambda_6}{\lambda_6^2 + \omega_1^2} \\[2mm]
\dfrac{B_3}{\lambda_1^2 + \omega_3^2} & \dfrac{B_3}{\lambda_2^2 + \omega_3^2} & \dfrac{B_3}{\lambda_5^2 + \omega_3^2} & \dfrac{B_3}{\lambda_6^2 + \omega_3^2} \\[2mm]
\dfrac{B_3 \lambda_1}{\lambda_1^2 + \omega_3^2} & \dfrac{B_3 \lambda_2}{\lambda_2^2 + \omega_3^2} & \dfrac{B_3 \lambda_5}{\lambda_5^2 + \omega_3^2} & \dfrac{B_3 \lambda_6}{\lambda_6^2 + \omega_3^2}
\end{bmatrix} . \tag{3.127}$$

Substituting into equation (3.124) yields

$$[1 \quad 1 \quad 1 \quad 1] + [\underline{C}_j \underline{\psi}(\lambda_1) \underline{C}_j \underline{\psi}(\lambda_2) \underline{C}_j \underline{\psi}(\lambda_5) \underline{C}_j \underline{\psi}(\lambda_6)] + \underline{K}_c \underline{\psi}(\lambda) = 0 . \tag{3.128}$$

The gain matrix $\underline{K}_c$ is found to be

$$K_c = [-0.155 \quad 0.096 \quad -1.78 \quad 0.087] \ . \tag{3.129}$$

Investigating the elements of $\underline{K}_c$, one finds that the first and third elements are negative, and therefore they reduce the stiffness of the system. On the other hand, the second and fourth elements are positive and therefore they introduce damping to the structure. The deflection response of the controlled bridge of this case is shown in Figure 3.32.

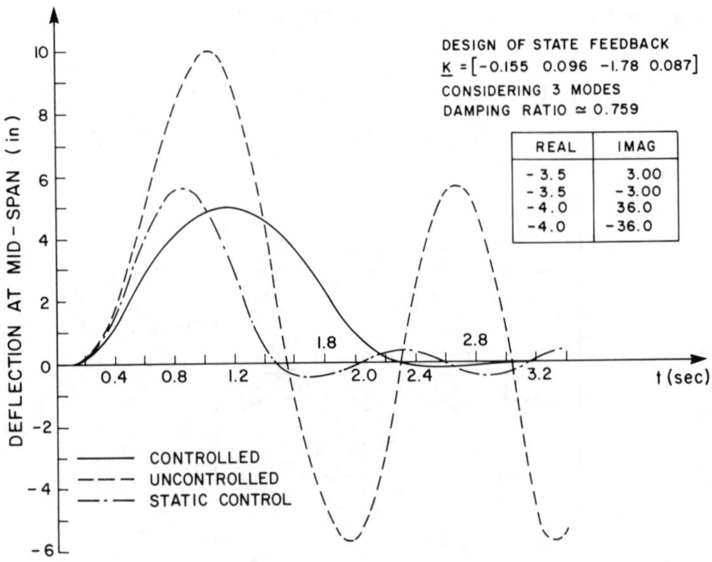

Figure 3.32 - Deflection Response Using an Observer and and State Feedback

3.4.3 *Summary and Conclusion*

A general description of the design of a control system by one of the modern design techniques and the application of this technique

to control a simple span bridge is given. The pole assignment
method proved to be straightforward and provides a design complying
with the required performance of the system. The advantage of
this method as compared with the trial and error (classical)
method and modal control method is obvious as the pole assignment
method was simple and general in the examples studied.

Although pole assignment method and modal control method
[3.16] have the same objective, which is to enforce direct changes
on the modes of the structure, the pole assignment method is
superior insofar as it enables one to change any number of modes
in any desired form.

The checking of the observability and/or controllability
of the open-loop and closed-loop systems is the main key for controlling these modes. Without checking these criteria, a design
of an output or a state feedback might be impossible if the closed-
loop system is uncontrollable. Therefore, checking the controllability of the open-loop and closed-loop system, before the design,
provides information on how well the structure can be controlled,
and if the locations of the control forces are useful or not.
A numerical example was presented in Section 3.4.2.3 to illustrate
this point, which was not apparent when using the modal control
method [3.16] for controlling the stiffness of only one mode.

The pole assignment method has also shown that using the
output variables as control variables may destabilize the system,
especially if the number of these variables is less than the order
of the system. Using an observer was useful in transforming the
output feedback into state feedback, and hence controlling the
stability of the system.

The results obtained in controlling the bridge in this
section confirmed the results developed in the previous section.
These results again are: the sensor must be placed at the control
force position and must measure the same type of response produced
by the control force in order to warrant the stability of the system.

Not much difference in response exists when considering seven modes
or three modes of vibration. Active feedback control was useful
in dampening the vibration of the structure.

3.5 ACTIVE STRUCTURAL CONTROL BY OPTIMAL CONTROL METHODS

It has been assumed in the previous sections that the supplied
control energy is unlimited. A comparison of the consumed energy
with the practical consumption is usually done after the process
of design. However, from the economical and system's response
point of view, one should strive for the best controlled response
warranting the consumption of a minimum amount of energy. The
problem then turns out to be an optimization problem, and optimal
control design methods must be used to solve the problem.

The optimal control problem is presented herein either
as a regulator problem, which may lead to a closed-loop control,
or as a tracking problem which may lead to a combination of closed-
loop and open-loop control. The interest in this section will
be focused on techniques which provide a feedback control.

It will be shown that regulator and tracking problems
are suitable for an application to optimal feedback control of
civil engineering structures. It will become apparent that the
tracking problems is the most suitable design method and provides
the best control response if and only if the applied disturbance
is known a priori. On the other hand, the regulator problem is
used whenever the disturbance applied to the structure is uncertain
or unpredictable.

3.5.1 *The Optimal Control Problem*

The dynamic behaviour of a system may be described by the following
state equation:

$$\dot{\underline{X}}(t) = \underline{A}(t)\underline{X}(t)+\underline{B}(t)\underline{U}(t)+\underline{d}(t) \; , \quad \underline{X}(t_0) = \underline{X} \; , \qquad (3.130)$$

in which $\underline{X} = \underline{X}(t)$ is an n-vector representing the state of the
system; $\underline{U} = \underline{U}(t)$ is an r-vector of the control variables; $\underline{A}(t)$
and $\underline{B}(t)$ are matrices of appropriate dimensions; and $\underline{d} = \underline{d}(t)$ is
an n-vector representing the disturbance.

The output or measurable variables to be controlled are
expressed as

$$\underline{Y}(t) = \underline{C}(t)\underline{X}(t) \; , \qquad (3.131)$$

in which $\underline{Y} = \underline{Y}(t)$ is an m-vector representing the measurable variables, and $\underline{C}(t)$ is a matrix of an appropriate dimension.

The design of optimal control devices is done in such a
way that certain specifications concerning the dynamic behaviour
of the controlled system are satisfied. Such specifications include requirements on the transient response, stability, steady
state error, and/or reduced parameter sensitivity. An objective
or cost function (performance index) is usually defined in terms
of these design specifications. The minimization of such a cost
function leads to an optimal control strategy. For the sake of
obtaining easily a closed form solution for the optimal control,
the objective function is chosen to be a quadratic function of
the form

$$J = \frac{1}{2} \, [\underline{Y}(t_f)-\underline{\alpha}]^T \underline{S}_0 \, [\underline{Y}(t_f)-\underline{\alpha}] + \frac{1}{2} \int_{t_0}^{t_f} \{[\underline{Y}(t)-\underline{\beta}(t)]^T \underline{Q}_0(t)[\underline{Y}(t)-\underline{\beta}(t)]$$

$$+ \, \underline{U}^T(t)\underline{R}(t)\underline{U}(t)\}dt \; , \qquad (3.132)$$

in which $\underline{S}_0$, $\underline{Q}_0(t)$, and $\underline{R}(t)$ are weighing matrices of appropriate
dimensions; $\underline{\alpha}$ is the desired steady state response; $\underline{\beta}(t)$ the desired
transient response; t_0 is the initial time; and t_f is the final
time of control.

The matrices $\underline{S}_0$, and $\underline{Q}_0(t)$ must be at least positive semi-definite, in order to ensure "a well-defined finite minimum for J" [3.14]. The matrix $\underline{R}(t)$ must be positive definite because $\underline{R}^{-1}(t)$ is required to exist for a finite solution. A justification of the above conditions imposed on the weighing matrices follows from the conditions which are sufficient to ensure at least a locally optimal solution, [3.14, p. 763]. The choice of the elements of the weighing matrices, however, is a major problem of optimal control theory. To ease this problem, these matrices are usually chosen to be diagonal with relatively large values for elements corresponding to variables for which a strong control of their magnitude is desired [3.13].

The minimization of the objective function, equation (3.132), subjected to equations (3.130) and (3.131) is referred to as a "tracking problem", since a control is desired such that the output variables track the vectors $\underline{\alpha}$ and $\underline{\beta}(t)$. If $\underline{\alpha}$, $\underline{\beta}(t)$, and $\underline{d}(t)$ are null vectors then the control problem is called a "regulator problem", since the steady state and transient responses are desired to be zero.

The necessary conditions for an optimal solution of the tracking or regulator problem may be obtained by using the calculus of variation which involves the concept of a Hamiltonian [3.13]. For the tracking problem, the Hamiltonian is defined as

$$H(\underline{X},\underline{U},\underline{\lambda}) = \frac{1}{2}(\underline{C}\,\underline{X} - \underline{\beta})^T \underline{Q}_0 (\underline{C}\,\underline{X} - \underline{\beta}) + \frac{1}{2}\underline{U}^T\underline{R}\,\underline{U} + \underline{\lambda}^T(\underline{A}\,\underline{X} + \underline{B}\,\underline{U} + \underline{d}) \, ,$$

$$(3.133)$$

in which $\underline{\lambda} = \underline{\lambda}(t)$ is an n-vector denoting the co-state variables or Lagrange multipliers.

For a minimum of J given by equation (3.132) and subjected to the constraints given by equations (3.130) and (3.131), it is necessary ([3.13,3.14]) that

$$\underline{\dot{X}} = \underline{A}\,\underline{X} + \underline{B}\,\underline{U} + \underline{d} \, , \qquad \underline{X}(t_0) = \underline{X} \, , \qquad\qquad (3.134)$$

$$\underline{\dot{\lambda}} = - \frac{\partial H}{\partial \underline{X}} = -\underline{C}^T Q_0 \underline{C}\ \underline{X} - \underline{A}^T \underline{\lambda} + \underline{C}^T Q_0 \underline{\beta}\ , \quad \underline{\lambda}(t_f) = \underline{C}^T S_0 [\underline{C}\ \underline{X}(t_f) - \underline{\alpha}]\ ,$$

$$(3.135)$$

$$\underline{g} = \frac{\partial H}{\partial \underline{U}} = \underline{R}\ \underline{U} - \underline{B}^T \underline{\lambda} = 0\ .$$

$$(3.136)$$

The optimal control is obtained from equation (3.136) as

$$\underline{U} = -\underline{R}^{-1} \underline{B}^T \underline{\lambda}\ .$$

$$(3.137)$$

Substituting equation (3.137) into equation (3.134),
equations (3.134) and (3.135) become

$$\underline{\dot{X}} = \underline{A}\ \underline{X} - \underline{B}\ \underline{R}^{-1} \underline{B}^T \underline{\lambda} + \underline{d}\ , \quad \underline{X}(t_0) = \underline{X}_0\ , \tag{3.138}$$

$$\underline{\dot{\lambda}} = -\underline{C}^T Q_0 \underline{C}\ \underline{X} - \underline{A}^T \underline{\lambda} + \underline{C}^T Q_0 \underline{\beta}\ , \quad \underline{\lambda}(t_f) = \underline{C}^T S_0 [\underline{C}\ \underline{X}(t_f) - \underline{\alpha}]\ .$$

$$(3.135)$$

For a regulator problem in which $\underline{\alpha} = \underline{\beta} = \underline{d} = 0$, equations
(3.138) and (3.135) become

$$\underline{\dot{X}} = \underline{A}\ \underline{X} - \underline{B}\ \underline{R}^{-1} \underline{B}^T \underline{\lambda}\ , \quad \underline{X}(t_0) = \underline{X}_0\ , \tag{3.139}$$

$$\underline{\dot{\lambda}} = -\underline{C}^T Q_0 \underline{C}\ \underline{X} - \underline{A}^T \underline{\lambda}\ , \quad \underline{\lambda}(t_f) = \underline{C}^T S_0 \underline{C}\ \underline{X}(t_f)\ . \tag{3.140}$$

Equations (3.138) and (3.135), or equations (3.139) and
(3.140) are called a two-point-boundary-value-problem (TPBVP).
That is so, because one of the boundary conditions is specified at
initial time and the other one at terminal time. An obvious difference between equations (3.138), (3.135) and equations (3.139),
(3.140) is that equations (3.138) and (3.135) are nonhomogeneous
vector differential equations, whereas equations (3.139) and (3.140)
are homogeneous differential equations. On the other hand, whereas
equations (3.138) and (3.135) provide control to track $\underline{\alpha}$, $\underline{\beta}$, and
$\underline{d}(t)$, the control provided by equations (3.139) and (3.140) is of
a general type and brings the system to zero state.

Since the objective in controlling a civil engineering
structure is likely to strive for zero transient and steady state
responses, it becomes obvious that the regulator problem is directly
applicable to the control of civil engineering structures, whereas
the tracking problem needs some trivial modifications. The application of both problems is shown later in Section 3.5.3.

3.5.2 *Solution of the Optimal Control Problem*

The simultaneous solution of equations (3.138) and (3.135) or equations (3.139) and (3.140) gives the optimal control strategy in
open-loop type, i.e., without feedback. There are several numerical
techniques available to obtain such a solution. Some of these
numerical techniques are based on the use of a transition matrix
[3.13], on gradient methods [3.23], on dynamic programming [3.24],
and other methods [3.25].

Closed-loop control is based on the feedback of the output or state variables. However, since the output feedback is also
subjected here to some restrictions (e.g., $[\underline{C}^T\underline{C}]^{-1}$ should exist),
the interest will be focused on using optimal state feedback con-
trol. In such cases, an observer [3.13,3.26] is required to estimate the state variables by using the available output.

To take, to a certain extent, advantage of the closed-
loop control, the solution of equations (3.138) and (3.135) is
assumed to be a combination of closed-loop and open-loop control.
That can be done by assuming the co-state variable $\underline{\lambda}(t)$ to be a
linear combination of the current state $\underline{X}(t)$ and the determined
state $\underline{q}(t)$ in the range t_0 to t_f. The variable $\underline{\lambda}(t)$ is assumed in
the form

$$\underline{\lambda}(t) = \underline{P}(t)\underline{X}(t) + \underline{q}(t) , \qquad\qquad (3.141)$$

in which $\underline{P} = \underline{P}(t)$ is an n×n symmetric matrix called "Riccati matrix";
and $\underline{q} = \underline{q}(t)$ is an n-vector which accounts for the desired responses
$\underline{\alpha}$ and $\underline{\beta}$ and the disturbance $\underline{d}(t)$.

The optimal control signal of the tracking problem is then obtained by substituting equation (3.141) into equation (3.137) yielding

$$\underline{U}(t) = -\underline{R}^{-1}\underline{B}^T P(t)\underline{X}(t) + \underline{U}_{EXT}(t) \, , \qquad (3.142)$$

in which $\underline{U}_{EXT}(t) = -\underline{R}^{-}\underline{B}^T q(t)$ is an external control [3.27] supplied to the system without dependence on the current state of the system. Hence, it is of the open-loop type.

Substituting equation (3.141) into equations (3.138) and (3.135) enables one to determine $\underline{P}(t)$ and $\underline{q}(t)$ from the following differential equations:

$$-\underline{\dot{P}} = \underline{P}\,\underline{A} + \underline{A}^T\underline{P} - \underline{P}\,\underline{B}\,\underline{R}^{-1}\underline{B}^T\underline{P} + \underline{C}^T\underline{Q}\,\underline{C} \, , \quad \underline{P}(t_f) = \underline{C}^T\underline{S}_0\underline{C} \, , \tag{3.143}$$

$$-\underline{\dot{q}} = \underline{P}\,\underline{d} + (\underline{A}^T - \underline{P}\,\underline{B}\,\underline{R}^{-1}\underline{B}^T)q - \underline{C}^T Q_0\underline{\beta} \, , \quad q(t_f) = -\underline{C}^T S_0\underline{\alpha} \, . \tag{3.144}$$

If $\underline{\alpha}$, $\underline{\beta}(t)$, and $\underline{d}(t)$ are null vectors, as in the case of the regulator problem, only the Riccati matrix differential equation, equation (3.143), must be solved for $\underline{P}(t)$, since $\underline{q}(t)$ in this case is a null vector. This is true because the TPBVP for regulator problems is homogeneous, and in order to satisfy the terminal conditions given by equation (3.140) one has to assume

$$\underline{\lambda}(t) = \underline{P}(t)\underline{X}(t) \, . \tag{3.145}$$

The optimal control of the regulator problem is then

$$\underline{U}(t) = -\underline{R}^{-1}\underline{B}^T\underline{P}(t)\underline{X}(t) \, , \tag{3.146}$$

which is only depending on the current state of the system. Hence, it is a closed-loop type.

Many numerical techniques are available for computing $\underline{P}$ and $\underline{q}$ [3.27,3.28]. For example, since $\underline{P}(t)$ is a symmetric matrix, only a set of $n(n+1)/2$ independent equations can be derived from equation (3.143). Using backward numerical integration, one can obtain $\underline{P}(t)$. Once $\underline{P}(t)$ has been determined, equation (3.144) is integrated backward to yield $\underline{q}(t)$.

One has to note that the preceding results are only applicable to linear, time-varying or invariant systems. For nonlinear systems one has to go through a process of linearization and iteration in order to determine the exact nominal state around which the system is linearized.

3.5.3 *Numerical Applications*

The bridge considered in this chapter is investigated here again, by optimal control techniques. For the sake of completeness, the equations of motion are given, considering the first three modes, as

$$\ddot{A}_1(t)+\omega_1^2 A_1(t) = \frac{2P}{mL}\sin\Omega_1 t - B_1[u(t)+c_1 A_1(t)+c_3 A_3(t)] \ , \qquad (3.147)$$

$$\ddot{A}_3(t)+\omega_3^2 A_3(t) = \frac{2P}{mL}\sin\Omega_3 t - B_3[u(t)+c_1 A_1(t)+c_3 A_3(t)] \ , \qquad (3.148)$$

in which $c_j = 2\ell j\pi/L \cos(j\pi a/L)$; $B_j = 4S\ell j\pi/mL^2 \cos(j\pi a/L)$; and S is the spring's constant. The second mode equation has not been written since it is uncontrolled.

Denominating the state variables as $x_1(t) = A_1(t)$, $x_2(t) = \dot{A}_1(t)$, $x_3(t) = A_3(t)$, and $x_4(t) = \dot{A}_2(t)$, equations (3.147) and (3.148) can be represented by the simulation diagram shown in Figure 3.33.

It is assumed that the observer has been designed as in Section 3.4.2.3 or in any other way [3.13]. The task of the designer is then to design the optimal control $\underline{U}^*$ as to guarantee a controlled response in a desirable fashion. From Figure 3.33, the activating signal $\underline{U}(t)$ follows as

$$\underline{U}(t) = \underline{R}(t) - \underline{B}_j[\underline{C}_j\underline{X}(t) + \underline{U}^*(t)] \ , \qquad (3.149)$$

in which $\underline{R}(t)$ is an input vector and $\underline{C}_j$ is given by

$$\underline{C}_j = [c_1 \quad 0 \quad c_3 \quad 0] \ . \tag{3.150}$$

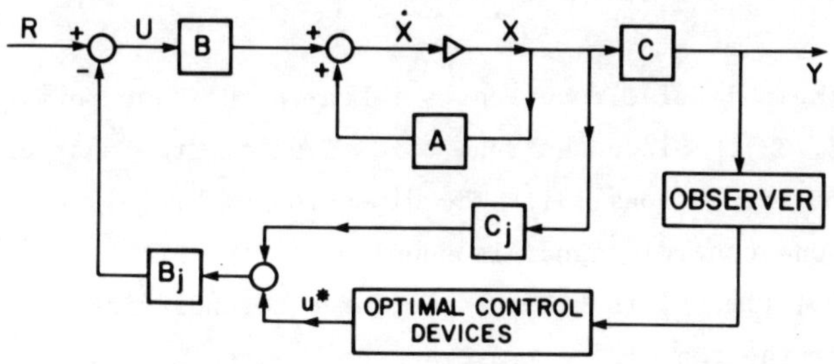

Figure 3.33 - Simulation Diagram for Optimal State Feedback

Equations (3.149) and (3.150) can now be written in the state form as

$$\dot{\underline{X}}(t) = A \ X(t) + \underline{B} \ \underline{U}(t) \ . \tag{3.151}$$

Substituting equation (3.149) into equation (3.151) one has

$$\dot{\underline{X}}(t) = \underline{A}_c \underline{X}(t) + \underline{B}_c \underline{U}^*(t) + \underline{d}(t) \ , \tag{3.152}$$

in which $\underline{A}_c = \underline{A} - \underline{B} \ \underline{B}_j \underline{C}_j$, $\underline{B}_c = -\underline{B} \ \underline{B}_j$, $\underline{d}(t) = \underline{B} \ \underline{R}(t)$, while $\underline{A}$, $\underline{B}$, $\underline{B}_j$, and $\underline{R}(t)$ are respectively given by

$$\underline{A} = \begin{bmatrix} 0 & 1 & 0 & 0 \\ -\omega_1^2 & 0 & 0 & 0 \\ 0 & 0 & 0 & 1 \\ 0 & 0 & -\omega_3^2 & 0 \end{bmatrix}, \quad \underline{B} = \begin{bmatrix} 0 & 0 \\ 1 & 0 \\ 0 & 0 \\ 0 & 1 \end{bmatrix}, \quad \underline{B}_j = \begin{bmatrix} B_1 \\ B_3 \end{bmatrix}, \quad \underline{R}(t) = \frac{2P}{mL} \begin{bmatrix} \sin \Omega_1 t \\ \sin \Omega_3 t \end{bmatrix} \ .$$

Again, the output is the difference in the rate of change of the two posts' positions. The output is expressed as

$$\underline{Y}(t) = \underline{C}\,\underline{X}(t) = \begin{bmatrix} 0 & \dfrac{c_1}{\ell} & 0 & \dfrac{c_3}{\ell} \end{bmatrix} \underline{X} \ .$$

The objective function is taken in the same form as in equation (3.132). Since the rank of $\underline{C}$ is unity, the matrices $\underline{S}_0$ and $\underline{Q}_0$ are of dimensions 1×1. The dimension of $\underline{R}(t)$ is also 1×1, since only one control signal is supplied to the structure. The final control time t_f is taken to be three seconds since the time required for the vehicle to traverse the bridge is L/v = 1.67 seconds. However, one may also find the minimum final control time by modifying the optimization problem [3.13,3.27].

3.5.3.1 Application of the Regulator Problem

It has been mentioned that the feedback control supplied by using the regulator problem is of closed-loop type and it does not depend on the characteristic of the applied disturbance. That means the vector $\underline{d}(t)$ will be omitted in equation (3.152). Applying the same steps as in Section 3.5.1 and 3.5.2, one obtains the Riccati matrix differential equation

$$-\underline{\dot{P}}(t) = \underline{P}\,\underline{A}_c + \underline{A}_c^T\underline{P} - \underline{P}\,\underline{B}_c\underline{R}^{-1}\underline{B}_c^T\underline{P} + \underline{C}^T\underline{Q}_0\underline{C} \ , \quad \underline{P}(t_f) = \underline{C}^T\underline{S}_0\underline{C} \ . \tag{3.153}$$

The Riccati matrix $\underline{P}(t)$ in equation (3.153) is of the dimension 4×4. However, due to its symmetry, only ten of its elements need to be computed. These elements are indicated in the following scheme:

$$\begin{bmatrix} P(1,1) & & Symmetric & \\ P(2,1) & P(2,2) & & \\ P(3,1) & P(3,2) & P(3,3) & \\ P(4,1) & P(4,2) & P(4,3) & P(4,4) \end{bmatrix} = \begin{bmatrix} P(1) & & Symmetric & \\ P(2) & P(3) & & \\ P(4) & P(5) & P(6) & \\ P(7) & P(8) & P(9) & P(10) \end{bmatrix} \tag{3.154}$$

To integrate backwards, the time variable in equation (3.153) is changed into $t = t_f - \tau$. The variable $\underline{V}(\tau) = \underline{P}(t_f - \tau)$ is used, and therefore equation (3.153) becomes

$$\underline{\dot{V}}(\tau) = \underline{V}\,\underline{A}_c + \underline{A}_c^T\underline{V} - \underline{V}\,\underline{B}_c\underline{R}^{-1}\underline{B}_c^T\underline{V} + \underline{C}^T\underline{Q}_0\underline{C} \ , \qquad \underline{V}(0) = \underline{C}^T\underline{S}_0\underline{C} \ . \tag{3.155}$$

The initial conditions for the ten differential equations derived from equation (3.155), using the arrangement shown in equation (3.154), are

$$v_1(0) = 0.0 \ , \qquad v_5(0) = 0.0 \ , \qquad v_9(0) = 0.0 \ ,$$

$$v_2(0) = 0.0 \ , \qquad v_6(0) = 0.0 \ , \qquad v_{10}(0) = S_0\left(\frac{c_3}{\ell}\right)^2 .$$

$$v_3(0) = S_0\left(\frac{c_1}{\ell}\right)^2 , \qquad v_7(0) = 0.0 \ ,$$

$$v_4(0) = 0.0 \ , \qquad v_8(0) = S_0\left(\frac{c_1 \times c_3}{\ell}\right), \tag{3.156}$$

Observing equations (3.155) and (3.156), one concludes that the optimal control solution is depending on the matrices $\underline{S}_0$, $\underline{Q}_0$, and $\underline{R}$. Therefore a number of trials must be performed in order to find the best values for these matrices yielding a reasonable controlled response and minimum energy consumption. The quality of the design is checked by investigating the controlled response of the structure under the effect of a prescribed disturbance. Therefore, equation (3.152) is used in order to find the effect of that disturbance on the state of the system in the presence of the assumed control.

Integrating the ten differential equations simultaneously, and using the initial conditions in equation (3.156), one is able to determine $\underline{V}(\tau)$ from t_0 to t_f. Reversing the order of the time, the original Riccati matrix $\underline{P}(t)$ is obtained. The optimal control is now calculated from

$$u^*(t) = -\underline{R}^{-1}\underline{B}_c^T\underline{P}(t)\underline{X}(t) = [(B_1P_2 + B_3P_7)x_1 + (B_1P_3 + B_3P_8)x_2 + (B_1P_5 + B_3P_9)x_3$$

$$+ (B_1P_8 + B_3P_{10})x_4]/R \ . \tag{3.157}$$

The current state $\underline{X}(t)$ is not known yet, and therefore one is unable to determine the active control before determining the current state. The current states are calculated by substituting equation (3.157) into equation (3.152) which yields

$$\dot{\underline{X}}(t) = [\underline{A}_c - \underline{B}_c \underline{R}^{-1} \underline{B}_c^T \underline{P}(t)]X(t) + \underline{d}(t) \ , \quad \underline{X}(t_0) = \underline{X}_0 \ . \quad (3.158)$$

Integrating equation (3.158) forward up to t_f, and considering zero initial conditions, yields the state $\underline{X}(t)$ from t_0 to t_f. These states can then be substituted into equation (3.157) for determining the active control response of the spring, or the ram's displacement. The ram's displacement must be within practical bounds, having also the cost of control in mind, etc., in order to make it acceptable. A simulation diagram is shown in Figure 3.34.

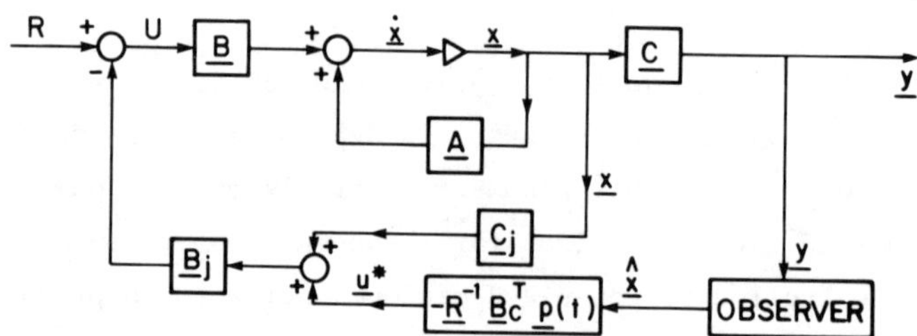

Figure 3.34 - Simulation Diagram of Regulator Problem

The deflection, velocity, and acceleration at mid-span of the bridge, respectively, are calculated from

$$y\left(\frac{L}{2}\ , \ t\right) = x_1(t) - x_3(t) \ , \qquad (3.159)$$

$$\dot{y}\left(\frac{L}{2}\ , \ t\right) = x_2(t) - x_4(t) \ , \qquad (3.160)$$

$$\ddot{y}\left(\frac{L}{2}\ , \ t\right) = \left[\dot{y}\left(\frac{L}{2}\ , \ t\right) - \dot{y}\left(\frac{L}{2}\ , \ t-h\right)\right]/h \ , \qquad (3.161)$$

in which h is time step of integration and is determined approximately as 0.01 of the time constant of the system.

The complete response of the deflection at mid-span, $y(L/2,t)$, and the ram's displacement, $u^*(t)$, for several trials using different values of $\underline{Q}_0$, $\underline{S}_0$, and $\underline{R}$ are shown in Figures 3.35 and 3.36, respectively. It can be deduced from Figure 3.35 that increasing the magnitude of $\underline{Q}_0$ has the effect of decreasing the deflection response of the system, but on the other hand, the energy consumption increases. The behaviour of system No. 4 is the best of those obtained in these trials since the effect of damping consists apparently in preventing oscillations and enforcing a decaying motion.

The deflection at mid-span, $y(L/2,t)$, ram's displacement, $u^*(t)$, and acceleration at mid-span, $\ddot{y}(L/2,t)$ for other trials are shown in Figures 3.37 to 3.39, respectively. The first observation to be mentioned is that by increasing $\underline{R}$, the active control $u^*(t)$ decreases, but the transient response increases. Another observation is that a significant increase in $\underline{Q}_0$ results in decreasing the transient response, but the damping effect on the free response becomes very weak. From observing the responses of the active control $u^*(t)$, and the acceleration $\ddot{y}(L/2,t)$ one concludes that systems No. 1 and 2 in Figure 3.37 are of an acceptable design.

3.5.3.2 Application of the Tracking Problem

A desirable objective in the control of civil engineering structures is to have a zero steady-state and fast decaying transient responses of the output variables. That objective can be expressed by setting $\underline{\alpha} = \underline{\beta}(t) = \underline{0}$ in equation (3.132). If the applied disturbance is known a priori and formulated in the state equation, e.g., equation (3.152), the co-state variable $\underline{\lambda}(t)$ must be assumed in the form of equation (3.141). This fact can further be certified by writing the two-point-boundary-value problem of this example, which is

$$\dot{\underline{X}} = \underline{A}_c \underline{X} - \underline{B}_c \underline{R}^{-1} \underline{B}_c^T \underline{\lambda} + \underline{d}(t) \ , \quad \underline{X}(t_0) = 0 \ , \tag{3.159}$$

$$\dot{\underline{\lambda}} = \underline{C}^T \underline{Q}_0 \underline{C} \ \underline{X} - \underline{A}_c \underline{\lambda} \ , \qquad \underline{\lambda}(t_f) = \underline{C}^T \underline{S}_0 \underline{C} \ \underline{X}(t_f) \ . \tag{3.160}$$

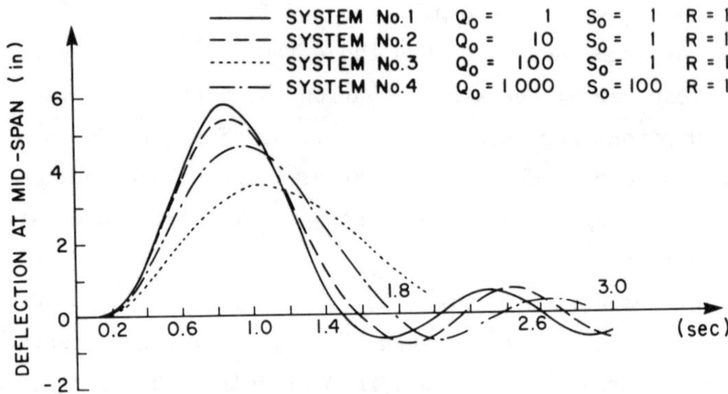

Figure 3.35 - Deflection Response using Regulator Control

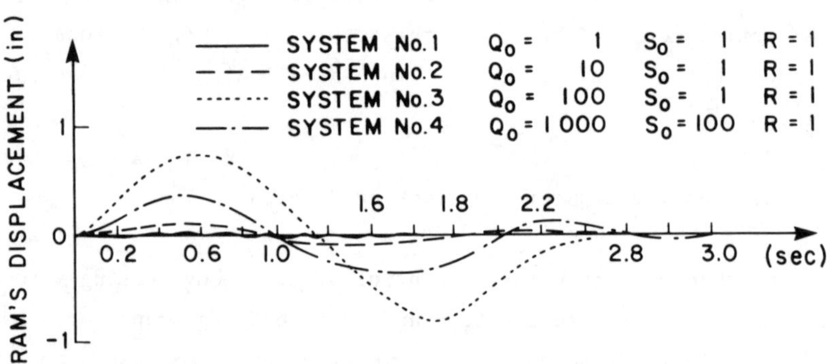

Figure 3.36 - Ram's Response using Regulator Control

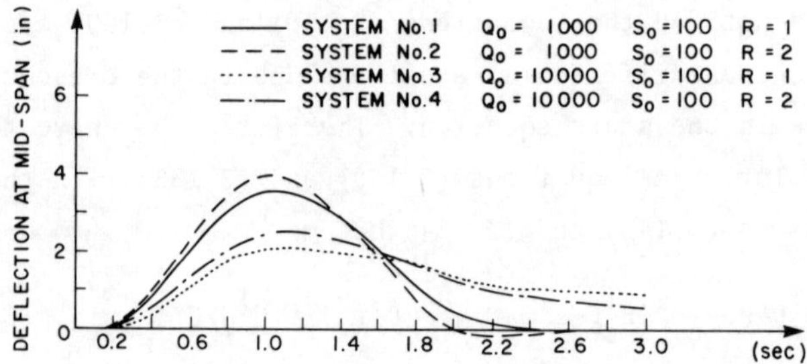

Figure 3.37 - Deflection Response using Regulator Control

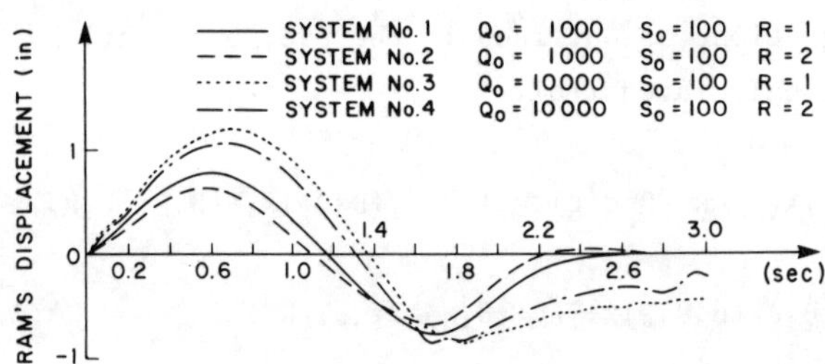

Figure 3.38 - Ram's Response using Regulator Control

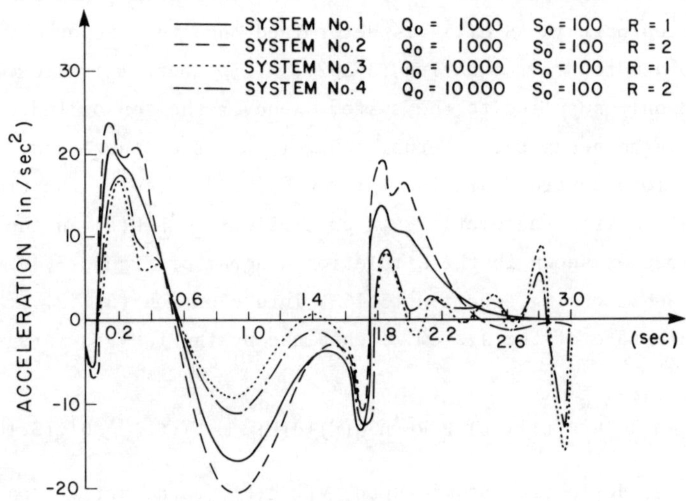

Figure 3.39 - Acceleration Response using Regulator Control

It is noticed that equations (3.159) and (3.160) are still nonhomogeneous differential equations due to the presence of the disturbance in the state equation. Therefore, they have the same type of solution as equations (3.138) and (3.135). In this example, equations (3.143) and (3.144) become

$$-\dot{\underline{P}}(t) = \underline{P}\,\underline{A}_c + \underline{A}_c^T\underline{P} - \underline{P}\,\underline{B}_c\underline{R}^{-1}\underline{B}_c^T\underline{P} + \underline{C}^T Q_0\underline{C} \ , \qquad \underline{P}(t_f) = \underline{C}^T S_0\underline{C} \ , \qquad (3.161)$$

$$-\dot{\underline{q}}(t) = \underline{A}_c^T\underline{q} - \underline{P}\,\underline{B}_c\underline{R}^{-1}\underline{B}_c^T\underline{q} + \underline{P}\,\underline{d} \ , \qquad\qquad \underline{q}(t_f) = \underline{0} \ . \qquad (3.162)$$

A solution to equations (3.161) and (3.162) is obtained by backward integration, as outlined in the previous section. The active control is calculated from

$$u^*(t) = -\underline{R}^{-1}\underline{B}_c^T\underline{P}(t)\underline{X}(t) - \underline{R}^{-1}\underline{B}_c^T\underline{q}(t) = [(B_1P_2+B_3P_7)x_1 + (B_1P_3+B_3P_8)x_2$$

$$+ (B_1P_5+B_3P_9)x_3 + (B_1P_8+B_3P_{10})x_4]/R + (B_1q_2+B_3q_4)/R \ . \qquad (3.163)$$

The closed-loop control part $[\underline{R}^{-1}\underline{B}_c^T\underline{P}(t)\underline{X}(t)]$ cannot be found until the current state $\underline{X}(t)$ has been determined. However, the open-loop part $[\underline{R}^{-1}\underline{B}_c^T\underline{q}(t)]$ is determined once the vector $\underline{q}(t)$ has been evaluated from equation (3.162). This control is stored [3.29] and only supplied to the system whenever the sensor indicates that the system needs the external control. On the other hand, the closed-loop control part is implemented by developing the time-varying gain matrix (automatic gain controller) $\underline{R}^{-1}\underline{B}_c^T\underline{P}(t)$ in the feedback loop as shown in the simulation diagram of Figure 3.40.

Substituting equation (3.163) into equation (3.152), the current state of the system is obtained by the forward integration of

$$\dot{\underline{X}}(t) = [\underline{A}_c - \underline{B}_c\underline{R}^{-1}\underline{B}_c^T\underline{P}(t)]\underline{X}(t) - \underline{B}_c\underline{R}^{-1}\underline{B}_c^T\underline{q}(t) + \underline{d}(t) \ , \qquad \underline{X}(t_0) = \underline{0} \ . (3.164)$$

The deflection at mid-span, $y(L/2,t)$, total active control $u^*(t)$, external active control, $u_{EXT}(t)$, and the acceleration at mid-span $\ddot{y}(L/2,t)$, for systems No. 1 and 3 of Figure 3.37, are shown in Figures 3.41, 3.42, 3.43 and 3.44, respectively. Comparing the deflection of the two systems in Figure 3.37 with that of Figure 3.44, one finds that the deflection is reduced in the tracking problem as compared to that in the regulator problem. However, the reduction is not significant.

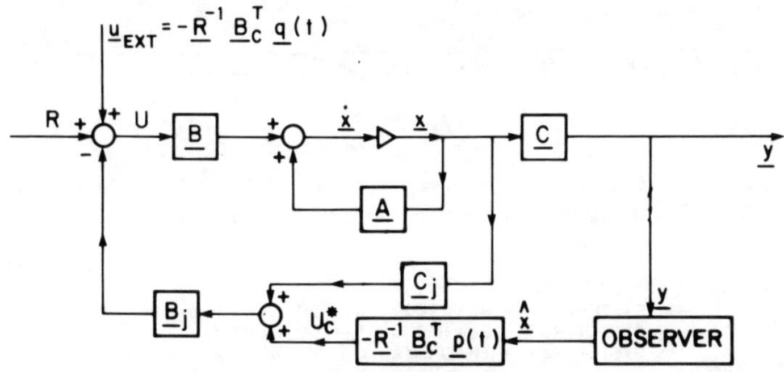

Figure 3.40 - Simulation Diagram of Tracking Problem

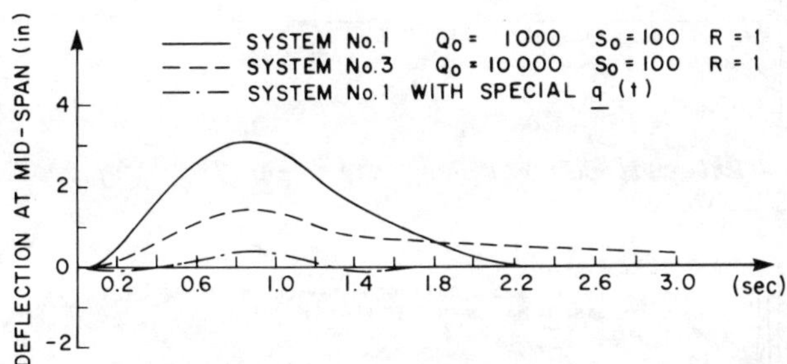

Figure 3.41 - Deflection Response using Tracking Control

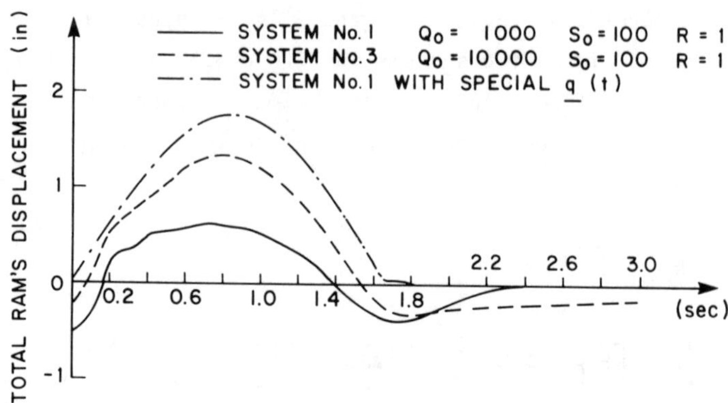

Figure 3.42 - Ram's Response using Tracking Control

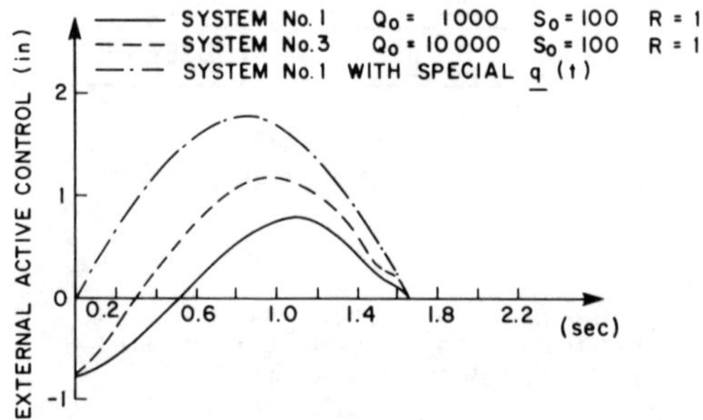

Figure 3.43 - External Control Response using Tracking Problem

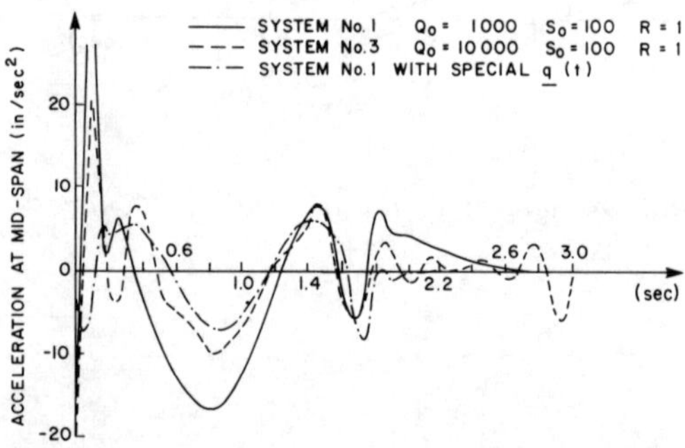

Figure 3.44 - Acceleration Response using Tracking Control

Observing equation (3.164), one concludes that a good design for the vector $\underline{q}(t)$ may lead to an excellent controlled response. In fact, one might eliminate completely the effect of the disturbance by equating $\underline{d}(t)$ with the term $[\underline{B}_c R^{-1} \underline{B}_c^T \underline{q}(t)]$ in equation (3.164). However, this leads to an inconsistent set of equations which is unsolvable. This is caused by the difference between the characteristic of the distrubance and the control moment. This fact can be investigated by looking at equations (3.147) and (3.148). It is noticed that the disturbance for each mode is both time-varying and distinct, whereas the control moment, its time-varying function, is the same for each mode. Therefore, it is impossible to find a control moment, $M(t)$, which would cancel the effect of the moving load without requiring that the control moment moves as the load does. That means $M(t)$ and $\delta(x-vt)$, in equation (3.2) must be functions of the same type [3.30].

However, it is possible to supply an external control which would eliminate the forcing function of only one of the modes. This becomes clearer by writing the two terms $d(t)$ and $[\underline{B}_c R^{-1} \underline{B}_c^T \underline{q}(t)]$ in detail as

$$\underline{d}(t) = [0 \quad 2P\sin\Omega_1 t \quad 0 \quad 2P\sin\Omega_3 t \quad]^T / mL \,, \tag{3.165}$$

$$\underline{B}_c R^{-1} \underline{B}_c^T \underline{q}(t) = [0 \quad B_1^2 q_2(t) + B_1 B_3 q_4(t) \quad 0 \quad B_1 B_3 q_2(t) + B_3^2 q_4(t)]^T / R \,. \tag{3.166}$$

Equating equations (3.165) and (3.166) term by term, one has

$$\frac{2P}{mL} \sin\Omega_1 t = \frac{B_1^2}{R} q_2(t) + \frac{B_1 B_3}{R} q_4(t) \,, \tag{3.167}$$

$$\frac{2P}{mL} \sin\Omega_3 t = \frac{B_1 B_3}{R} q_2(t) + \frac{B_3^2}{R} q_4(t) \,. \tag{3.168}$$

A unique solution for $q_2(t)$ and $q_4(t)$ from equations (3.167) and (3.168) is not possible as mentioned before. However,

one may satisfy the single equation (3.167) in order to eliminate
the forcing function of the first mode, which is well-known to contribute predominantly to the response. Assuming $q_4(t)$ to be zero,
the function $q_2(t)$ is evaluated from equation (3.167) as

$$q_2(t) = \frac{2PR}{mLB_1^2}\,\sin\Omega_1 t\ . \tag{3.169}$$

The state equation, equation (3.164), can now be written
in the following modified form:

$$\dot{\underline{X}}(t) = [\underline{A}_c - \underline{B}_c\,\underline{R}^{-1}\underline{B}_c^T\underline{P}(t)]\underline{X}(t) +
\begin{bmatrix}
0 \\
0 \\
0 \\
\dfrac{2P}{mL}\left(\sin\Omega_3 t - \dfrac{B_3}{B_1}\sin\Omega_1 t\right)
\end{bmatrix} \tag{3.170}$$

in which the first mode is unloaded, whereas the third mode has
become subjected to a modified forcing function.

The previous results could not be realized by Schorn
[3.31], in his example as he dealt with one mode only. He designed
the control system to eliminate the forcing function of the first
mode, and therefore he obtained zero controlled response. But the
effect on the higher order modes could not be verified as they had
been neglected. That emphasizes the conclusion obtained in the
previous sections that as many modes as possible should be included
in the design.

The deflection at mid-span, $y(L/2,t)$, the total active
control, $u^*(t)$, external active control, $u_{EXT}(t)$, and the acceleration at mid-span, $\ddot{y}(L/2,t)$, for the above special case, are shown,
respectively, in Figures 3.41, 3.42, 3.43 and 3.44, where the
above system is denoted by "system with special $\underline{q}(t)$". As compared
with systems No. 1 and 3 in the same figures, it can be noticed
that the best behaviour is that obtained for the above system.
However, the system with special $\underline{q}(t)$ requires a larger control
energy than the other two systems.

3.5.4　*Summary and Conclusions*

The optimal control problem was presented in Section 3.5 either as
a regulator problem or as a tracking problem. When feedback control strategy was the concern, it was taken into consideration
that the regulator problem provides a closed-loop control and the
tracking problem provides a combination of closed-loop and open-
loop control. Both problems have been applied to the optimal control of the bridge considered in the previous sections. It has been
shown that if the disturbance is known a priori and included in
the state equation, then the tracking problem is the most suitable
design method and usually provides the best controlled response.
Since, in general, the a priori form of the disturbance is known
for the civil engineering structures, it is recommended to use the
tracking optimal control problem. The external control would then
be designated to eliminate the expected disturbance, whereas the
closed-loop control takes care of any fluctuation of the actual
disturbance about the expected one. On the other hand, for structures, whose disturbance (forcing function) cannot safely be
predicted, the closed-loop control provides the active stiffness
and damping for a proper control of such structures.

It has been shown that in order to obtain an excellent
controlled response by using the tracking problem, the function
representing control force and disturbance must be of the same
type. Also, it has been shown that as many vibrational modes as
possible must be included in the design.

As a general conclusion, it can be stated that the optimal
control law depends mainly on appropriately specifying weighing
matrices and the solving of a Riccati matrix, if feedback control
was preferred. It became obvious from the previous examples that the

two main problems in optimal control are the appropriate specification of the weighing matrices, which may require a long trial
process, and the solving of a Riccati differential equation, which
becomes cumbersome for an equation of a high order. These two
difficulties lead to the conclusion that the use of optimal feedback
control should better be restricted to low order systems or to
structures whose equations of motion can be approximated by a
limited number of modes.

3.6   STUDY OF SECONDARY EFFECTS

In the previous sections, the bridge was represented by an idealized-
smooth simple beam loaded with a constant load which is moving
with constant speed. The effect of inertia coming from the moving
load, the effect of the normal force caused by the control mechanism,
and the effect of the uneveness of the bridge deck on the controlled
response, have been neglected. To check the effect of these factors
(inertia; normal force; uneveness) on the controlled response of
the structure obtained in the previous sections, these factors
will now be added to the same control system designed previously
disregarding such effects. It is believed that studying the influence of these factors in this way gives better information than
designing a new control system accounting for these factors. The
optimal control systems designed and recommended in Section 3.4.3
are therefore used in the following analysis. The structure's
equations of motion will only differ from the ones previously used
insofar as the factors mentioned above are included. To obtain
good information on the influence of these factors, the effect of
each of them will be studied separately.

In reality, a bridge is subjected to many kinds of moving
loads. In order to make this study feasible and applicable, the
controlled response of the bridge under different kinds of moving
loads compared with its uncontrolled response will be investigated.

The kinds of loads that will be taken into account are a harmonic
load moving with constant speed, a distributed load moving with
constant speed, a constant load moving with accelerated speed, and
a constant load moving with decelerated speed. An investigation
of the effect of these types of loading enables one to judge
whether the design of the suggested control mechanism of Section
3.4.3.1 remains acceptable or not. It has been shown in Section
3.4 that the tracking problem provides excellent results under the
condition that the applied forcing function is known a priori.
Unfortunately, this condition is not satisfied for bridge structures.
Therefore, one has to use the closed-loop control which provides
the damping and stiffness needed for controlling the structure in
the presence of the uncertainty regarding the applied load which
may have to be expected.

3.6.1 *Inertia Effect of the Moving Load*

By considering the inertia of the moving load, the equation of
motion of the system shown in Figure 3.45 is obtained as

$$EI \frac{\partial^2 y}{\partial x^4} + m \frac{\partial^2 y}{\partial t^2} = \left[P - \frac{P}{g} \frac{\partial^2 y}{\partial t^2} \right] \delta(x-vt) + M(t)\delta'(x-a) - M(t)\delta'(x-L+a) \; ,$$

$$(3.171)$$

in which g is the acceleration due to gravity.

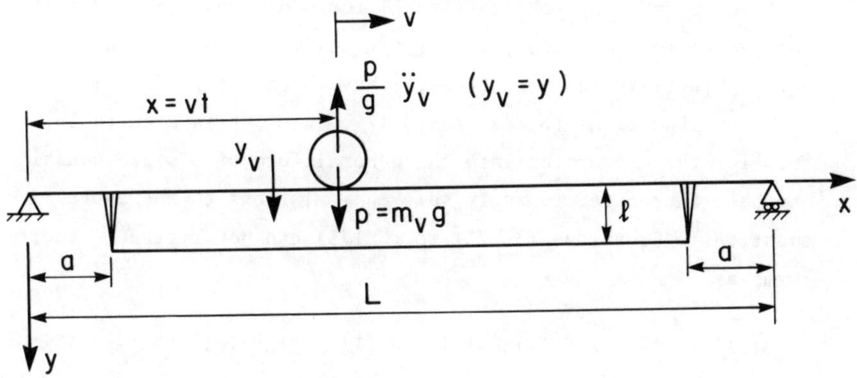

Figure 3.45 - Inertia Force of Moving Mass

A solution for equation (3.171) is assumed in the form of

$$y(x,t) = \sum_{j=1}^{\infty} \sin \frac{j\pi x}{L} A_j(t) , \qquad (3.172)$$

in which the function $\sin(j\pi x/L)$ was chosen to satisfy the boundary conditions.

Substituting equation (3.172) into equation (3.171), and applying the sine integral transformation [3.34], one obtains, for three modes, the following equations:

$$\ddot{A}_1(t)+\omega_1^2 A_1(t) = \frac{2P}{mL} \sin\Omega_1 t - \frac{2P}{mLg} \sin\Omega_1 t[\sin\Omega_1 t\ddot{A}_1(t)+\sin\Omega_2 t\ddot{A}_2(t)$$

$$+ \sin\Omega_3 t\ddot{A}_3(t)]-B_1[c_1 A_1(t)+c_3 A_3(t)+u(t)] , \qquad (3.173)$$

$$\ddot{A}_2(t)+\omega_2^2 A_2(t) = \frac{2P}{mL} \sin\Omega_2 t - \frac{2P}{mLg} \sin\Omega_2 t[\sin\Omega_1 t\ddot{A}_1(t)+\sin\Omega_2 t\ddot{A}_2(t)$$

$$+ \sin\Omega_3 t\ddot{A}_3(t)] , \qquad (3.174)$$

$$\ddot{A}_3(t)+\Omega_3^2 A_3(t) = \frac{2P}{mL} \sin\Omega_3 t - \frac{2P}{mLg} \sin\Omega_3 t[\sin\Omega_1 t\ddot{A}_1(t)+\sin\Omega_2 t\ddot{A}_2(t)$$

$$+ \sin\Omega_3 t\ddot{A}_3(t)]-B_3[c_1 A_1(t)+c_3 A_3(t)+u(t)] , \qquad (3.175)$$

in which $B_j = (4S\ell j\pi/mL^2)\cos(j\pi a/L)$ for odd modes; $c_j = (2\ell\pi j/L)\cos(j\pi a/L)$ for odd modes; $\omega_j^2 = (j\pi/L)^4 EI/m$; $\Omega_j = j\pi v/L$; and $M(t) = S[u(t)+\ell y'(a,t)-\ell y'(L-a,t)]$ have been used.

Equations (3.173) to (3.175), are not in a suitable form for the conversion into the general form of a state equation. One has, therefore, to apply some manipulations to the above equations. Equations (3.173) to (3.175) can be written in shorter forms as

$$(1+a_{11})\ddot{A}_1(t)+a_{12}\ddot{A}_2(t)+a_{13}\ddot{A}_3(t)+\omega_1^2 A_1(t) = a_1-B_1[c_1 A_1(t)+c_3 A_3(t)+u(t)] ,$$

$$(3.176)$$

$$a_{21}\ddot{A}_1(t)+(1+a_{22})\ddot{A}_2(t)+a_{23}\ddot{A}_3(t)+\omega_2^2 A_2(t) = a_2 , \qquad (3.177)$$

$$a_{31}\ddot{A}_1(t)+a_{32}\ddot{A}_2(t)+(1+a_{33})\ddot{A}_3(t)+\omega_3^2 A_3(t) = a_3-B_3[c_1 A_1(t)+c_3 A_3(t)+u(t)] , \qquad (3.178)$$

in which $a_{ij} = (2P/mLg)\sin(\Omega_i t)\sin(\Omega_j t)$; and $a_i = (2P/mL)\sin\Omega_i t$.

The above equations can also be brought into a matrix form as

$$\underline{A}^* \begin{bmatrix} \ddot{A}_1(t) \\ \ddot{A}_2(t) \\ \ddot{A}_3(t) \end{bmatrix} + \begin{bmatrix} \omega_1^2 A_1(t) \\ \omega_2^2 A_2(t) \\ \omega_3^2 A_3(t) \end{bmatrix} = \begin{bmatrix} a_1 \\ a_2 \\ a_3 \end{bmatrix} - \begin{bmatrix} B_1 \\ 0 \\ B_3 \end{bmatrix} [c_1 A_1(t) + c_3 A_3(t) + u(t)] \qquad (3.179)$$

in which the matrix $\underline{A}^*$ is given by

$$\underline{A}^* = \begin{bmatrix} (1+a_{11}) & a_{12} & a_{13} \\ a_{21} & (1+a_{22}) & a_{23} \\ a_{31} & a_{32} & (1+a_{33}) \end{bmatrix} . \qquad (3.180)$$

The inverse of the matrix $\underline{A}^*$ is assumed to be in the form of

$$\underline{A}^{*^-} = \frac{1}{DET} \begin{bmatrix} b_{11} & b_{12} & b_{13} \\ b_{21} & b_{22} & b_{23} \\ b_{31} & b_{32} & b_{33} \end{bmatrix} , \qquad (3.181)$$

in which b_{ij} is the element (i,j) of $Adj(\underline{A}^*)$, and DET is the determinant of $\underline{A}^*$.

Multiplying equation (3.179) by $\underline{A}^{*^{-1}}$, one obtains

$$\begin{bmatrix} \ddot{A}_1(t) \\ \ddot{A}_2(t) \\ \ddot{A}_3(t) \end{bmatrix} + \underline{A}^{*^{-1}} \begin{bmatrix} \omega_1^2 A_1(t) \\ \omega_2^2 A_2(t) \\ \omega_3^2 A_3(t) \end{bmatrix} = \underline{A}^{*^{-1}} \begin{bmatrix} a_1 \\ a_2 \\ a_3 \end{bmatrix} - \underline{A}^{*^{-1}} \begin{bmatrix} B_1 \\ 0 \\ B_3 \end{bmatrix} [c_1 A_1(t) + c_3 A_3(t) + u(t)] . \qquad (3.182)$$

Equation (3.182) can now be written in the first order state equation form. Considering $x_1 = A_1$, $x_2 = \dot{A}_1$, $x_3 = A_2$, $x_4 = \dot{A}_2$, $x_5 = A_3$, and $x_6 = \dot{A}_3$, equation (3.182) becomes

$$\dot{x}_1(t) = x_2 ,$$

$$\dot{x}_2(t) = \frac{1}{DET} [(b_{11}a_1 + b_{12}a_2 + b_{13}a_3) - (B_1 b_{11} + B_3 b_{13})$$
$$(c_1 x_1 + c_3 x_5 + u) - (\omega_1^2 b_{11} x_1 + \omega_2 b_{12} x_3$$
$$+ \omega_3^2 b_{13} x_5] ,$$

$$\dot{x}_3(t) = x_4 ,$$

$$\dot{x}_4(t) = \frac{1}{DET} [(b_{21}a_1 + b_{22}a_2 + b_{23}a_3) - (B_1 b_{21} + B_3 b_{23})$$
$$(c_1 x_1 + c_3 x_5 + u) - (\omega_1^2 b_{21} x_1 + \omega_2 b_{22} x$$
$$+ \omega_3^2 b_{23} x_5)] ,$$

$$\dot{x}_5(t) = x_6 ,$$

$$\dot{x}_6(t) = \frac{1}{DET} [(b_{31}a_1 + b_{32}a_2 + b_{33}a_3) - (B_1 b_{31} + B_3 b_{33})$$
$$(c_1 x_1 + c_3 x_5 + u) - (\omega_1^2 b_{31} x_1 + \omega_2^2 b_{32} x_3$$
$$+ \omega_3^2 b_{33} x_5)] ,$$

$$(3.183)$$

Equation (3.183) represents a time-varying linear system, since the variable b_{ij} is time varying and multiplied by $\underline{X}(t)$. Observing the second mode's equation, it becomes apparent that the second mode has become coupled with the other modes, which was not the case when the inertia effect was neglected.

3.6.1.1 Uncontrolled Response

The uncontrolled structure's response, including the inertia effect, is obtained by assuming $B_1 = B_3 = 0$, in equation (3.183). The uncontrolled response of deflection at mid-span is plotted in Figure 3.46 together with the uncontrolled response, neglecting the inertia effect. The uncontrolled response of acceleration at

mid-span for both, considering or neglecting the inertia effect,
is also plotted in Figure 3.47. It is obvious from the figures
that the effect of inertia on the deflection and acceleration
responses is noticeable and should be included in the design.

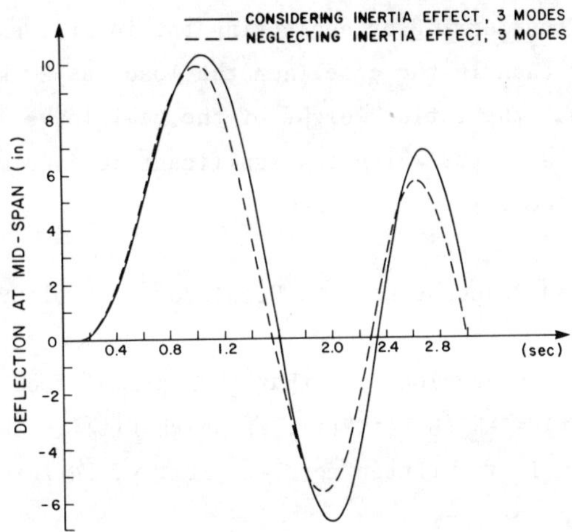

Figure 3.46 - Uncontrolled Response of Deflection at Mid-Span

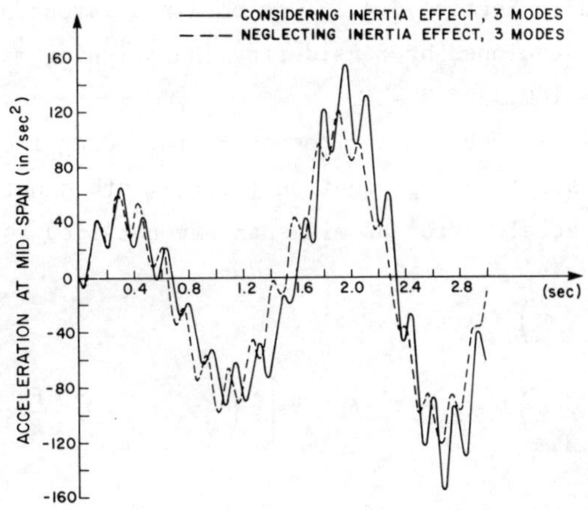

Figure 3.47 - Uncontrolled Response of Acceleration, 20 Kips

Since inertia depends on the mass of the vehicle, another investigation for a heavier vehicle, which weighs 30 kips instead of 20 kips, was carried out. The deflection and acceleration at mid-span, for this case, are shown respectively in Figures 3.48 and 3.49. It is noticed that the difference between the uncontrolled response considering or neglecting the inertia effect is relatively bigger than in the case when the load was 20 kips. This indicates that the ratio "weight of the vehicle versus weight of the bridge" is a factor which is significant for the magnitude of the inertia effect.

3.6.1.2 Controlled Response using a Regulator Control System

It has been shown, in Section 3.4, that the optimal feedback control law is $u(t) = -\underline{R}^{-1}\underline{B}_c^T\underline{P}(t)X(t)$, in which $\underline{P}(t)$ is the Riccati matrix. This formula is written more explicitly, in this section as,

$$u(t) = [(B_1P_2+B_3P_7)x_1+(B_1P_3+B_3P_8)x_2+(B_1P_5+B_3P_9)x_5+(B_1P_8+B_3P_{10})x_6]/R \ .$$

$$(3.184)$$

The Riccati matrix $\underline{P}(t)$ used here is the same one which was recommended in Section 3.4.3.1 to obtain a favourable design. This matrix is developed by considering the weighing matrices $\underline{Q}_0 = 1000$, $\underline{S}_0 = 100$, and $R = 1$.

The active controlled response, including inertia effect, is obtained by substituting equation (3.184) into equation (3.183). Deflection and acceleration at mid-span are obtained as

$$y\left(\frac{L}{2}, t\right) = x_1(t)-x_5(t) \ , \qquad\qquad (3.185)$$

$$\ddot{y}\left(\frac{L}{2}, t\right) = \dot{x}_2(t)-\dot{x}_6(t) \simeq \left[\dot{y}\left(\frac{L}{2}, t\right) - \dot{y}\left(\frac{L}{2}, t-h\right)\right]/h \ ,$$

$$(3.186)$$

in which h is the time step of integration.

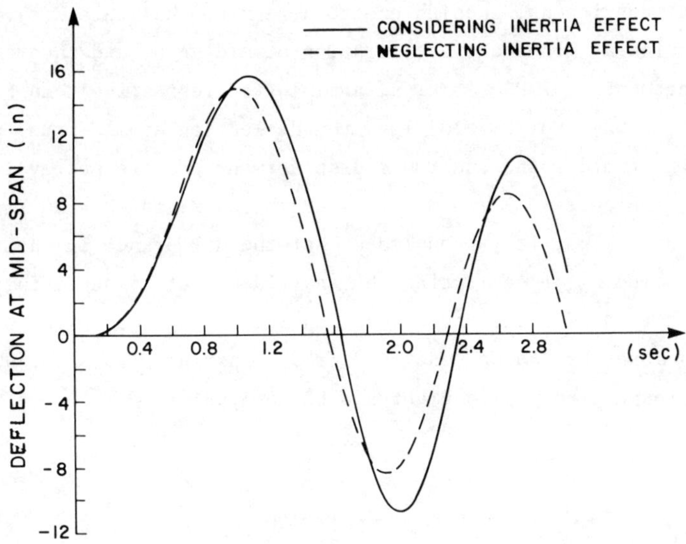

Figure 3.48 - Uncontrolled Response of Deflection, P = 30 Kips

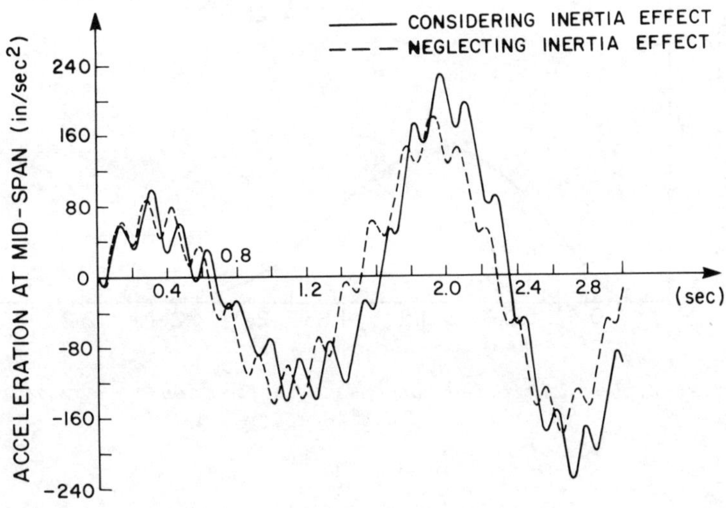

Figure 3.49 - Uncontrolled Response of Acceleration, P = 30 Kips

Integrating equation (3.183) forward, considering zero
initial conditions, enables one to determine the state $\underline{X}(t)$. The
deflection at mid-span, acceleration at mid-span, and the ram's
displacement, for P = 20 kips, are plotted respectively in Figures
3.50 to 3.52. For P = 30 kips, the deflection at mid-span, acceleration at mid-span, and ram's displacement are, respectively,
shown in Figures 3.53 to 3.55. In both cases, for P = 20 kips and
for P = 30 kips, it is concluded that the difference in the response
when including or neglecting the inertia effect is negligible. However, that difference increases when increasing the weight of the
moving load, as can be concluded from comparing Figures 3.50 to
3.52, respectively with Figures 3.53 to 3.55.

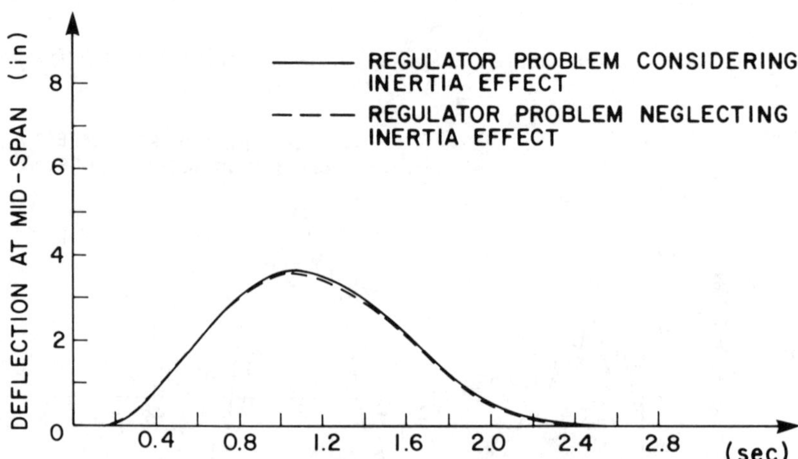

*Figure 3.50 - Controlled Response of Deflection by Regulator
Control, P = 20 Kips*

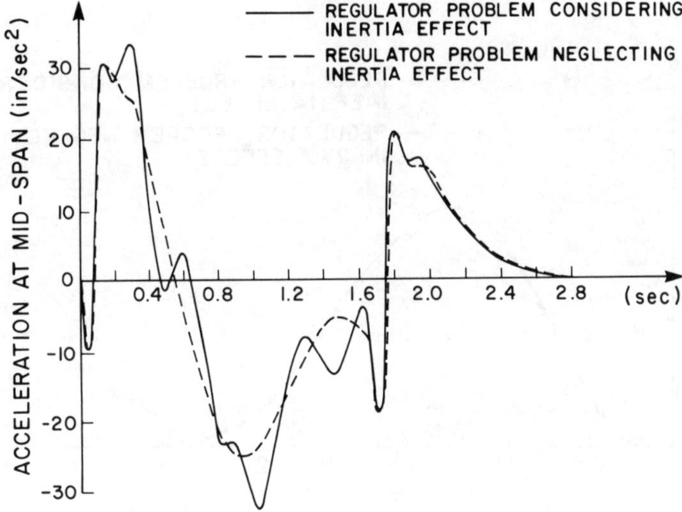

Figure 3.51 - Controlled Response of Acceleration by Regulator Control, P = 20 Kips

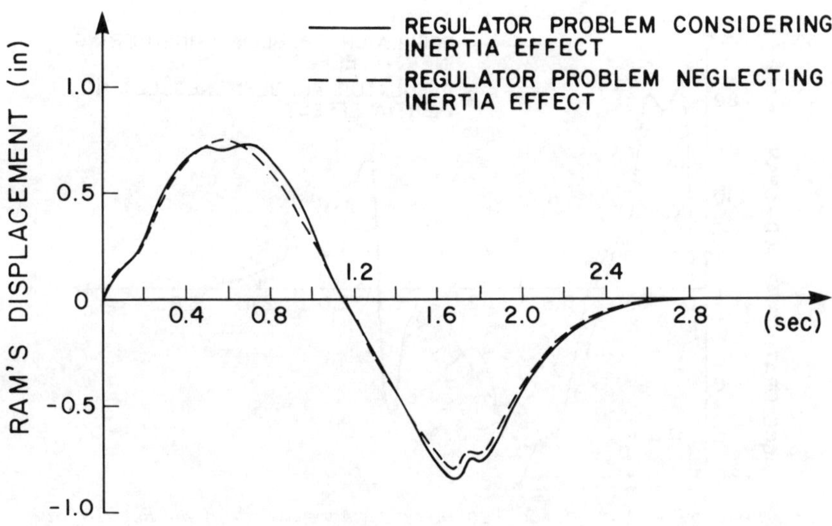

Figure 3.52 - Active Control Response by Regulator Control, P = 20 Kips

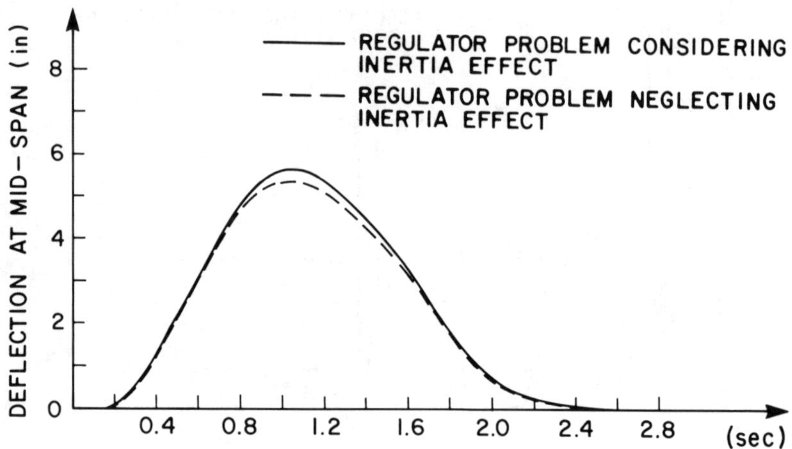

Figure 3.53 - *Controlled Response of Deflection by Regulator Control, P = 30 Kips*

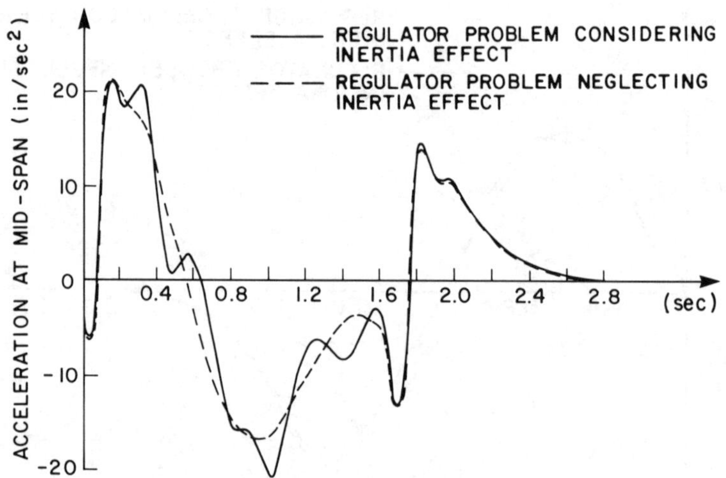

Figure 3.54 - *Controlled Response of Acceleration by Regulator Control, P = 30 Kips*

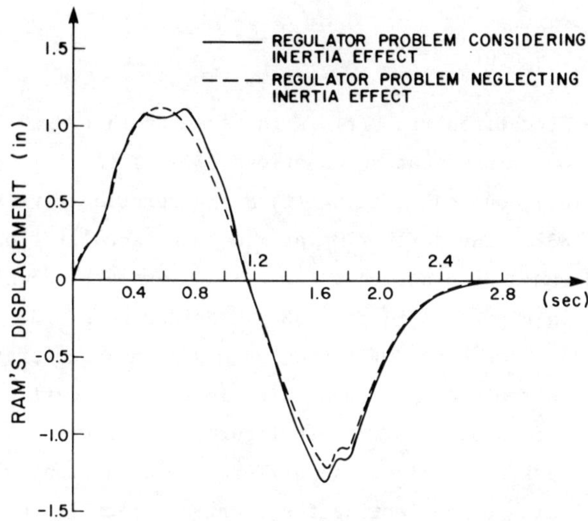

Figure 3.55 - Active Control Response by Regulator Control,
  P = 30 Kips

3.6.1.3 Controlled Response using a Tracking Control System

It has been shown in Section 3.4.3.2 that the optimal control law
provided by using the tracking method is expressed as
$u(t) = -\underline{R}^{-1}\underline{B}_c^T\underline{P}(t)\underline{X}(t) - \underline{R}^{-1}\underline{B}_c^T\underline{q}(t)$, in which $\underline{q}(t)$ is a vector which
provides the external control. The above formula is expressed as

$$u(t) = [(B_1P_2+B_3P_7)x_1(t)+(B_1P_3+B_3P_8)x_2(t)+(B_1P_5+B_3P_9)x_5(t)$$

$$+ (B_1P_8+B_3P_{10})x_6(t)]/R + (B_1q_2+B_3q_4)/R . \qquad (3.187)$$

The vector $\underline{q}(t)$ used here is the special $\underline{q}(t)$ which
eliminates the forcing function of the first mode, i.e., $q_2(t)$ and
$q_4(t)$ in equation (3.184) are given by

$$q_2(t) = \frac{2PR}{mLB_1^2} \sin\Omega_1 t \ , \tag{3.188}$$

$$q_4(t) = 0.0 \ . \tag{3.189}$$

The Riccati matrix $\underline{P}(t)$ is the same as that used in the previous section. Substituting equations (3.187) to (3.189) into equations (3.183), one can obtain $\underline{X}(t)$ by integrating forward equations (3.183). The deflection at mid-span, acceleration at mid-span, and ram's displacement, for P = 20 kips are shown, respectively, in Figures 3.56 to 3.58. Considering P = 30 kips, the controlled response of deflection at mid-span, acceleration at mid-span, and ram's displacement are plotted, respectively, in Figures 3.59 to 3.61. From these figures, one concludes that, also for the tracking problem, the difference in the controlled responses including or neglecting the inertia effect is not significant. The controlled response for P = 30 kips is also shown in Figures 3.59 to 3.61, using a control system designed to track a load of P = 20 kips only. The significant difference in the controlled responses, of this case and the case in which the control tracks P = 30 kips, indicates that one has to know the exact applied load in order to control the structure properly. However, good results may still be obtained if the forcing function is known at least statistically.

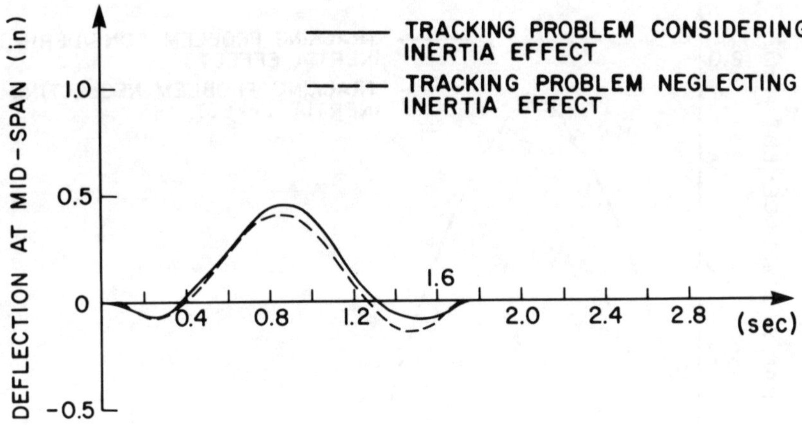

Figure 3.56 - Controlled Response of Deflection by Tracking Control, P = 20 Kips

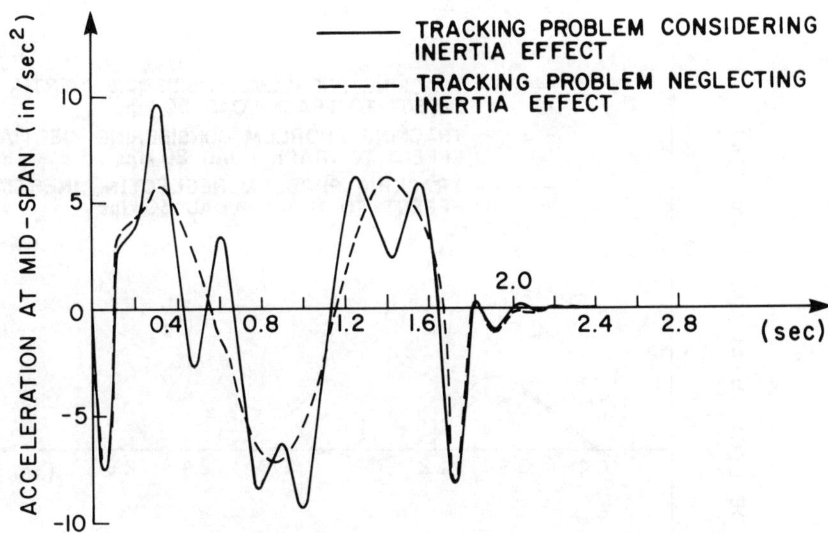

Figure 3.57 - Controlled Response of Acceleration by Tracking Control, P = 20 Kips

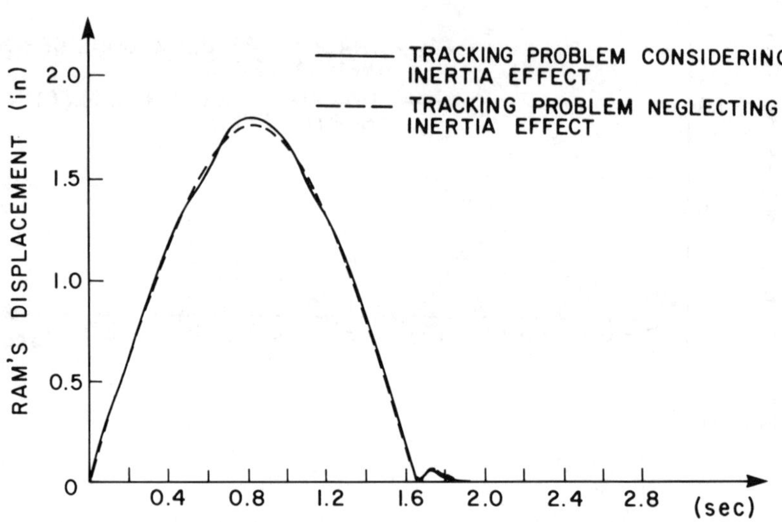

Figure 3.58 - Active Control Response by Tracking Control,
P = 20 Kips

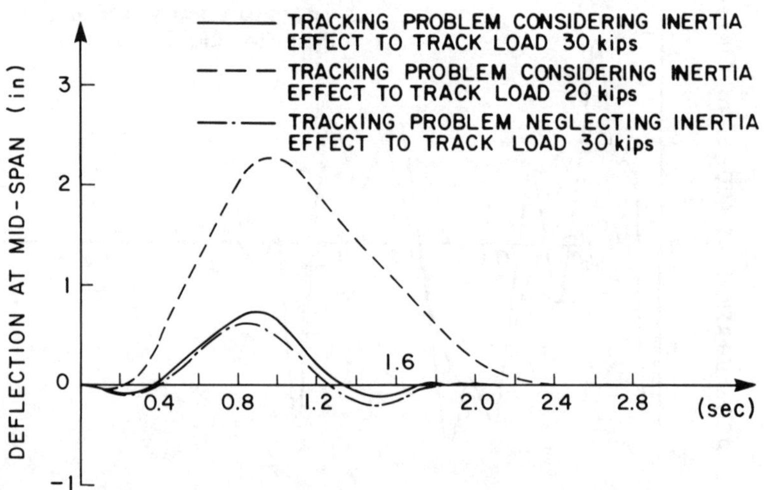

Figure 3.59 - Controlled Response of Deflection by Tracking
Control, P = 30 Kips

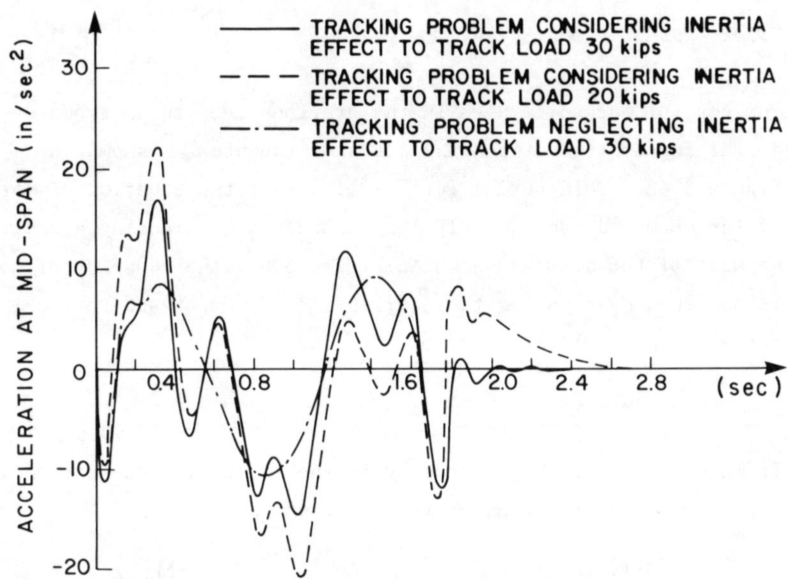

Figure 3.60 - Controlled Response of Acceleration by Tracking Control, P = 30 Kips

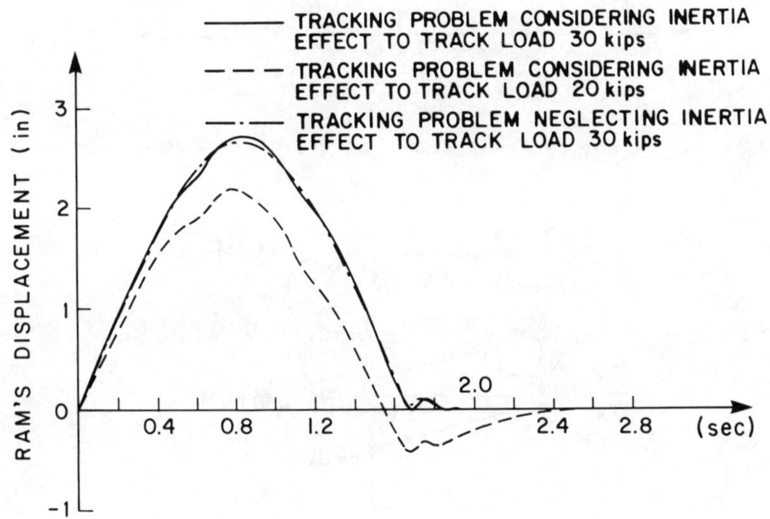

Figure 3.61 - Active Control Response by Tracking Control, P = 30 Kips

3.6.2 *Normal Force Effect*

The control mechanism used in the previous sections introduces a
normal force N(t) between the posts of the beam as shown in
Figure 3.62. This normal force will affect the equation of motion
of the beam and subsequently its response. Considering a small
segment of the beam as shown in Figure 3.63, the equation of motion
is obtained by studying the equilibrium of moment and vertical
forces. The equilibrium of moment reads

$$-dM + V \cdot dx + N \frac{\partial y}{\partial x} dx = 0 \ , \qquad (3.190)$$

in which V is the shear force, y is the deflection at section x,
N is the normal force and M is the moment at section x.

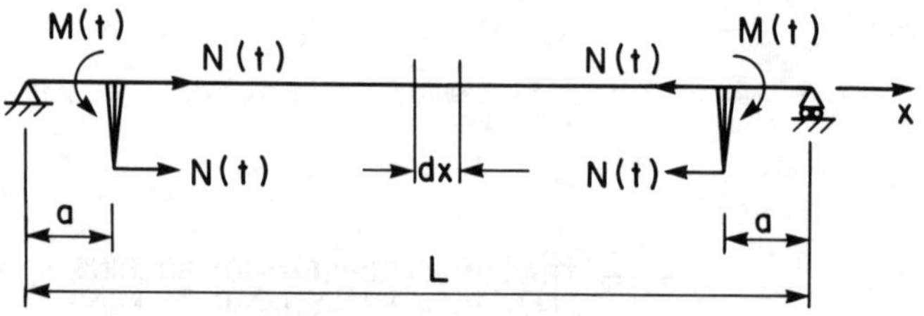

Figure 3.62 - *Normal Force Caused by the Control Mechanism*

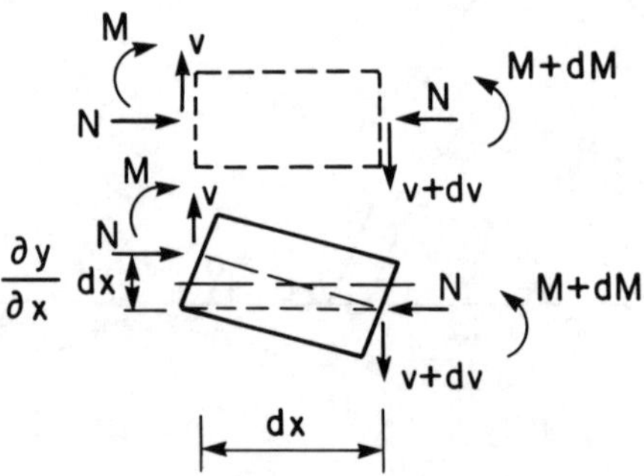

Figure 3.63 - *Free-Body Diagram Considering Normal Forces*

Using the principle of d'Alembert, one has

$$m \frac{\partial^2 y}{\partial t^2} dx - dV = 0 \;, \tag{3.191}$$

in which m is the mass per unit length, and t is the time.

Using equations (3.190) and (3.191), and since $M(x) = -EI(\partial^2 y/\partial x^2)$, one arrives at [3.35, 3.36]

$$EI \frac{\partial^4 y}{\partial x^4} + N(t) \frac{\partial^2 y}{\partial x^2} + m \frac{\partial^2 y}{\partial t^2} = 0 \;, \tag{3.192}$$

which describes the equation of motion of free vibration.

The equation of motion of the controlled structure, shown in Figure 3.62 is then given by

$$EI \frac{\partial^4 y}{\partial x^4} + N(t) \frac{\partial^2 y}{\partial x^2} + m \frac{\partial^2 y}{\partial t^2} = P\delta(x-vt)+M\delta'(x-a)-M(t)\delta'(x-L+a) \;, \tag{3.193}$$

in which it is assumed that N(t) is constant between the posts
at any instant.

Equation (3.193) is only valid at sections between x = a
and x = L-a. Therefore, an approximate solution is assumed in the
form of

$$y(x,t) = \sum_{j=1}^{\infty} \sin \frac{j\pi x}{L} A_j(t) \;, \tag{3.194}$$

in which the function sin (jπx/L) was chosen to satisfy the boundary
conditions and the compatibility conditions at x = a and x = L-a.

The normal force N(t) is evaluated for three modes as

$$N(t) = S[c_1 A_1(t)+c_3 A_3(t)+u(t)] \tag{3.195}$$

in which S is the stiffness of the spring; $c_j = (2\ell j\pi/L)\cos(j\pi a/L)$
for odd modes.

Substituting equation (3.195) into equation (3.193) and
applying the sine integral transformation, one obtains, after some
manipulations, for three modes

$$\ddot{A}_1 + \omega_1^2 A_1 - D_1[c_1 A_1 + c_3 A_3 + u]A_1 = \frac{2P}{mL}\sin\Omega_1 t - B_1[c_1 A_1 + c_3 A_3 + u] \, , \qquad (3.196)$$

$$\ddot{A}_2 + \omega_2^2 A_2 - D_2[c_1 A_1 + c_3 A_3 + u]A_2 = \frac{2P}{mL}\sin\Omega_2 t \, , \qquad (3.197)$$

$$\ddot{A}_3 + \omega_3^2 A_3 - D_3[c_1 A_1 + c_3 A_3 + u]A_3 = \frac{2P}{mL}\sin\Omega_3 t - B_3[c_1 A_1 + c_3 A_3 + u] \, , \qquad (3.198)$$

in which $D_j = (j^2\pi^2 S)/(mL^2)$.

Invesitgating equations (3.196) to (3.198) one concludes
that these equations have become simultaneous nonlinear differential
equations. That was not the case when the normal force was neglected. The design of a control law for nonlinear systems is very
cumbersome. The analysis in this section will show how significant
the error using a control system designed for an approximate linear
system, neglecting the normal force, and applying that control
system on the actual nonlinear system, considering the normal force,
may be.

3.6.2.1 Passive Control Response

The passive controlled response is obtained by considering $u(t) = 0$
in equations (3.196) and (3.198). Passive control response of
deflection at mid-span, and of acceleration at mid-span, considering
and neglecting the normal force, are shown respectively in
Figures 3.64 and 3.65. From these figures, it is concluded that
the normal force has a slight effect on the passive controlled
response.

3.6.2.2 Active Control Response using a Regulator Control System

Denominating $x_1 = A_1$, $x_2 = \dot{A}_1$, $x_3 = A_2$, $x_4 = \dot{A}_2$, $x_5 = A_3$, and
$x_6 = \dot{A}_3$, equations (3.196) to (3.198) can be written in the state
form as

$$\dot{x}_1(t) = x_2 ,$$

$$\dot{x}_2(t) = -\omega_1^2 x_1 + D_1[c_1 x_1 + c_3 x_5 + u]x_1 + \frac{2P}{mL}\sin\Omega_1 t$$

$$- B_1[c_1 x_1 + c_3 x_5 + u] ,$$

$$\dot{x}_3(t) = x_4 ,$$

$$\dot{x}_4(t) = -\omega_2^2 x_3 + D_2[c_1 x_1 + c_3 x_5 + u]x_3 + \frac{2P}{mL}\sin\Omega_2 t ,$$

$$\dot{x}_5(t) = x_6 ,$$

$$\dot{x}_6(t) = -\omega_3^2 x_5 + D_3[c_1 x_1 + c_3 x_5 + u]x_5 + \frac{2P}{mL}\sin\Omega_3 t$$

$$- B_3[c_1 x_1 + c_3 x_5 + u] .$$

$$(3.199)$$

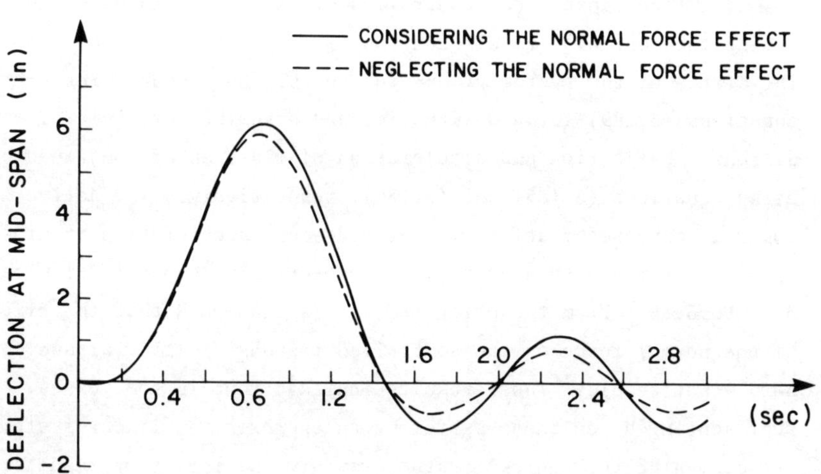

Figure 3.64 - Passive Controlled Response of Deflection

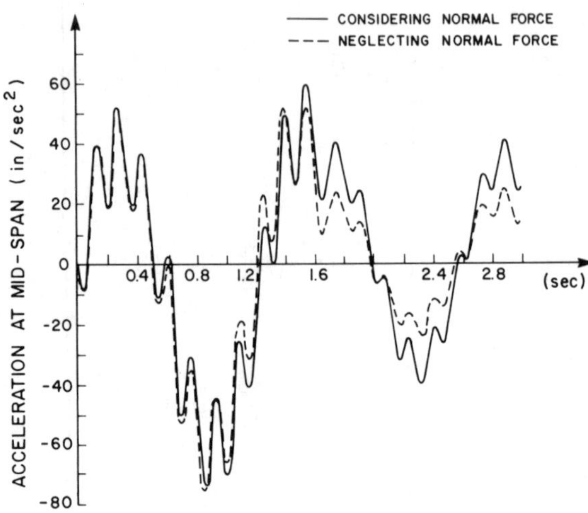

Figure 3.65 - Passive Controlled Response of Acceleration

The closed-loop controlled system is obtained by substituting the control u(t) of equation (3.184) into equations (3.199).
The states of the system at any instant are obtained by integrating
equations (3.199) forward with time and assuming zero initial conditions. Deflection and acceleration of mid-span are obtained by
using equations (3.185) and (3.186), respectively. The active
control response of deflection at mid-span, acceleration at mid-
span, and ram's displacement are, respectively, shown in Figures
3.66 to 3.68. From these figures, it is concluded that the effect
of the normal force on the controlled response is trivial and
hardly noticeable. These results indicate that in some cases,
representing a nonlinear system by an approximated linear system
for designing the control system, may give satisfactory results.
Since designing a control for nonlinear systems is difficult and
time consuming, it is recommended to design the control for an
approximate linear system and then check the effect of neglecting
the nonlinearity on the controlled response [3.37].

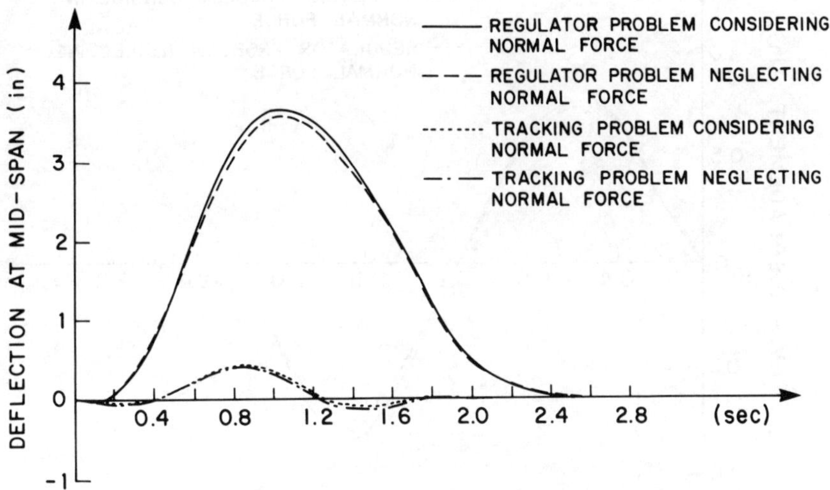

Figure 3.66 - Active Controlled Response of Deflection by Regulator and Tracking Control

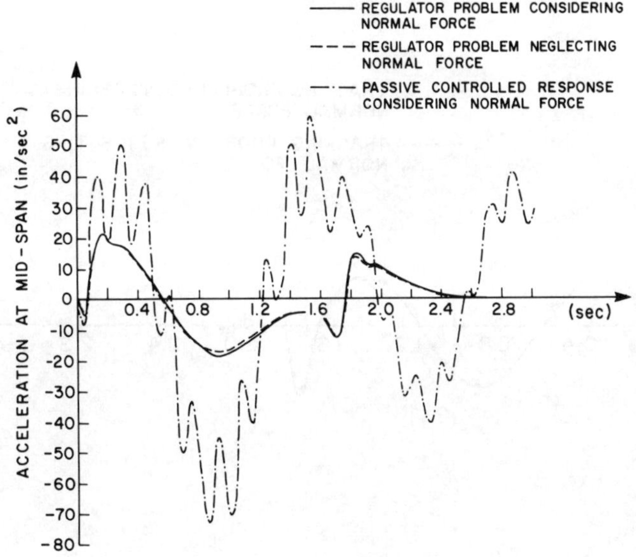

Figure 3.67 - Active Controlled Response of Acceleration by Regulator Control

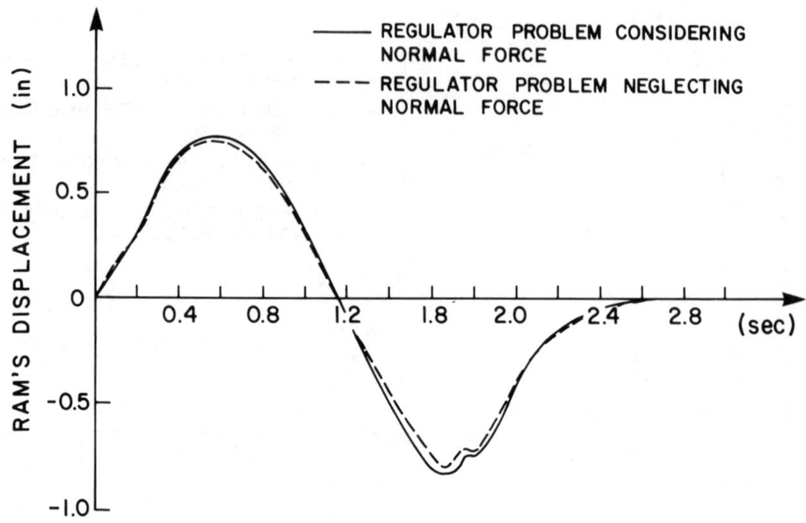

Figure 3.68 - Active Control Response by Regulator Control

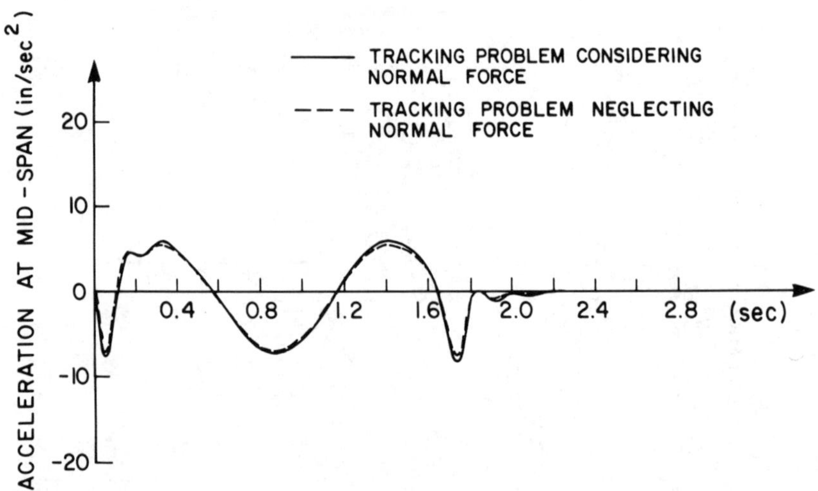

Figure 3.69 - Active Controlled Response of Acceleration by Tracking Control

3.6.2.3 Active Control Response using a Tracking Control System

Substituting equations (3.187), (3.189) into equations (3.199)
will result in a structure controlled by a closed-loop and a special
open-loop control. Deflection at mid-span and acceleration at
mid-span are plotted, respectively, in Figures 3.66 to 3.69. It is
also concluded in this case that the effect of the normal force on
the controlled response is negligible. Again, these results are
attributed to the fact that the closed-loop control is found to be
a function of the system's actual response and not to be a predetermined control.

3.6.3 *Effect of Uneveness of the Bridge Deck*

In the previous section, it has been assumed that the lateral displacement of the vehicle is the same as that of the bridge. In
such a case, the vehicle is represented by a constant force (unsprung weight) equal to the vehicle's weight, as shown in Figure
3.70. However, for an accurate analysis one has to consider the
vehicle as an independent dynamic system, which may be represented
by an undamped O.D.O.F. system (sprung mass), as shown in Figure
3.71. In this case, the load applied to the bridge is given by

$$P(t) = K[v_1(t)-y(\overline{x},t)]+m_v g \ , \qquad (3.200)$$

in which K is the stiffness of the tires, m_v is the mass of the
vehicle, x is the position of the vehicle from the left support,
g is the acceleration due to gravity, and $v_1(t)$ is the vertical
displacement of the vehicle.

Consider now the bridge being in its undeformed state
as shown in Figure 3.72. Let the uneveness of the bridge surface
be represented by the function r(x). The applied load on the
bridge is then given by

$$P(t) = K[v_1(t)-r(\overline{x})]+m_v g \ . \qquad (3.201)$$

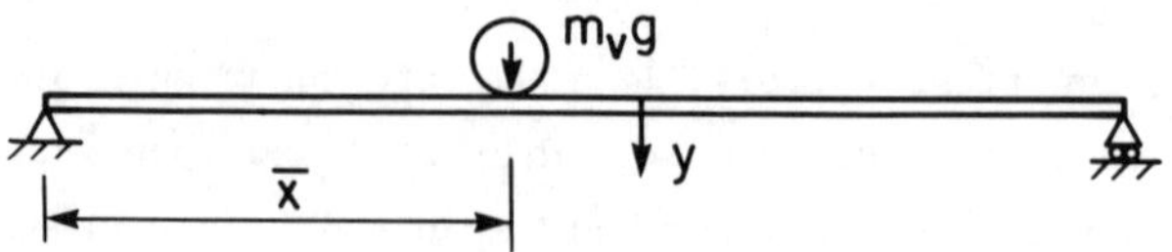

Figure 3.70 - Smooth Bridge Deck with Moving Load

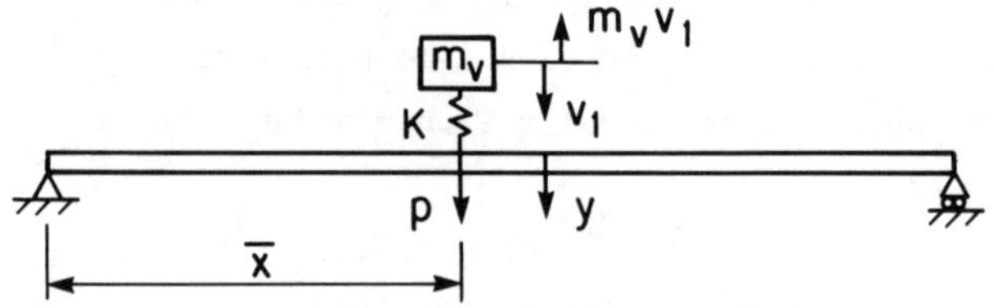

Figure 3.71 - Smooth Bridge Deck with Sprung Mass

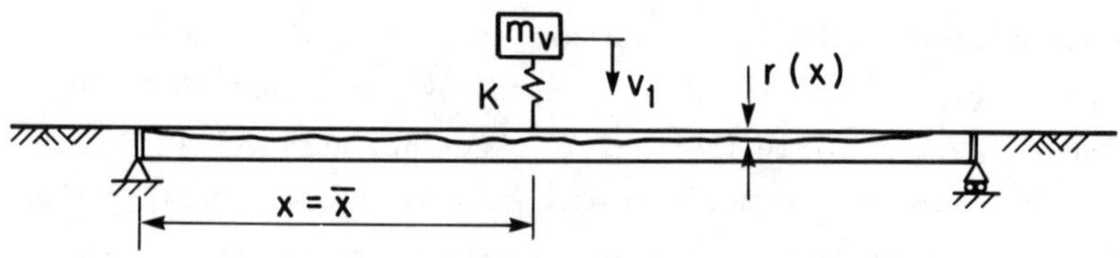

Figure 3.72 - Undeformed Irregular Bridge's Deck

The applied load, taking in addition the deflection of
the bridge into account, Figure 3.73, is

$$P(t) = K[v_1(t)-r(\overline{x})-y(\overline{x},t)]+m_v g \; . \tag{3.202}$$

Therefore the equations of vertical motion of the vehicle
and the bridge are

$$m_v \ddot{v}_1(t)+K[v_1(t)-r(\overline{x})-y(\overline{x},t)] = 0 \; , \tag{3.203}$$

$$EI \frac{\partial^4 y}{\partial x^4} + m \frac{\partial^2 y}{\partial t^2} = K[v_1(t)-r(\overline{x})-y(\overline{x},t)]\delta(x-\overline{x})+m_v g\delta(x-\overline{x})$$
$$+M(t)\delta'(x-a)-M(t)\delta'(x-L+a) \; . \tag{3.204}$$

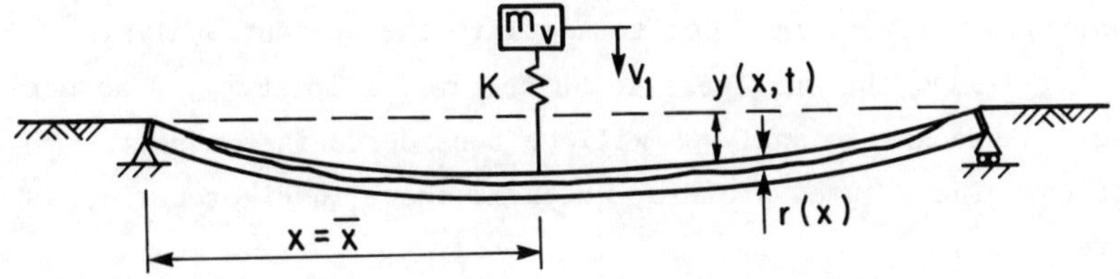

Figure 3.73 - Deformed Irregular Bridge's Deck

A solution is assumed in the form of

$$y(x,t) = \sum_{j=1}^{\infty} \sin\frac{j\pi x}{L} A_j(t) \ . \tag{3.205}$$

Substituting equation (3.205) into equation (3.204), and applying the sine integral transformation, one obtains, for three modes, the following relationships:

$$m_v\ddot{v}_1(t) + K[v_1(t) - r(\bar{x}) - A_1\sin\Omega_1 t - A_2\sin\Omega_2 t - A_3\sin\Omega_3 t] = 0 \ , \tag{3.206}$$

$$\ddot{A}_1 + \omega_1^2 A_1 = \frac{2K}{mL}[v_1(t) - r(\bar{x}) - A_1\sin\Omega_1 t - A_2\sin\Omega_2 t - A_3\sin\Omega_3 t]\sin\Omega_1 t$$

$$+ \frac{2m_v g}{mL}\sin\Omega_1 t - B_1[c_1 A_1 + c_3 A_3 + u(t)] \ , \tag{3.207}$$

$$\ddot{A}_2 + \omega_2^2 A_2 = \frac{2K}{mL}[v_1(t) - r(\bar{x}) - A_1\sin\Omega_1 t - A_2\sin\Omega_2 t - A_3\sin\Omega_3 t]\sin\Omega_2 t$$

$$+ \frac{2m_v g}{mL}\sin\Omega_2 t \ , \tag{3.208}$$

$$\ddot{A}_3 + \omega_3^2 A_3 = \frac{2K}{mL}[v_1(t) - r(\bar{x}) - A_1\sin\Omega_1 t - A_2\sin\Omega_2 t - A_3\sin\Omega_3 t]\sin\Omega_3 t$$

$$+ \frac{2m_v g}{mL}\sin\Omega_3 t - B_3[c_1 A_1 + c_3 A_3 + u(t)] \ , \tag{3.209}$$

in which $\bar{x} = vt$ is the position of the vehicle measured from the left support, $\Omega_j = j\pi v/L$; and B_j, c_j have previously been defined.

The uneveness of the bridge deck is in general a random quantity. However, in order to simplify the present analysis, one may assume the uneveness to be deterministic, e.g., a harmonic wave. A stochastic analysis will be considered in the next section. The assumed profile, which is shown in Figure 3.74, is expressed as

$$r(x) = h \left(1 - \cos \frac{2\pi x}{\ell_1} \right), \tag{3.210}$$

in which h and ℓ_1 are explained in Figure 3.74.

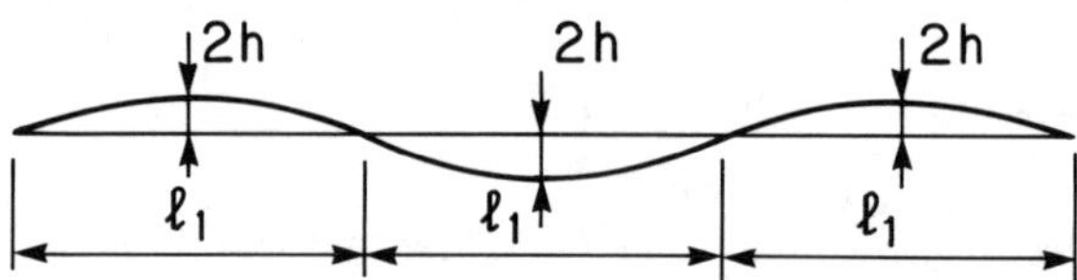

Figure 3.74 - Uneveness Profile of Bridge's Deck

Denominating $x_1 = A_1$, $x_2 = \dot{A}_1$, $x_3 = A_2$, $x_4 = \dot{A}_2$, $x_5 = A_3$, $x_6 = \dot{A}_3$, $x_7 = v_1$, and $x_8 = \dot{v}_1$, equations (3.206) to (3.209) can thus be written in the state form as

$$\dot{x}_1(t) = x_2,$$

$$\dot{x}_2(t) = -\omega_1^2 x_1 + \frac{2K}{mL} [x_7 - r(vt) - x_1 \sin\Omega_1 t - x_3 \sin\Omega_2 t - x_5 \sin\Omega_3 t]\sin\Omega_1 t$$
$$+ \frac{2m_v g}{mL} \sin\Omega_1 t - B_1 [c_1 x_1 + c_5 x_5 + u(t)] ,$$

$$\dot{x}_3(t) = x_4 ,$$

$$\dot{x}_4(t) = -\omega_2^2 x_3 + \frac{2K}{mL} [x_7 - r(vt) - x_1 \sin\Omega_1 t - x_3 \sin\Omega_2 t - x_5 \sin\Omega_3 t]\sin\Omega_2 t$$
$$+ \frac{2m_v g}{mL} \sin\Omega_2 t ,$$

$$\dot{x}_5(t) = x_6 ,$$

$$\dot{x}_6(t) = -\omega_3^2 x_5 + \frac{2K}{mL}[x_7 - r(vt) - x_1\sin\Omega_1 t - x_3\sin\Omega_2 t - x_5\sin\Omega_3 t]\sin\Omega_3 t$$

$$+ \frac{2m_v g}{mL}\sin\Omega_3 t - B_3[c_1 x_1 + c_3 x_5 + u(t)] \ ,$$

$$\dot{x}_7(t) = x_8 \ ,$$

$$\dot{x}_8(t) = -\frac{K}{m_v}[x_7 - r(vt) - x_1\sin\Omega_1 t - x_3\sin\Omega_2 t - x_5\sin\Omega_3 t] \ ,$$

in which $r(vt) = h[1-\cos(2\pi vt/\ell_1)]$; $K = 20$ kips/in; $m_v = 20/g$ kips-mass; $\ell_1 = 4$ ft.

    The above equations will be denoted by

$$\dot{\underline{X}}(t) = f(\underline{X}, u, \underline{C}) \ , \tag{3.211}$$

in which $\underline{X}$ is the state vector of dimension 8×1, and $\underline{C}$ is a time-varying vector.

3.6.3.1 Uncontrolled Response

The uncontrolled response is obtained by setting $B_1 = B_3 = 0$ in equations (3.211). Integrating equations (3.211) and considering zero initial conditions, one can determine the state of the structure at any instant. The deflection and acceleration at mid-span are determined from equations (3.185) and (3.186), respectively. These responses for $h = 0.25''$ are shown, respectively in Figures 3.75 and 3.76. For $h = 0.75''$, the uncontrolled responses of deflection and acceleration at mid-span are shown in Figures 3.77 and 3.78, respectively. These responses are compared with the case of $h = 0$ for sprung and unsprung masses. Investigating these figures, one concludes that, in general, the effect of unevenness on the response is very significant. There is a remarkable difference in the acceleration between considering and neglecting the unevenness of the bridge deck. The amplitude of acceleration gets greater by increasing the unevenness depth. For example, the largest amplitude of the acceleration, considering an uneveness with h = 0.25", is 600 in/sec, whereas for h = 0.75", the largest amplitude is 1250 in/sec^2. These two amplitudes due to uneveness are compared with the 120 in/sec^2 for unsprung mass, and the 180 in/sec^2 for sprung mass, when the uneveness was not present. The comparison shows that there is a dangerous effect due to an uneveness in the bridge deck on the human comfort and the safety of the structure from the fatigue point of view.

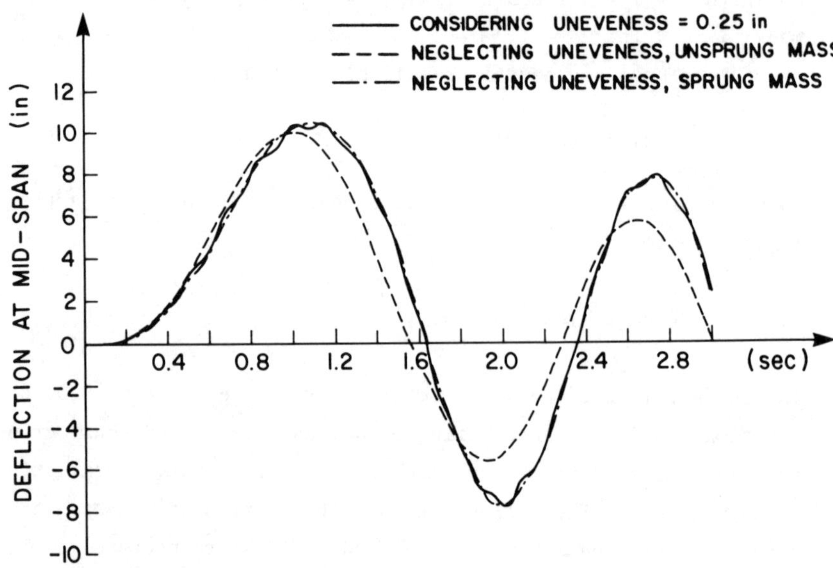

Figure 3.75 - Uncontrolled Response of Deflection, h = 0.25 in.

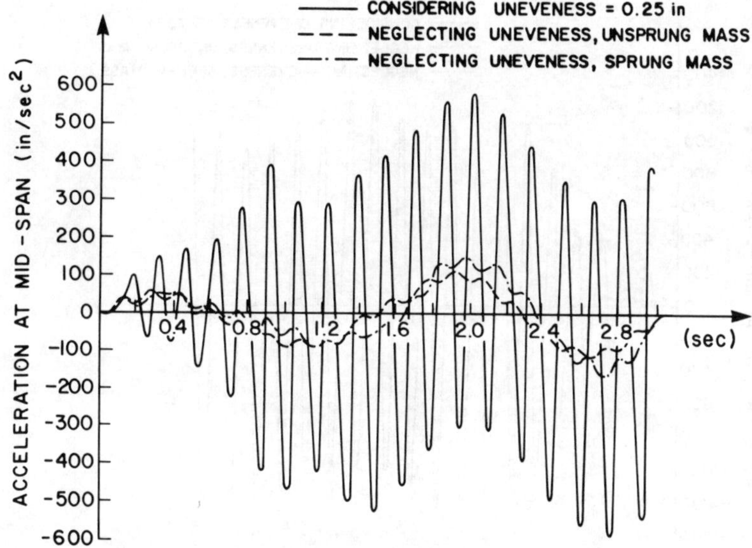

Figure 3.76 - Uncontrolled Response of Acceleration, h = 0.25 in.

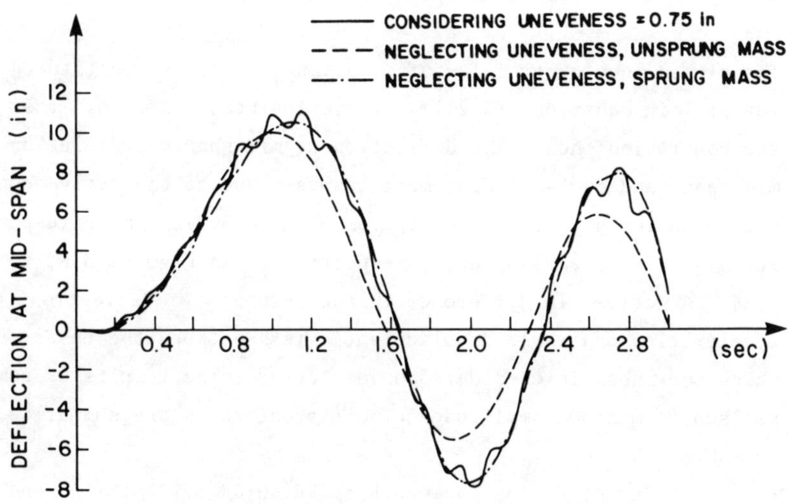

Figure 3.77 - Uncontrolled Response of Deflection, h = 0.75 in.

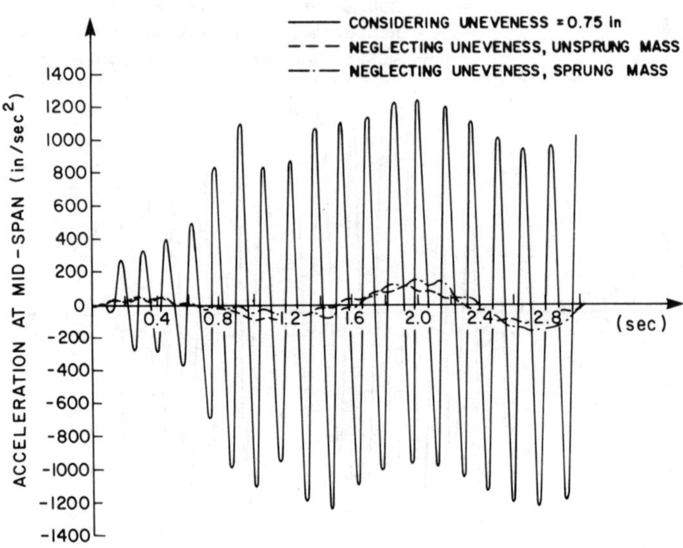

Figure 3.78 - Uncontrolled Response of Acceleration, h = 0.75 in.

3.6.3.2 Active Control Response Using a Regulator Control System

The closed-loop control law of equation (3.184) is substituted
for u(t) in equations (3.211). The calculated states are then
the controlled ones. The deflection at mid-span, acceleration at
mid-span, and ram's displacement response due to the active control
are shown for h = 0.25", in Figures 3.79 to 3.81. These responses
for h = 0.75" are also shown, respectively, in Figures 3.82 to
3.84. A noticeable difference in the response of deflection
acceleration and ram's displacement exists between the cases in
which uneveness is considered or neglected. The results of comparison of maximum amplitudes of different cases are summarized
in Table 3.5.

Investigating Table 3.5 and Figures 3.77 - 3.84, one
arrives at some conclusions: Although the controller has damped
the vibration, the difference in the transient response between

considering and neglecting uneveness is still noticeable. The ram
is oscillating very fast as compared with the cases of neglected
uneveness. The maximum amplitude of acceleration for h = 0.25"
is much smaller than for h = 0.75". These two amplitudes are
much larger than those obtained when the uneveness was neglected.

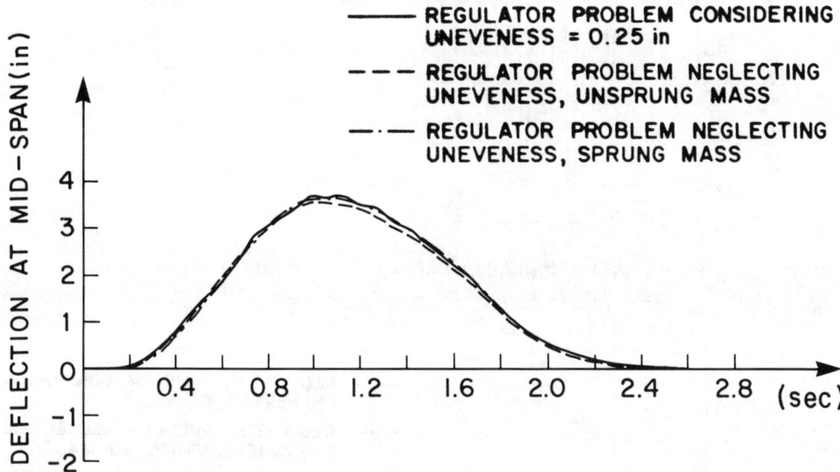

*Figure 3.79 - Controlled Response of Deflection by Regulator
Control, h = 0.25 in.*

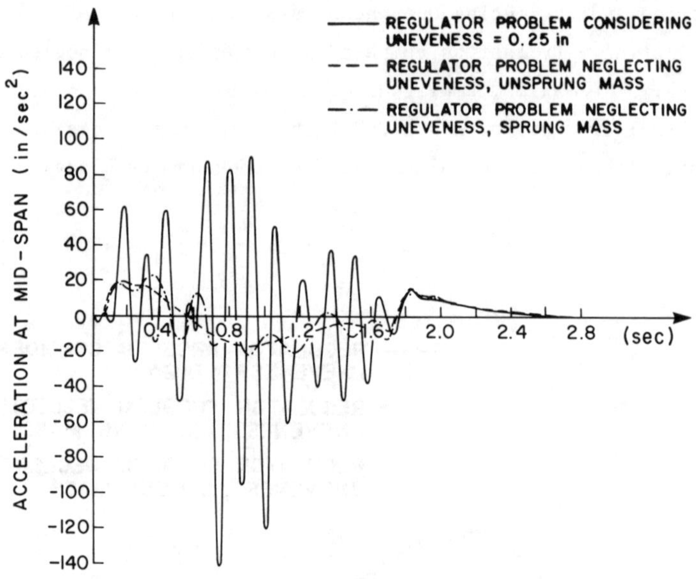

Figure 3.80 - Controlled Response of Acceleration by Regulator Control, h = 0.25 in.

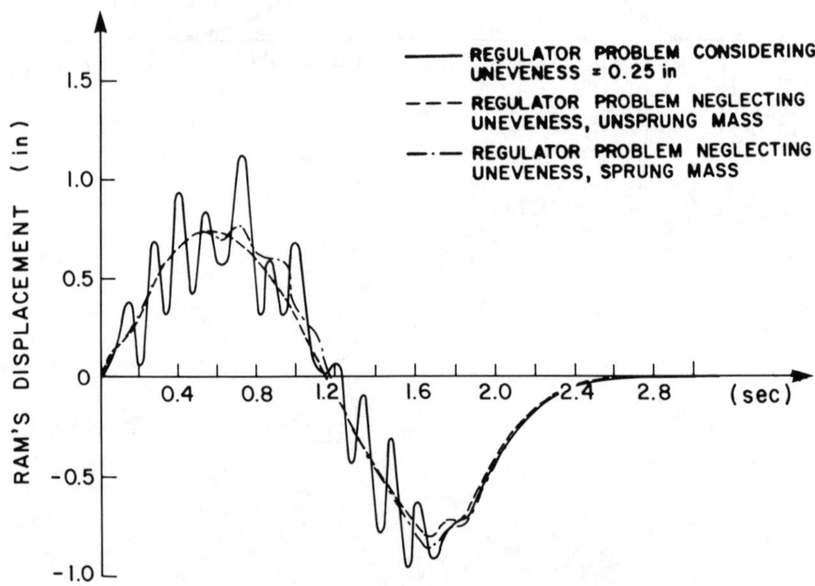

Figure 3.81 - Active Control Response by Regulator Control, h = 0.25 in.

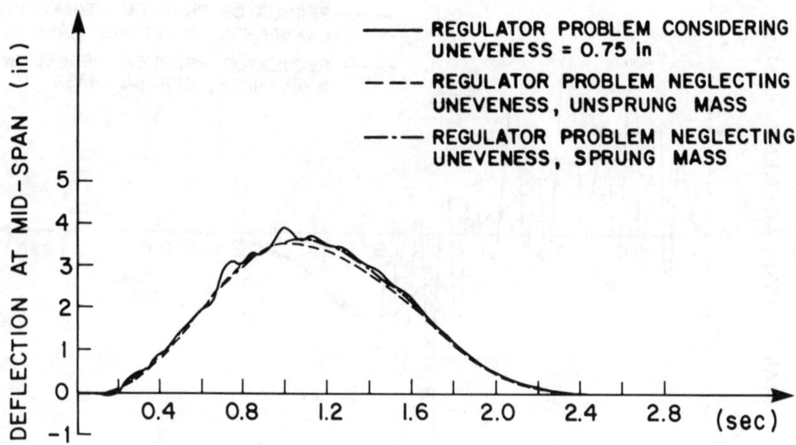

Figure 3.82 - Controlled Response of Deflection by Regulator Control, h = 0.75 in.

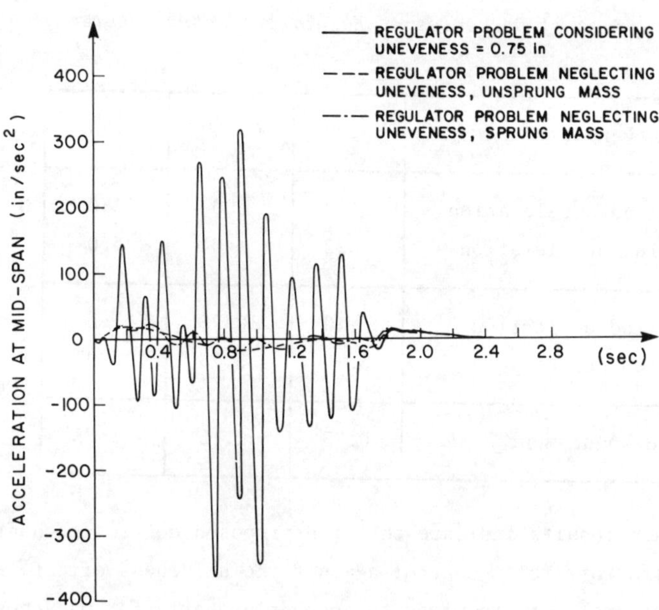

Figure 3.83 - Controlled Response of Acceleration by Regulator Control, h = 0.75 in.

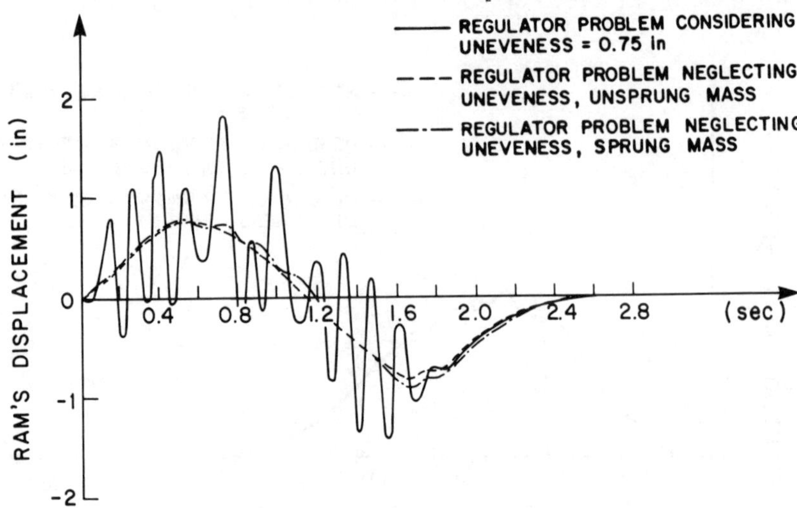

Figure 3.84 - Active Control Response by Regulator Control,
h = 0.75 in.

Table 3.5 - Effect of Uneveness on Regulator Control System

Uneveness =	0.0"	0.25"	0.75"
uncontrolled acceleration	120	600	1250
controlled acceleration	22	140	370
uncontrolled deflection	10	10.25	11.0
controlled deflection	3.50	3.75	3.95
ram's displacement	0.75	1.15	1.85

Such results indicate that the proposed control mechanism
is not satisfactory for a control against the uneveness effect, and
a new control system is required to account properly for uneveness.
The fast oscillation of the ram needs further investigation to

confirm whether the ram's response is feasible or not. These
two questions are recommended for future research.

3.6.3.3 Active Control Response using a Tracking Control System

The active controlled response using a tracking control is obtained
by substituting equation (3.187) for u(t) in equations (3.211). The
deflection at mid-span, acceleration at mid-span, and ram's displacement for both cases h = 0.25" and h = 0.75" are shown, respectively,in Figures 3.85 to 3.90. The maximum amplitudes of deflection, acceleration, and ram's displacement as compared with the
uncontrolled ones are listed in Table 3.6.

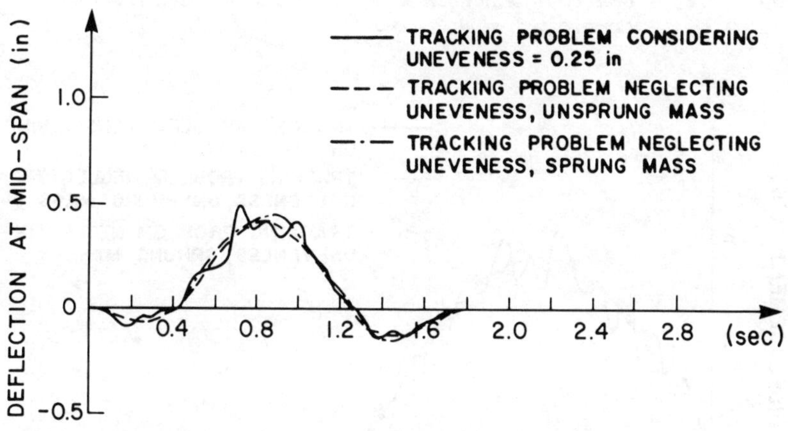

*Figure 3.85 - Controlled Response of Deflection by Tracking
       Control, h = 0.75 in.*

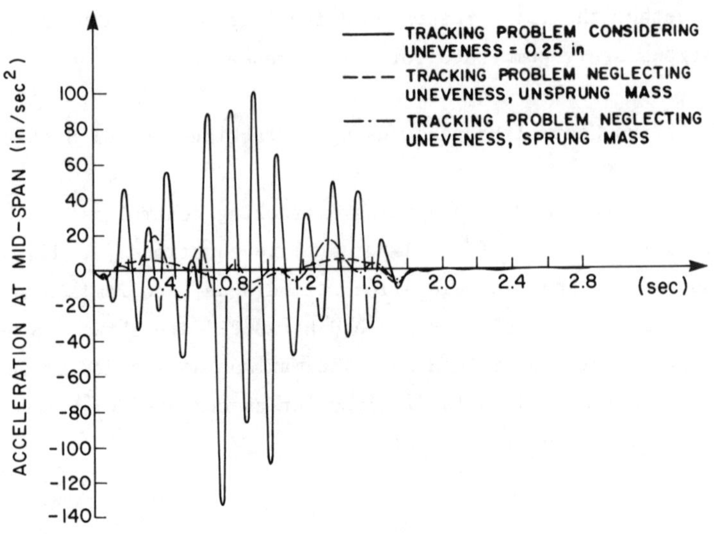

Figure 3.86 - *Controlled Response of Acceleration by Tracking Control, h = 0.25 in.*

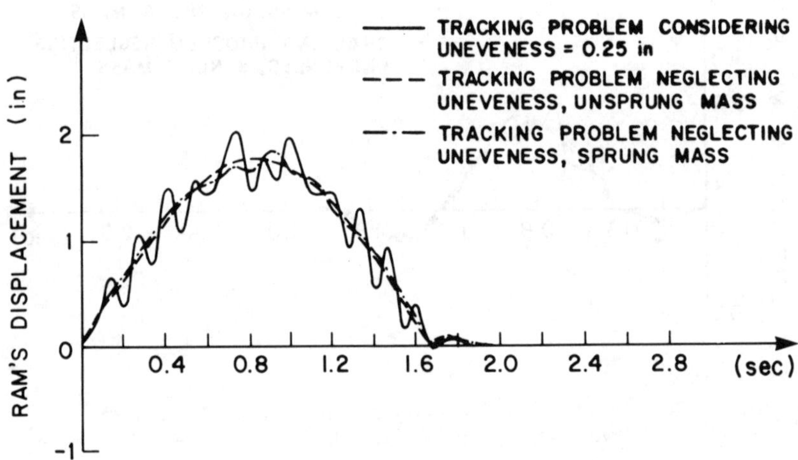

Figure 3.87 - *Active Control Response by Tracking Control, h = 0.25 in.*

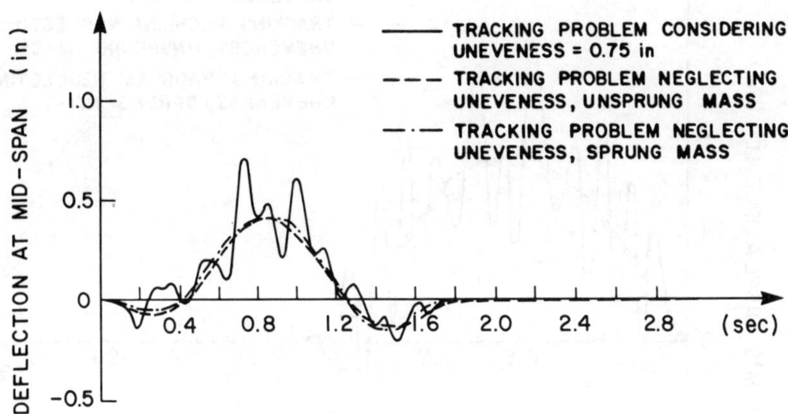

Figure 3.88 - Controlled Response of Deflection by Tracking Control, h = 0.75 in.

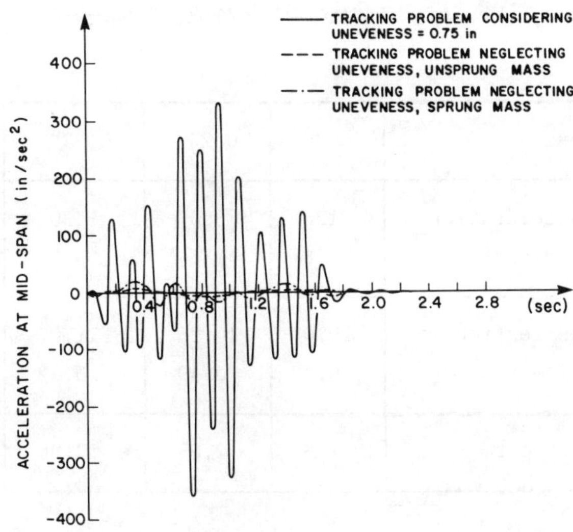

Figure 3.89 - Controlled Response of Acceleration by Tracking Control, h = 0.75 in.

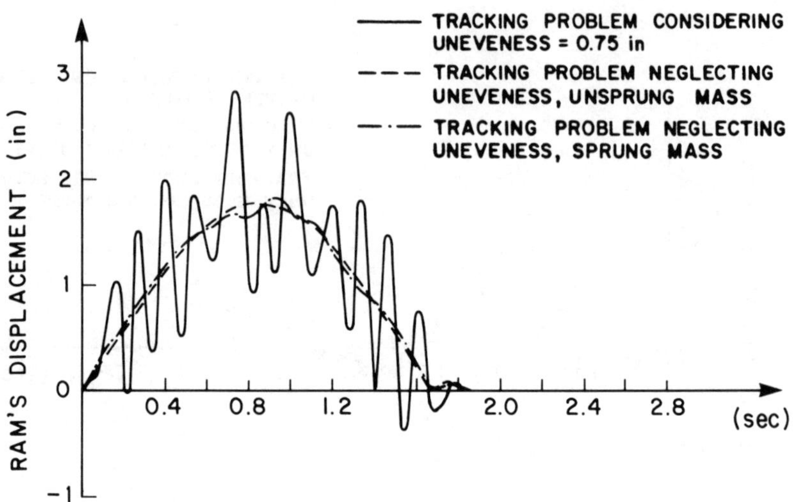

Figure 3.90 - Active Control Response by Tracking Control,
h = 0.75 in.

Table 3.6 - Effect of Uneveness on Tracking Control System

Uneveness =	0.0"	0.25"	0.75"
uncontrolled acceleration	120	600	1250
controlled acceleration	8	133	355
uncontrolled deflection	10.0	10.25	11.0
controlled deflection	0.40	0.49	0.72
ram's displacement	1.80	2.05	2.90

From Table 3.6, it is concluded again that the effect of the uneveness is significant, though the controller could dampen the vibration. The difference in the controlled acceleration response between using regulator and tracking control in the presence of uneveness is not noticeable. Tracking control, therefore, was not useful in this case since it, approximately yields the same control as the regulator control. However, the tracking control reduces the deflection of the bridge to a greater extent than the regulator control. From the human comfort and fatigue point of view, the acceleration response is more important than deflection response. Therefore, one concludes that the control designed with neglecting the uneveness is not satisfactory to control against the uneveness effect, though the control has reduced the uncontrolled acceleration response by about 75 percent.

3.6.4 *Stochastic Control Against Uneveness*

In this section, a control against the random uneveness of the bridge deck is developed. The statistical properties of the irregularities of the bridge deck will be used to find a stochastic optimal control law. A comparison between the controlled and uncontrolled response shall be given.

3.6.4.1 Stochastic Control Problem

The dynamic behaviour of a linear system subjected to a random disturbance can be described by the following state differential equation:

$$\dot{\underline{X}}(t) = \underline{A}(t)\underline{X}(t)+\underline{B}(t)\underline{U}(t)+\underline{d}(t) \; ; \quad \underline{X}(t_0) = \underline{X}_0 \; , \qquad (3.212)$$

in which $\underline{X}_0$ is an n-vector stochastic process independent of $\underline{d}(t)$ and $\underline{A}(t)$ and $\underline{B}(t)$ are matrices of approximate dimensions.

The output of the system can be expressed as

$$\underline{Y}(t) = \underline{C}(t)\underline{X}(t) \ . \tag{3.213}$$

In order to design a control law accounting for the random disturbance, a model is designed so that the output of that model is similar in its statistics to the statistics of the disturbance $\underline{d}(t)$. It has been a general practice [3.13,3.38] to let the model be driven by white noise. The equation of motion of the model (filter) can be written in a matrix form as

$$\underline{\dot{Z}}(t) = \underline{A}_f(t)\underline{Z}(t)+\underline{W}(t) \ ; \quad \underline{Z}(t_0) = \underline{Z}_0 \ , \tag{3.214}$$

in which $\underline{Z}(t)$ is a p-vector representing the state variables of the model; $\underline{W}(t)$ is a p-vector representing the white noise; $\underline{Z}_0$ is a p-vector stochastic process; and $\underline{A}_f(t)$ is a matrix of appropriate dimension.

The relation between the actual disturbance, $\underline{d}(t)$, and the model's variables, $\underline{Z}(t)$, can be expressed as

$$\underline{d}(t) = \underline{C}_f(t)\underline{Z}(t) \tag{3.215}$$

in which $\underline{C}_f(t)$ is a matrix of dimension n×p.

Equations (3.212), (3.214) and (3.215) can be combined to yield one single system given by

$$\underline{\dot{X}}_c = \underline{A}_c(t)\underline{X}_c(t)+\underline{B}_c(t)\underline{U}(t)+\underline{W}_c(t) \ , \quad \underline{X}_c(t_0) = \begin{bmatrix} \underline{X}_0 \\ \underline{Z}_0 \end{bmatrix} \ , \tag{3.216}$$

in which $\underline{X}_c(t) = [\underline{X}^T(t)\underline{Z}^T(t)]^T$, and $\underline{A}_c(t)$, $\underline{B}_c(t)$, $\underline{W}_c(t)$ are given by

$$\underline{A}_c(t) = \begin{bmatrix} \underline{A}(t) & \underline{C}_f(t) \\ 0 & \underline{A}_f(t) \end{bmatrix} \ , \tag{3.217}$$

$$\underline{B}_c(t) = [\underline{B}^T(t) \quad \underline{0}^T]^T \ , \tag{3.218}$$

$$\underline{W}_c(t) = [\underline{0} \quad \underline{W}^T(t)]^T \ . \tag{3.219}$$

The measurable output can also be expressed as

$$\underline{Y}(t) = [\underline{C}(t) \quad \underline{0}] \begin{bmatrix} \underline{X}(t) \\ \underline{Z}(t) \end{bmatrix} . \tag{3.220}$$

To find an optimal control law, one has to specify an objective function. In the presence of random variables, an objective function is considered as a function of the average response and control energy [3.32,3.33]. Using a quadratic objective function, one may use

$$j = E \left\{ \frac{1}{2} \int_{t_0}^{t_f} [\underline{Y}^T(t)\underline{Q}_0(t)\underline{Y}(t) + \underline{U}^T(t)\underline{R}(t)\underline{U}(t)]dt + \frac{1}{2} \underline{Y}^T(t_f)\underline{S}_0\underline{Y}(t_f) \right\} , \tag{3.221}$$

in which E denotes an average or expectation; $\underline{Q}_0(t)$ is a positive definite weighing matrix; $\underline{R}(t)$ is a positive definite weighing matrix; and $\underline{S}_0$ is a non-negative weighing matrix.

By means of equation (3.220), one can rewrite equation (3.221) as

$$J = E \left\{ \frac{1}{2} \int_{t_0}^{t_f} [\underline{X}^T(t)\underline{Q}(t)\underline{X}(t) + \underline{U}^T(t)\underline{R}(t)\underline{U}(t)]dt + \frac{1}{2} \underline{X}^T(t_f)\underline{S} \, \underline{X}(t_f) \right\} , \tag{3.222}$$

in which $\underline{S}$ and $\underline{Q}(t)$ are non-negative matrices given by

$$\underline{Q}(t) = \underline{C}^T(t)\underline{Q}_0(t)\underline{C}(t) , \tag{3.223}$$

$$\underline{S} = \underline{C}^T(t_f)\underline{S}_0\underline{C}(t_f) . \tag{3.224}$$

In the presence of white noise in equation (3.216), and considering the objective function in equation (3.222), the optimal control law is given by [3.13,3.27,3.38]

$$\underline{U}^*(t) = -\underline{R}^{-1}(t)\underline{B}_c^T\underline{P}(t)\underline{X}_c(t) , \tag{3.225}$$

in which $\underline{P}(t)$ is a Riccati matrix of size (n+p).

The Riccati matrix can be determined by solving the Riccati matrix differential equation which is given by

$$-\dot{\underline{P}}(t) = \underline{P}(t)\underline{A}_c(t) + \underline{A}_c^T(t)\underline{P}(t) ,$$

$$-\underline{P}(t)\underline{B}_c(t)\underline{R}^-(t)\underline{B}_c^T(t)\underline{P}(t) + \underline{Q}_c , \qquad \underline{P}(t_f) = \underline{S}_c , \qquad (3.226)$$

in which $\underline{Q}_c$ and $\underline{S}_c$ are obtained as

$$\underline{Q}_c(t) = \begin{bmatrix} \underline{Q}(t) & 0 \\ 0 & 0 \end{bmatrix} , \qquad (3.227)$$

$$\underline{S}_c = \begin{bmatrix} \underline{S} & 0 \\ 0 & 0 \end{bmatrix} . \qquad (3.228)$$

The matrix $\underline{P}$ is partitioned into the following form

$$\underline{P}(t) = \begin{bmatrix} \underline{P}_{11}(t) & \underline{P}_{12}(t) \\ \underline{P}_{21}(t) & \underline{P}_{22}(t) \end{bmatrix} , \qquad (3.229)$$

in which $\underline{P}_{11}(t)$ is of dimension n×n, $\underline{P}_{12}(t)$ is of dimension n×p, $\underline{P}_{21}(t) = \underline{P}_{12}^T(t)$, and $\underline{P}_{22}(t)$ is of dimension p×p.

Substituting equations (3.217), (3.218), (3.227) and (3.228) in equation (3.226) one has

$$-\dot{\underline{P}}_{11}(t) = \underline{P}_{11}\underline{A}(t) + \underline{A}^T(t)\underline{P}_{11} - \underline{P}_{11}\underline{B}(t)\underline{R}^{-1}(t)\underline{B}^T(t)\underline{P}_{11} + \underline{Q}(t) ,$$

$$\underline{P}_{11}(t_f) = \underline{S} , \qquad (3.230)$$

$$-\dot{\underline{P}}_{12}(t) = \underline{P}_{11}\underline{C}_f(t) + \underline{A}^T(t)\underline{P}_{12} + \underline{P}_{12}\underline{A}_f(t) - \underline{P}_{11}\underline{B}(t)\underline{R}^{-1}(t)\underline{B}^T(t)\underline{P}_{12} ,$$

$$\underline{P}_{12}(t_f) = 0 , \qquad (3.231)$$

$$-\dot{\underline{P}}_{21}(t) = -\dot{\underline{P}}_{12}^T , \qquad (3.232)$$

$$-\dot{\underline{P}}_{22}(t) = \underline{P}_{21}\underline{C}_f(t) + \underline{P}_{22}\underline{A}_f(t) + \underline{C}_f^T(t)\underline{P}_{12} + \underline{A}_f^T(t)\underline{P}_{22} - \underline{P}_{21}\underline{B}(t)\underline{R}^{-1}(t)\underline{B}^T(t)\underline{P}_{12} ,$$

$$\underline{P}_{22}(t_f) = 0 . \qquad (3.233)$$

Observing equations (3.230) to (3.233), one finds that $\underline{P}_{11}(t)$ is independent of $\underline{P}_{21}$, $\underline{P}_{12}$, and $\underline{P}_{22}$. The quantity $\underline{P}_{11}$ can be determined by means of $\underline{A}(t)$, $\underline{B}(t)$, $\underline{R}(t)$, $\underline{Q}(t)$, and $\underline{S}$. On the other hand, $\underline{P}_{12}(t)$ and $\underline{P}_{22}(t)$ are dependent on $\underline{P}_{11}(t)$.

The optimal control in equation (3.225) can now be written as

$$\underline{U}^*(t) = -\underline{R}^{-1}(t)\underline{B}^T(t)\underline{P}_{11}(t)\underline{X}(t) - \underline{R}^{-1}(t)\underline{B}^T(t)\underline{P}_{12}(t)\underline{Z}(t) \ . \qquad (3.234)$$

As can be deduced from equation (3.234), the optimal control is a combination of closed-loop control since it depends on $\underline{X}(t)$, and open-loop control which depends on $\underline{Z}(t)$. The open-loop control is then dependent on the assumed model and the statistics of the disturbance. If the model equation and the white noise intensity represent closely the actual disturbance, one may then control efficiently against the random disturbance.

In structural control terminology, one may call the foregoing problem "the stochastic tracking control problem" as compared with the deterministic tracking control problem presented in Sections 3.4.

To check the controlled response of the structure, the variance of the system is calculated from

$$\dot{\underline{K}}(t) = [\underline{A}_c - \underline{B}_c\underline{R}^{-1}\underline{B}_c^T\underline{P}]\underline{K}(t) + \underline{K}(t)[\underline{A}_c - \underline{B}_c\underline{R}^{-1}\underline{B}_c^T\underline{P}]^T + \underline{V}(t) \ , \qquad (3.235)$$

in which $\underline{K}(t)$ is the variance matrix of size $(n+p)$, and $\underline{V}(t)$ is the white noise intensity.

The initial conditions of equation (3.235) are obtained from

$$\underline{K}(t_0) = \underline{E}[\underline{X}_c(t_0)\underline{X}_c^T(t_0)] = \begin{bmatrix} \underline{K}_{xx}(t_0) & 0 \\ 0 & \underline{K}_{zz}(t_0) \end{bmatrix} \ , \qquad (3.236)$$

in which $\underline{K}_{xx}(t_0)$ is the initial condition of the structural response; and $\underline{K}_{zz}(t_0)$ is the initial condition of the model's response.

The variance of the uncontrolled response can be deter-
mined from equation (3.235) by considering $\underline{P}(t) = \underline{0}$. The magnitude
of the objective function at the optimal control can be evaluated
from [3.13]

$$J_{min} = \text{Tr}\left[\underline{P}(t_0)\underline{K}(t_0) + \int_{t_0}^{t_f} \underline{P}(t)\underline{V}(t)dt\right] , \qquad (3.237)$$

in which $\text{Tr}[---]$ is the trace operator, obtained from the summation
of diagonal elements of a matrix.

For the uncontrolled response, the objective function
measures the structural response and can be obtained from

$$J = \text{Tr}\left[\int_{t}^{t_f} \underline{Q}_c(t)\underline{K}(t)dt\right] . \qquad (3.238)$$

The variance of the applied control can also be determined
from equation (3.225) as

$$\underline{K}_{uu}(t) = \underline{R}^{-1}\underline{B}_c^T\underline{P}\ \underline{K}(t)\underline{P}\ \underline{B}_c\underline{R}^{-1} . \qquad (3.239)$$

This variance can be used for comparing the selected
trials of control instead of using equation (3.237). The stochastic
tracking control problem may be represented by the block diagram
of Figure 3.91.

3.6.4.2 Numerical Application

The unevenness $r(\bar{x})$ is considered to be a Gaussian random variable
of zero mean and variance σ^2. The mean square value of the bridge
unevenness can be assumed [3.39,3.40] as

$$E[r(vt_1)r(vt_2)] = \sigma^2 \exp\left(-\frac{|t_1-t_2|}{\theta}\right) , \qquad (3.240)$$

in which $\sigma^2 = 0.475$ sq.in.; and θ is the time constant $= 0.7759$ sec.

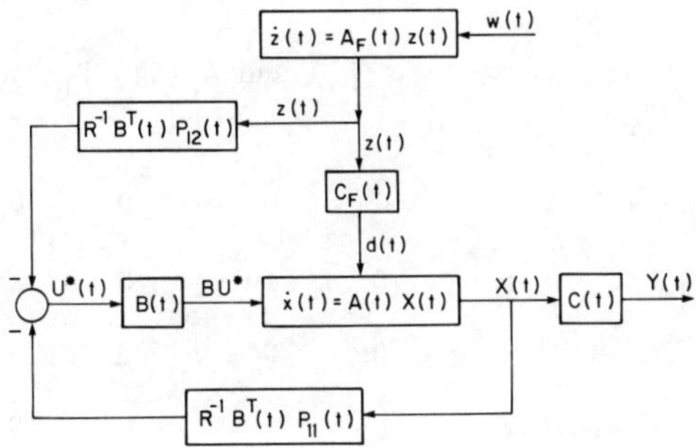

Figure 3.91 - Stochastic Tracking Control Problem

The uneveness r(vt) can be modelled by the output of the following first-order differential equation [3.38],

$$\dot{r}(vt) = -\frac{1}{\theta}\,r(vt) + W(vt) \ , \tag{3.241}$$

in which W(vt) is a white noise of intensity μ.

In general, the co-variance function of r(vt) is given by [3.38]

$$R_{rr}(t_1,t_2) = \left(\sigma^2 - \frac{\mu\theta}{2}\right) e^{-(t_1+t_2)/\theta} + \frac{\mu\theta}{2}\,e^{-|t_1-t_2|/\theta} \ . \tag{3.242}$$

Thus, from equations (3.240) and (3.242), the white noise intensity is obtained from

$$\mu = \frac{2\sigma^2}{\theta} \ , \tag{3.243}$$

from which $\mu = 1.22436$ sq in/sec.

The mean square value of the white noise can then be written as

$$E[W(vt_1)W(vt_2)] = \mu\delta(vt_1-vt_2) \ . \tag{3.244}$$

By introducing $x_5(t) = r(vt)$, equation (3.241) can be augmented with equation (3.212) to yield

$$\dot{\underline{X}}_c(t) = \underline{A}_c(t)\underline{X}_c(t) + \underline{B}_c\underline{U}(t) + \underline{W}_c(t) \ , \qquad\qquad (3.245)$$

in which $\underline{X}_c(t) = [x_1\ x_2\ x_3\ x_4\ x_5]^T$, and $A_c(t)$, B_c, $W_c(t)$ are, respectively, given by

$$\underline{A}_c(t) = \begin{bmatrix} 0 & 1 & 0 & 0 & 0 \\ (-\omega_1^2 - B_1C_1C - \bar{K}\sin\Omega_1^2 t) & 0 & \bar{K}\sin\Omega_1 t & 0 & -\bar{K}\sin\Omega_1 t \\ 0 & 0 & 0 & 1 & 0 \\ \dfrac{K}{m_v}\sin\Omega_1 t & 0 & -\dfrac{K}{m_v} & 0 & \dfrac{K}{m_v} \\ 0 & 0 & 0 & 0 & -\beta \end{bmatrix}, \qquad (3.246)$$

where $\beta = 1/\theta$,

$$\underline{B}_c = [0\ \ -B_1\ \ 0\ \ 0\ \ 0]^T \ , \qquad\qquad (3.247)$$

$$\underline{W}_c(t) = [0\ \ p\sin\Omega_1 t\ \ 0\ \ 0\ \ W(vt)]^T \ . \qquad\qquad (3.248)$$

The disturbance vector $\underline{W}_c(t)$ can be decomposed into a deterministic part and a stochastic part as follows:

$$\underline{W}_c(t) = \underline{W}_{c1}(t) + \underline{W}_{c2}(t) \ , \qquad\qquad (3.249)$$

in which $\underline{W}_{c1}(t)$ and $\underline{W}_{c2}(t)$ are given by

$$\underline{W}_{c1}(t) = [0\ \ p\sin\Omega_1 t\ \ 0\ \ 0\ \ 0]^T \ , \qquad\qquad (3.250)$$

$$\underline{W}_{c2}(t) = [0\ \ 0\ \ 0\ \ 0\ \ W(vt)]^T \ . \qquad\qquad (3.251)$$

To control against the deterministic disturbance $\underline{W}_{c1}(t)$, one may use the method presented in references [3.30] and [3.41]. To control against the random disturbance $\underline{W}_{c2}(t)$, the present approach will be used. The optimal control, $\underline{U}^*(t)$, can be found if an objective function similar to equation (3.222) is specified. To determine the proper weighing matrices, one has to perform some trials. These weighing matrices will be of the form

$$\underline{Q}_c = \rho_1 \begin{bmatrix} 1 & 1 & 1 & 1 & 0 \\ 1 & 2 & 1 & 1 & 0 \\ 1 & 1 & 1 & 1 & 0 \\ 1 & 1 & 1 & 1 & 0 \\ 0 & 0 & 0 & 0 & 0 \end{bmatrix} , \qquad (3.252)$$

$$\underline{R} = [\rho_2], \qquad (3.253)$$

$$\underline{S}_c = \rho_3 \begin{bmatrix} 1 & 1 & 1 & 1 & 0 \\ 1 & 2 & 1 & 1 & 0 \\ 1 & 1 & 1 & 1 & 0 \\ 1 & 1 & 1 & 1 & 0 \\ 0 & 0 & 0 & 0 & 0 \end{bmatrix} , \qquad (3.254)$$

in which ρ_1, ρ_2, and ρ_3 are weighing factors.

A Riccati matrix is obtained from solving equation (3.226). However, only $\underline{P}_{11}$ and $\underline{P}_{12}$ need to be determined since the optimal control is given by equation (3.234). Determination of $\underline{P}_{11}(t)$ and $\underline{P}_{12}(t)$ will be done through backward integration of equations (3.230) and (3.231). Thus, one has $(4 \times 5/2 + 4)$ first-order differential equations. Matrix $\underline{P}_{11}(t)$ represents 10 first-order differential equations and $\underline{P}_{12}(t)$ represents four differential equations.

In equation (3.231), $\underline{A}_f$ and $\underline{C}_f$ are given by

$$\underline{A}_f = [-\beta] , \qquad (3.255)$$

$$\underline{C}_f = [0 \quad -\overline{K}\sin\Omega_1 t \quad 0 \quad \frac{K}{m_v}]^T . \qquad (3.256)$$

To check the efficiency of the designed control, the variance of the controlled deflection and acceleration of the bridge is compared with the uncontrolled variance. The initial conditions of the whole system is considered to be of zero variance, e.g.,

$$\underline{K}(t_0) = \underline{0} . \tag{3.257}$$

Using equation (3.235), one may calculate the variance matrix $\underline{K}(t)$, in which $\underline{V}(t)$ is obtained from

$$E[W_{C_2}(t_1)W_{C_2}^T(t_2)] = \underline{V}(t)\delta(t_1-t_2) , \tag{3.258}$$

$$\underline{V}(t) = \begin{bmatrix} 0 & 0 & 0 & 0 & 0 \\ 0 & 0 & 0 & 0 & 0 \\ 0 & 0 & 0 & 0 & 0 \\ 0 & 0 & 0 & 0 & 0 \\ 0 & 0 & 0 & 0 & \mu \end{bmatrix} . \tag{3.259}$$

However, since the deterministic disturbance is the source which causes the random disturbance, the matrix $\underline{V}(t)$ becomes

$$\underline{V}(t) = \begin{bmatrix} 0 & 0 & 0 & 0 & 0 \\ 0 & p^2 \sin \Omega^2 t & 0 & 0 & 0 \\ 0 & 0 & 0 & 0 & 0 \\ 0 & 0 & 0 & 0 & 0 \\ 0 & 0 & 0 & 0 & \mu \end{bmatrix} . \tag{3.260}$$

The variance of the deflection and acceleration at mid-span are calculated from

$$\sigma_{x_1}^2(t) = \underline{K}_{x_1,x_1}(t) , \tag{3.261}$$

$$\sigma^2_{\ddot{x}_1}(t) = \frac{[K_{\dot{x}_1,\dot{x}_1}(t)-K_{\dot{x}_1,\dot{x}_1}(t-h)]}{h} \ , \tag{3.262}$$

in which h is the time step of integration.

Considering the numerical data in Section 3.2, the variances of the deflection are plotted in Figure 3.92 and the variances of the acceleration are given in Figure 3.93 for both the controlled and uncontrolled responses. From these figures one may conclude that the best control for the unevenness of the bridge deck has been obtained.

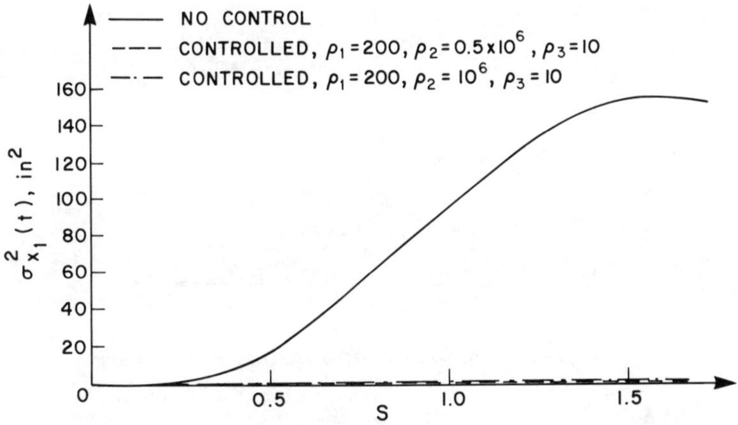

Figure 3.92 - Variances of Bridge Deflections

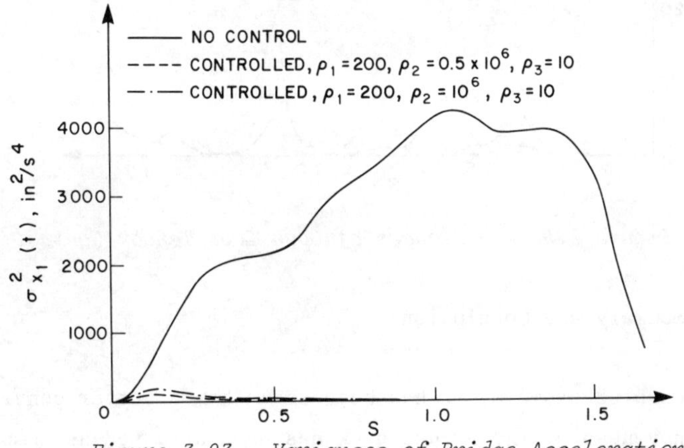

Figure 3.93 - Variances of Bridge Accelerations

The variance of the total tendon control is given in Figure 3.94 and that for the open-loop control is shown in Figure 3.95. One may realize that the developed control force is within the range of pactical applications. In fact, by investigating Figures 3.94 and 3.95, one may compute the maximum variance for the total control force, which is approximately 97.65 kips^2, and the maximum variance of the open-loop control force, which is approximately 2.343 kips^2. It is evident that both values are within the practical limits.

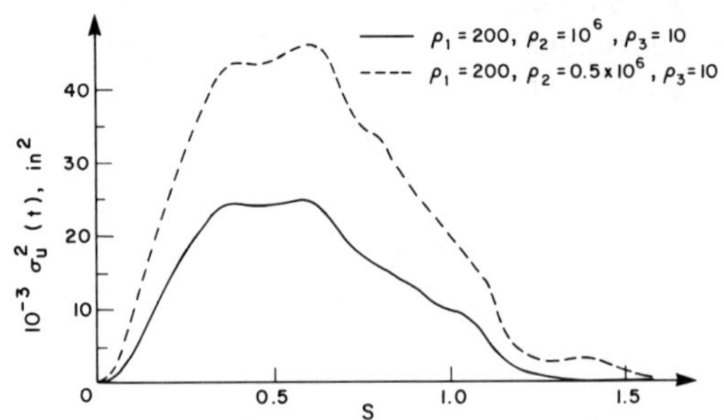

Figure 3.94 - Variances of Total Tendon Control

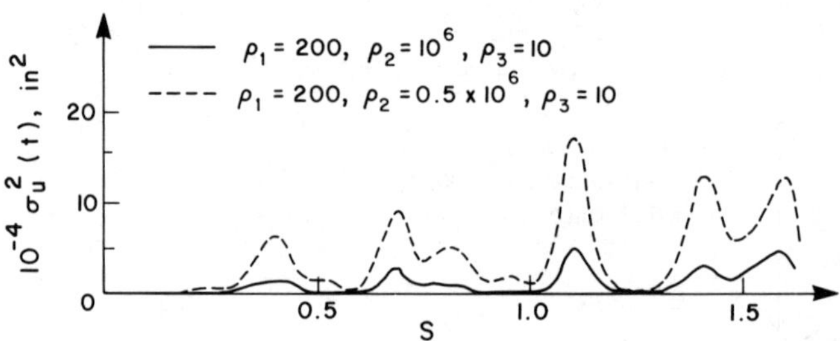

Figure 3.95 - Variances of Open-Loop Tendon Control

3.6.4.3 Summary and Conclusions

In the presented example, it has been shown how a better control against uneveness of the bridge deck has been attained as compared with that of Section 3.6.3. These results put the researchers in

a better position for controlling efficiently flexible civil
engineering structures, which are usually subjected to random
loading. However, the problem of proper modelling for the random
disturbance still needs further investigation in order to enable
the designers to find the best model [3.42].

3.6.5 *Effect of Various Moving Loads*

The previous study was only devoted to the case of a bridge under
a concentrated load, which is moving with constant speed. In
reality, a bridge is subjected to different types of moving loads,
which may even have random character. In order to make this study
as close to reality and applicable in practice as possible, some
moving loads which may act on the structure and its controlled
response due to these loads are examined in this section. One has
to notice that only the regulator control system, neglecting
secondary effects, is used here due to the uncertainty about the
moving loads.

3.6.5.1 Pulsating Force Moving with Constant Speed

This problem has been of interest in connection with the vibration
of railway bridges due to the passage of steam locomotives [3.5].
The unbalanced weight on the driving wheels produces a harmonic
force. The equation of motion of the controlled structure, in
this case is

$$EI \frac{\partial^4 y}{\partial x^4} + m \frac{\partial^2 y}{\partial t^2} = P\sin\Omega_p t\delta(x-vt)+M(t)\delta'(x-a)-M(t)\delta'(x-L+a) \quad , \quad (3.263)$$

in which Ω_p is the frequency of the pulsating force.

Resonance occurs if the frequency Ω_p equals the frequency
of any of the vibrational modes. Assuming that Ω_p is equal to the
first mode frequency, ω_1, the three mode equations are then obtained
after some manipulation as

$$\ddot{A}_1 + \omega_1^2 A_1 = \frac{2P}{mL} \sin\omega_1 t \sin\Omega_1 t - B_1 [c_1 A_1 + c_3 A_3 + u(t)] \ ,$$

$$\ddot{A}_2 + \omega_2^2 A_2 = \frac{2P}{mL} \sin\omega_1 t \sin\Omega_2 t \ , \qquad\qquad (3.264)$$

$$\ddot{A}_3 + \omega_3^2 A_3 = \frac{2P}{mL} \sin\omega_1 t \sin\Omega_3 t - B_3 [c_1 A_1 + c_3 A_3 + u(t)] \ .$$

Transforming equations (3.264) into the state form, and
substituting equation (3.184) for $u(t)$ in equations (3.264), one
is able to determine the controlled state of the structure. The
uncontrolled response is obtained by considering $B_1 = B_3 = 0$. A
comparison between the controlled and uncontrolled responses of
deflection and acceleration at mid-span, respectively, is given
in Figures 3.96 and 3.97. It is shown that the effect of resonance
is pronounced for the uncontrolled response of deflection and
acceleration, however, the controller could stabilize the structure
against the resonance effect. The improved controlled response is
ascribed to introducing active damping and stiffness to the controlled structure. This indicates that the designed regulator
control system can also control satisfactorily against harmonic
moving loads. The controller response is also shown in Figure
3.98, which indicates the feasibility of the active control energy
used.

3.6.5.2 Uniform Load Moving with Constant Speed

This case corresponds to the load caused by a train moving with a
constant speed. The train is assumed to be 300 feet long and
moving with a speed 60 ft/sec. Therefore, the train takes 5 seconds
to traverse the bridge. A good time estimate to accomplish the
control is assumed to be 6 seconds, i.e., t_f = 6 sec. A Riccati
matrix must therefore be found by integrating Riccati's equation
over 6 seconds and not over 3 seconds, as it has been done in the
previous section. However, the results obtained indicate that
the Riccati matrix for both cases is the same.

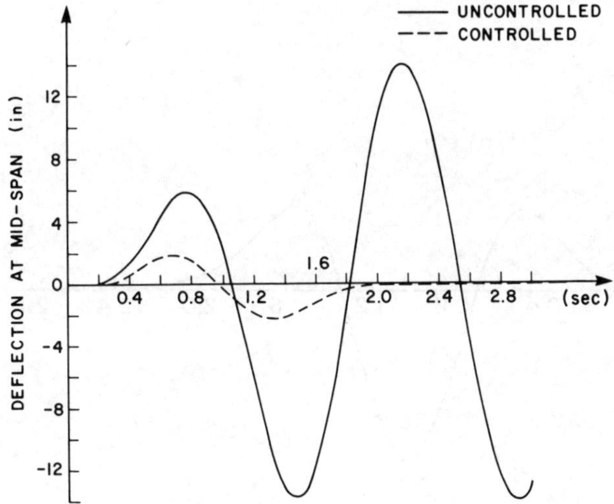

Figure 3.96 - Deflection Response due to Harmonic Loading

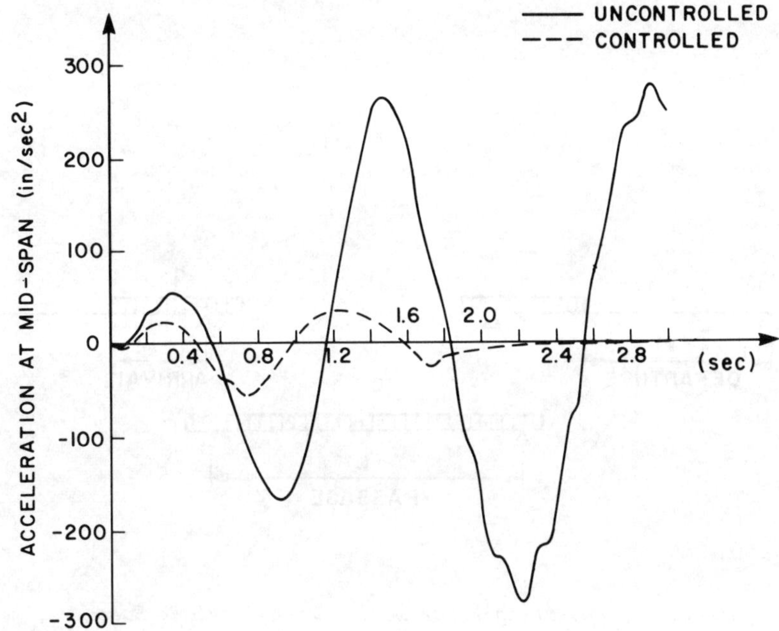

Figure 3.97 - Acceleration Response due to Harmonic Loading

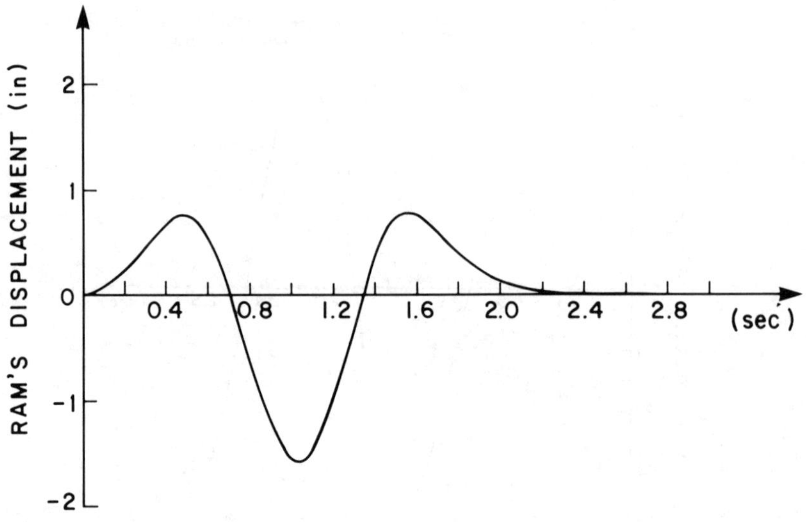

Figure 3.98 - Active Control Response due to Harmonic Loading

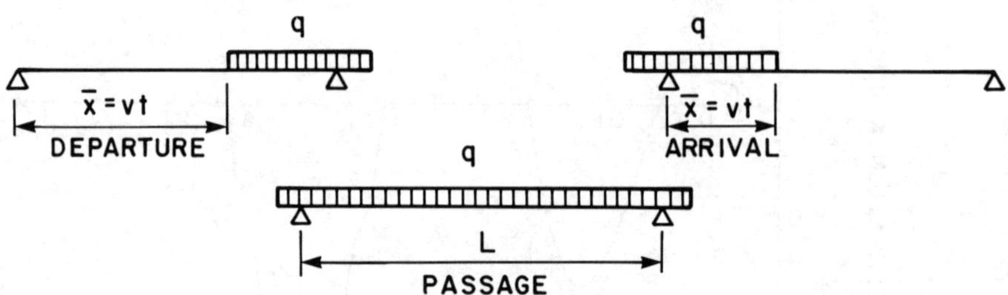

Figure 3.99 - Pulsating Force Moving with Constant Speed

The equations of motion of the forced response are divided into three portions as shown in Figure 3.99. One part corresponds to the arrival of the train until it covers the overall span of the bridge. The second part corresponds to the period of time in which the load covers the overall span. The last part is valid during the departure of the train. The interval of time for each of the three portions is 1.67 seconds and their equations of motion are, respectively, given by [3.34] ,

$$EI \frac{\partial^4 y}{\partial x^4} + m \frac{\partial^2 y}{\partial t^2} = q[1-H(x-vt)]+M(t)\delta'(x-a)-M(t)\delta'(x-L+a) , \qquad (3.265)$$

$$EI \frac{\partial^4 y}{\partial x^4} + m \frac{\partial^2 y}{\partial t^2} = q+M(t)\delta'(x-a)-M(t)\delta'(x-L+a) , \qquad (3.266)$$

$$EI \frac{\partial^4 y}{\partial x^4} + m \frac{\partial^2 y}{\partial t^2} = qH(x-vt)+M(t)\delta'(x-a)-M(t)\delta'(x-L+a) , \qquad (3.267)$$

in which $H(x)$ is the Heaviside unit function which is unity for $x \geq 0$, and zero for $x < 0$; and q is the magnitude of the load per unit length, assumed to be 50 lb/ft.

Applying an integral transformation to each one of equations (3.265) to (3.267), one is able to determine the mode's equations. The three modes' equations of equation (3.265) are

$$\ddot{A}_1+\omega_1^2 A_1 = \frac{2q}{m\pi} (1-\cos\Omega_1 t)-B_1[c_1 A_1+c_3 A_3+u(t)] , $$

$$\ddot{A}_2+\omega_2^2 A_2 = \frac{2q}{2m\pi}(1-\cos\Omega_2 t) , \qquad\qquad (3.268)$$

$$\ddot{A}_3+\omega_3^2 A_3 = \frac{2q}{3m\pi} (1-\cos\Omega_3 t)-B_3[c_1 A_1+c_3 A_3+u(t)] , $$

in which $0 \leq t \leq 1.67$.

The first three modes of equation (3.266) are obtained as

$$
\begin{aligned}
\ddot{A}_1+\omega_1^2 A_1 &= \frac{4q}{m\pi} - B_1[c_1A_1+c_3A_3+u(t)] \ , \\[2mm]
\ddot{A}_2+\omega_2^2 A_2 &= 0 \ , \\[2mm]
\ddot{A}_3+\omega_3^2 A_3 &= \frac{4q}{3m\pi} - B_3[c_1A_1+c_3A_3+u(t)] \ ,
\end{aligned}
\qquad (3.269)
$$

in which $1.67 \leq t \leq 3.33$.

$$
\begin{aligned}
\ddot{A}_1+\omega_1^2 A_1 &= \frac{2q}{m\pi}(\cos\Omega_1 t+1)-B_1[c_1A_1+c_3A_3+u(t)] \ , \\[2mm]
\ddot{A}_2+\omega_2^2 A_2 &= \frac{2q}{2m\pi}(\cos\Omega_2 t-1) \ , \\[2mm]
\ddot{A}_3+\omega_3^2 A_3 &= \frac{2q}{3m\pi}(\cos\Omega_3 t+1)-B_3[c_1A_1+c_3A_3+u(t)] \ ,
\end{aligned}
\qquad (3.270)
$$

which are valid for $3.33 \leq t \leq 5$.

The controlled response is obtained by substituting equation (3.184) for $u(t)$ in equations (3.268) to (3.270). The uncontrolled response is obtained by assuming $B_1 = B_3 = 0$. A comparison between the controlled and uncontrolled acceleration at mid-span is plotted in Figure 3.101. Investigating Figures 3.100 and 3.101, one concludes that the controller, in this case too, was able to dampen the vibration and to reduce the transient responses of deflection and acceleration. The control energy needed to accomplish the control was also feasible as shown in Figure 3.102.

3.6.5.3 Concentrated Load Moving with Decelerated Speed

This situation occurs if there is a traffic sign at the end of the bridge, which causes the vehicle to decrease its velocity. It is assumed that the load is moving with speed v, and then stops at the far end of the bridge. Furthermore, it is assumed that the motion is uniformly decelerated. The deceleration is obtained as

$$
a = (v_f^2-v_0^2)/2d \ ,
\qquad (3.271)
$$

in which v_f is the final speed, v_0 is the original speed, and d
is the distance.

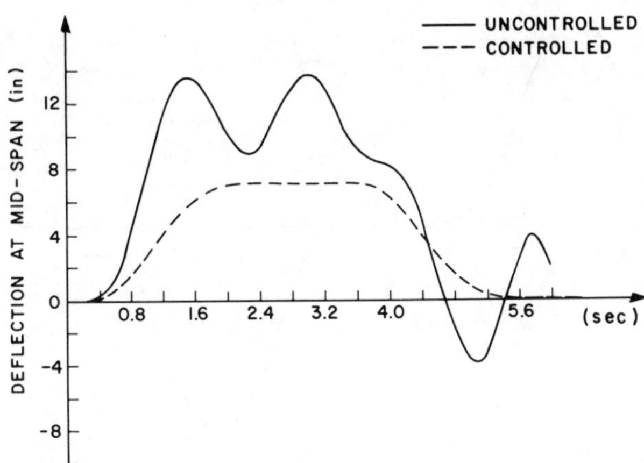

Figure 3.100 - Deflection Response due to Moving Train Load

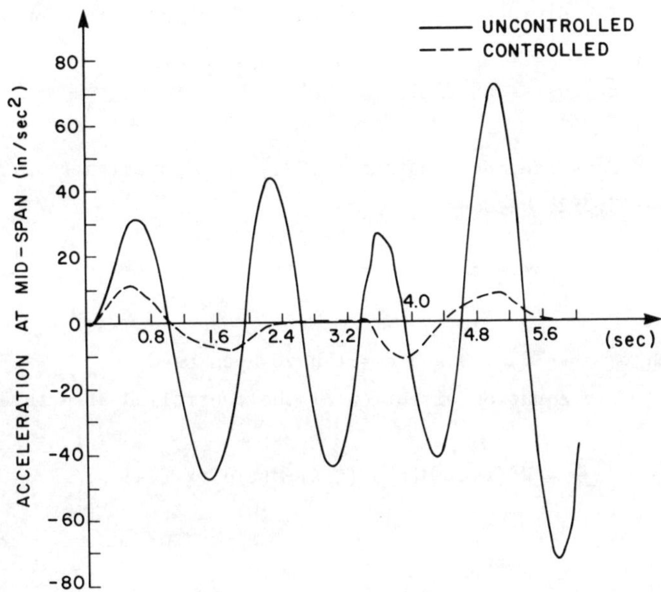

Figure 3.101 - Acceleration Response due to Moving Train Load

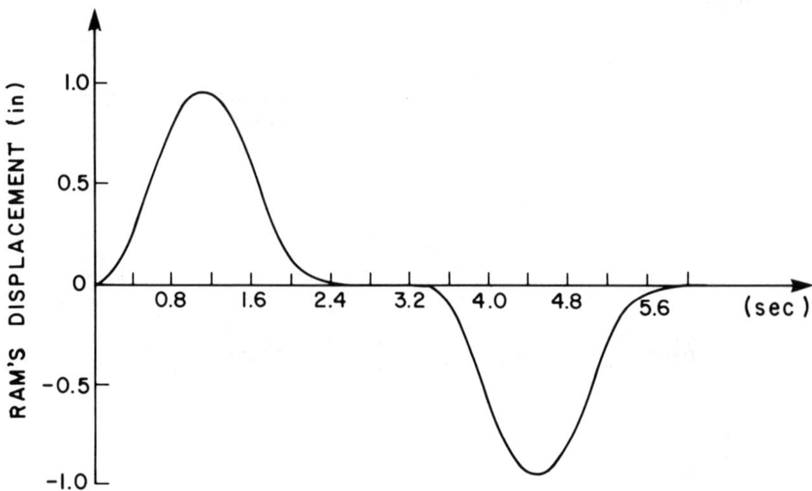

Figure 3.102 - Active Control Response due to Moving Train Load

The distance, at any instant, from the left support is

$$\bar{x} = v_0 t + at^2/2 \ . \tag{3.272}$$

Substituting equation (3.271) into equation (3.272), equation (3.272) becomes

$$\bar{x} = vt - \frac{v^2 t^2}{4L} \ , \tag{3.273}$$

in which $v_0 = v$, $v_f = 0$ and $d = L$ have been used.

The equation of motion of the controlled structure is

$$EI \frac{\partial^4 y}{\partial x^4} + m \frac{\partial^2 y}{\partial t^2} = P\delta(x-\bar{x}) + M(t)\delta'(x-a) + M(t)\delta'(x-L+a) \ . \tag{3.274}$$

Applying the integral transformation, the three modes' equations are

$$\ddot{A}_1 + \omega_1^2 A_1 = \frac{2P}{mL} \sin \frac{\bar{x}\pi}{L} - B_1[c_1A_1 + c_3A_3 + u(t)] \ , \left.\begin{array}{l} \\ \\ \end{array}\right\}$$

$$\ddot{A}_2 + \omega_2^2 A = \frac{2P}{mL} \sin \frac{2\bar{x}\pi}{L} \ , \qquad\qquad (3.275)$$

$$\ddot{A}_3 + \omega_3^2 A_3 = \frac{2P}{mL} \sin \frac{3\bar{x}\pi}{L} - B_3[c_1A_1 + c_3A_3 + u(t)] \ ,$$

in which $\bar{x}$ is given by equation (3.273).

The uncontrolled response is obtained by considering $B_1 = B_3 = 0$. The controlled response is obtained by substituting equation (3.184) for $u(t)$ in equations (3.275). Since the vehicle traverses the bridge in 3.34 seconds, the controller was offered 5 seconds, $t_f = 5$, to achieve the control. However, the Riccati matrix was found to be the same as for $t_f = 3$ seconds.

A comparison between the controlled and uncontrolled responses of deflection, and acceleration at mid-span is shown, respectively,in Figures 3.103 and 3.104. It can be concluded that the controller has dampened the vibration and reduced the transient response. The ram's displacement response is also shown in Figure 3.105, which indicates the feasibility of control energy consumption.

3.6.5.4 Concentrated Load Moving with Accelerated Speed

This situation occurs in practice if there is a traffic sign before the vehicle enters the bridge. The vehicle starts its motion with zero speed. It is assumed that the motion is uniformly accelerated, and that the vehicle has reached the speed v at the far end of the bridge. The distance from the left support is then

$$\bar{x} = \frac{v^2 t^2}{4L} \ . \qquad\qquad (3.276)$$

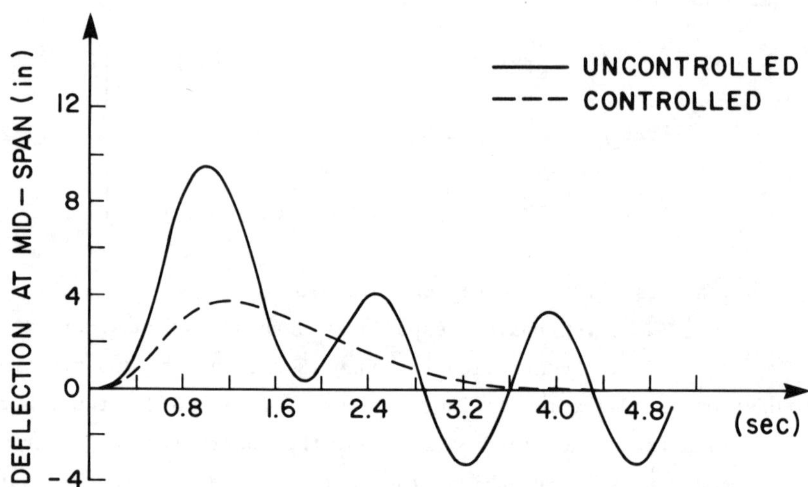

Figure 3.103 - Deflection Response due to Decelerated Motion

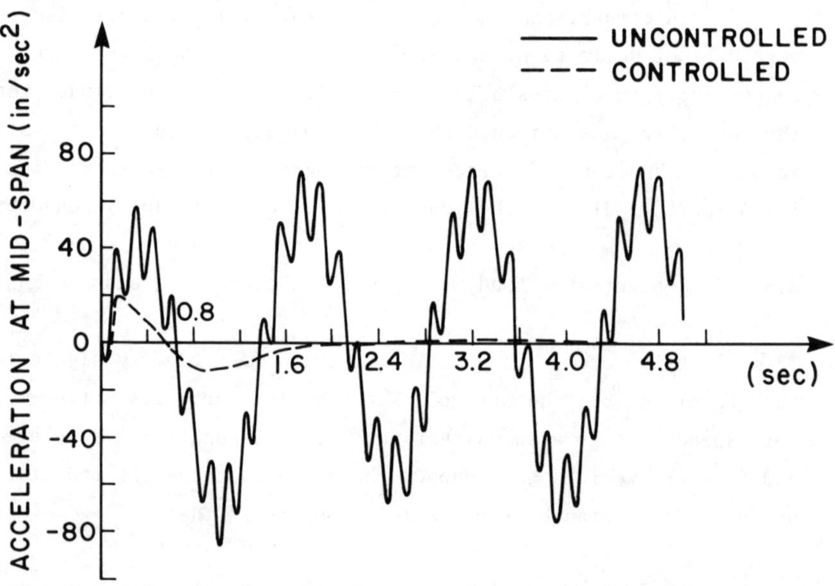

Figure 3.104 - Acceleration Response due to Decelerated Motion

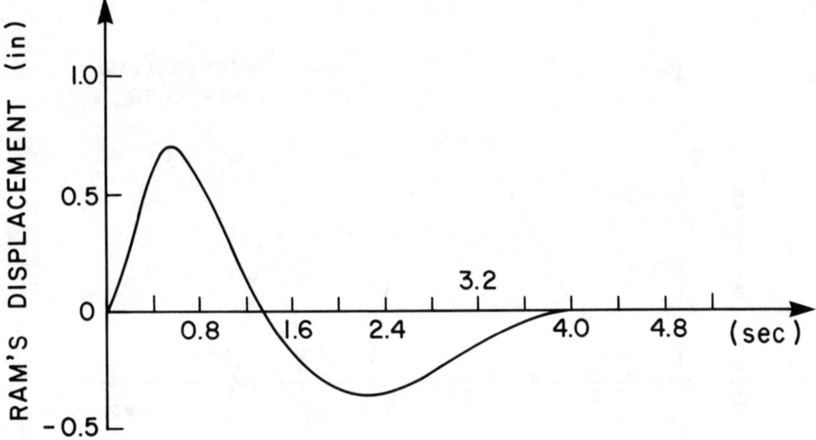

Figure 3.105 - Active Control Response due to Decelerated Motion

The three modes' equations for this case are the same as equations (3.275) except that $\bar{x}$ is now to be substituted by equation (3.276). A comparison between the controlled and uncontrolled responses of deflection and acceleration at mid-span, respectively, is shown in Figures 3.106 and 3.107. It can be concluded that the controller has provided the required control to dampen the vibration and to reduce the transient response. Comparing the uncontrolled responses of Figures 3.103 and 3.104 with those of Figures 3.106 and 3.107, one realizes that the effect of the decelerated motion on the response of the bridge is more severe than that of the accelerated motion. Therefore, it is recommended to post the traffic signals at the entrance of the bridge rather than at its exit, if the bridge is uncontrolled. The ram's displacement of the controller is displayed in Figure 3.108, showing its feasibility.

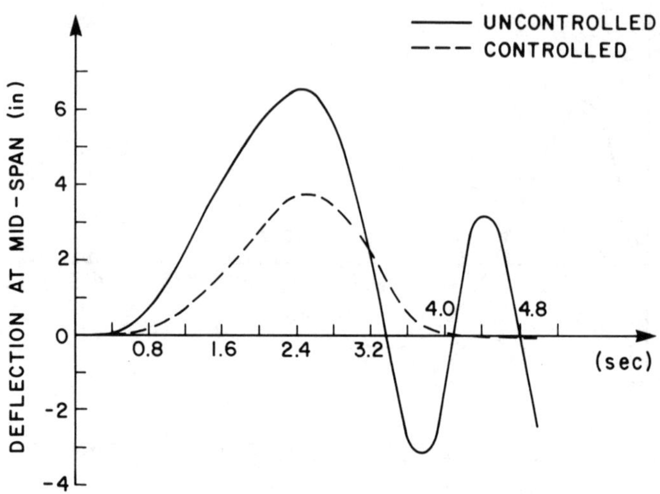

Figure 3.106 - Deflection Response due to Accelerated Motion

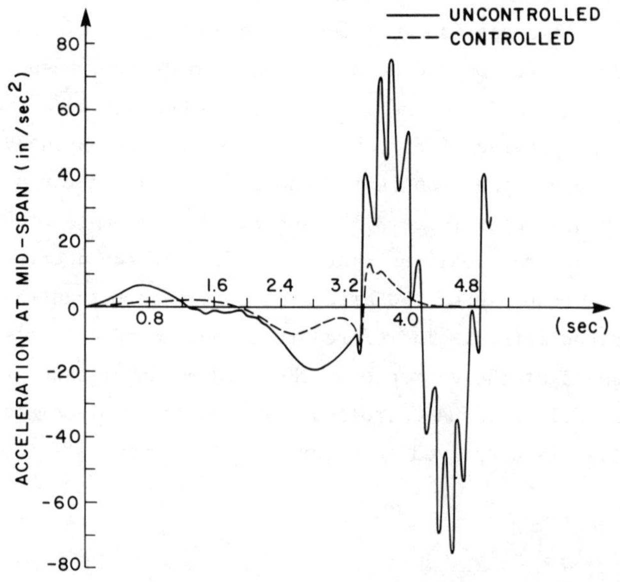

Figure 3.107 - Acceleration Response due to Accelerated Motion

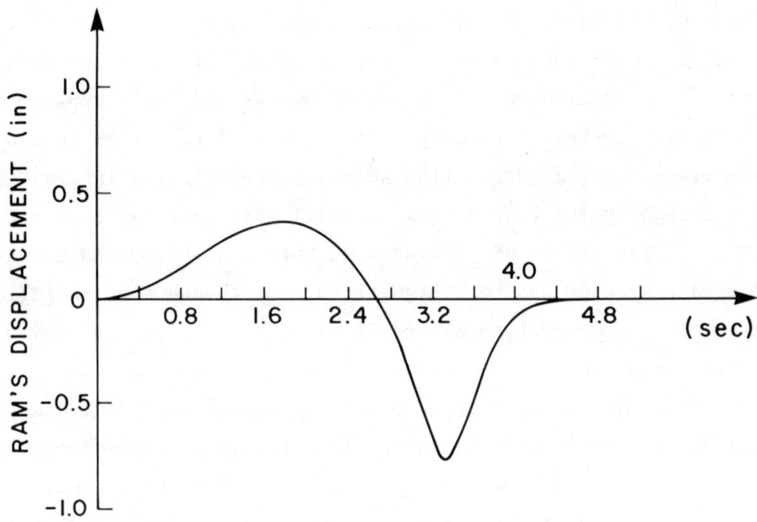

Figure 3.108 - Active Control Response due to Accelerated Motion

3.6.6 *Summary and Conclusions*

In this section, the effect of some factors, which have been neglected in the previous sections, were investigated. The effect of
different types of moving loads on the controlled response has been
analyzed.

The inertia effect on the uncontrolled response was
noticeable, especially when the ratio "vehicle's weight versus
bridge's weight" was increased. This leads to the conclusion that
the inertia effect should be included in the design of bridges.

It has been found, however, that the difference in the
controlled response between considering or neglecting inertia was
negligible for both regulator and tracing controlled systems. This
shows how effective the designed control system was which depends
on the feedback of the actual structural response.

The normal force caused by the control mechanism had a
slight effect on the passive control response. However, the effect

of the normal force on the active controlled response was trivial
and hardly noticeable for both regulator and tracking controlled
systems. A conclusion may then be drawn as follows: representing
a nonlinear system (including normal force effect) with an approximate linear system (neglecting normal force effect), in order to
ease designing the control system, will give reasonable results.

The effect of uneveness on both controlled and uncontrolled responses was very significant. Although the controlled
response of acceleration was equal to only 25 percent of the uncontrolled response, the designed control system was not effective
to warrant the desired response satisfying human comfort and safety
with respect to fatigue failure. As a result of uneveness, the ram
was oscillating very fast. However, it has been shown that by
proper modelling of the uneveness, one could design a stochastic
control capable of dampening out the bridges vibration.

By studying the effect of different moving loads on
the regulator controlled system, it has been found that the controller could dampen the vibration, and was able to reduce the
transient response as compared with the uncontrolled response.

Investigating the effect of an accelerated motion on the
bridge behaviour, it was found that posting the traffic signals at
the exit of the bridges should be avoided as the effect of decelerated motion on the uncontrolled response of bridges is harmful.

3.7 OPTIMAL CONTROL AND SENSORS LOCATIONS

In Section 3.3, it has been shown that in order to let the estimated
state variables converge to the actual states in a very short time,
the observer poles must possess greater negative real parts than
the structure poles. With this assumption, the observer requires
a high gain which in turn increases its sensitivity to any observation noise. It is necessary to design an optimal observer recon-
ciling the speed of accurate estimation and the desire to keep the
consumed energy low.

The optimal design methods available in the literature use the stochastic optimal control theory [3.13, 3.38], which depends on assuming the variance of observation noise and the variance of the applied disturbance. The observer gain is obtained after solving a Riccati matrix equation. The optimal observer gain is derived considering several trials concerning assumed variances. For high order structural systems this method is not practical because of the large dimension of the Riccati matrix equation.

In this section, the optimal design of an observer using the pole assignment method is given. It shall be shown that the optimal observer is not only the one which is characterized by an optimal gain, but also by the optimal location of the sensors and choice of their types, the optimal location of the control forces and their optimal design. It is also shown that the suitable type of sensor and its location differs from one structure to another, depending on the natural frequencies and mode shapes of the structure. This result is in contrast to the general belief that the acceleration sensor is the one which should preferably be used in controlling structures.

3.7.1 *The Optimal Observer*

The observer used for the system defined in equation (3.212) has the form [3.13, 3.38]

$$\dot{\underline{Z}} = \underline{F}\,\underline{Z} + \underline{B}_c\,\underline{Y} + \underline{B}\underline{U} \; , \quad \underline{Z}_0 \; , \tag{3.277}$$

in which $\underline{Z}$ is the estimated state vector given by the observer; $\underline{Y}$ is the measurement vector; $\underline{Z}_0$ is the initial observer's state vector; $\underline{B}_c$ is the gain matrix of the observer; and $\underline{F}$ is the observer's state matrix of dimension n×n.

The measurement vector $\underline{Y}$ is given by

$$\underline{Y} = \underline{C}\,\underline{X} + \underline{d}_2 \; , \tag{3.278}$$

in which $\underline{C}$ is a matrix of dimension m×n, which represents the number, locations, and types of sensors; and $\underline{d}_2$ is the observation noise vector of order m×1.

The errors between the actual states $\underline{X}$ and the estimated states $\underline{Z}$ are espressed as

$$\underline{e} = \underline{X} - \underline{Z} \ . \tag{3.279}$$

From equations (3.212) and (3.277) to (3.279), one obtains

$$\underline{\dot{e}} = \underline{F}\ \underline{e} + (\underline{d}_1 - \underline{B}_c\underline{d}_2) \ , \qquad \underline{e}_0 = \underline{X}_0 - \underline{Z}_0 \ , \tag{3.280}$$

where

$$\underline{F} = \underline{A} - \underline{B}_c\underline{C} \ . \tag{3.281}$$

For the error $\underline{e}$ to decay with time, it is obvious from equation (3.280) that $\underline{F}$ must be asymptotically stable. Moreover, in order to minimize the time required for convergence towards the estimation, the eigenvalues of $\underline{F}$ must have larger negative real parts than the poles of the controlled structure, which demands a high gain. However, $\underline{B}_c$ should be as small as possible in order to minimize the observation noise effect. One way of satisfying these various conditions is by using optimal control theory in terms of which the problem is formulated as follows:

$$J_1 = \frac{1}{2} \int_{t_0}^{t_f} \underline{e}^T\underline{e}\,dt = \text{minimum} \ , \tag{3.282}$$

$$\underline{\dot{e}} = (A - \underline{B}_c C)\underline{e} + \underline{d}_1 - \underline{B}_c\underline{d}_2 \ , \qquad \underline{e}_0 \ , \tag{3.283a}$$

$$(\underline{e}^T\underline{e})_{t_f} = 0 \ , \tag{3.283b}$$

where t_f is the unspecified terminal time to be minimized, and $\underline{e}(t_0)$ are the initial estimation errors.

The available design methods [3.38] depend on guessing the variances of $\underline{d}_1$ and $\underline{d}_2$ in order to solve for the Riccati matrix $\underline{Q}$ from

$$\underline{Q} = \underline{V}_1 - \underline{Q}\underline{C}^T\underline{V}_2\underline{C}\underline{Q} + \underline{Q}\underline{A}^T + \underline{A}\underline{Q} \; , \quad \underline{Q}_0 \; , \tag{3.284}$$

in which $\underline{V}_1(t)$ is the intensity of $\underline{d}_1$, $\underline{V}_2(t)$ is the intensity of $\underline{d}_2$, and $\underline{Q}(t)$ is the Riccati matrix of order n×n.

The optimal gain matrix, $\underline{B}_c$, for the *assumed* variances $\underline{V}_1$ and $\underline{V}_2$, and for the available measurements, $\underline{C}\,\underline{X}$, is obtained from

$$\underline{B}_c = \underline{Q}\underline{C}^T\underline{V}_2^{-1} \; . \tag{3.285}$$

For high order structural systems, finding the optimal observer, using equations (3.284) and (3.285), is a very lengthy and complicated process. An easier and faster method for time-invariant structural systems is given in the next section.

3.7.2 *The Design Method*

As stated by equation (3.277) and shown in Figure 3.109, the forcing terms for the observer is represented by $(\underline{B}_c\underline{Y})$ and $(\underline{B}\underline{U})$. The optimal observer is the one which consumes only reasonable amounts of energy for securing low noise. This can be achieved when all variables involved in $(\underline{B}_c\underline{Y})$ and $(\underline{B}\underline{U})$ are optimal. In other words, it is necessary to aim at the optimal location of the sensors, and to optimize the type of sensors, the observer gain matrix, the control forces and their locations in order to obtain an optimal observer. An easy and fast approach consists in using the pole-assignment method. This method is illustrated below.

The eigenvalues of the observer matrix $\underline{F}$ in equation (3.281) are determined from

$$\Delta(s) = |\underline{I}_n s - \underline{A} + \underline{B}_c\underline{C}| = 0 \; , \tag{3.286}$$

in which $\underline{I}_n$ is the identity matrix of order n×n, and s are the eigenvalues.

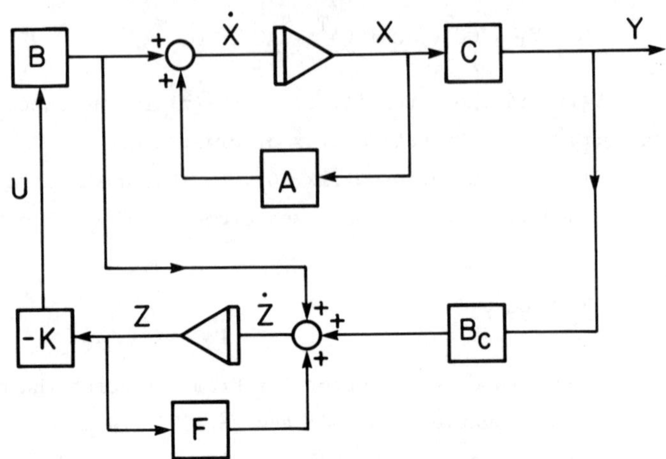

Figure 3.109 - Block Diagram for the Observer

Equation (3.286) can be written as

$$\Delta(s) = \left\lfloor \underline{I}_n s - \underline{A} \right\rvert \left\lvert \underline{I}_n + \underline{\phi}(s)\underline{B}_c\underline{C} \right\rvert = 0 , \tag{3.287}$$

in which $\underline{\phi}(s)$ is the inverse of $(\underline{I}_n s - \underline{A})$.

Since the observer poles are, in general, different than the structure poles, the first determinant in equation (3.287) is not zero for the observer poles. Therefore, the observer poles must satisfy the condition

$$\left\lvert \underline{I}_n + \underline{\phi}(s)\underline{B}_c\underline{C} \right\rvert = 0 . \tag{3.288}$$

Equation (3.288) can be written, using the determinant identity, as

$$\left\lvert \underline{I}_m + \underline{C}\phi(s)\underline{B}_c \right\rvert = 0 , \tag{3.289}$$

in which $\underline{I}_m$ is the identity matrix of dimension m×m.

By specifying the desirable observer poles in equation (3.289), one can determine the optimal gain matrix $\underline{B}_c$ for each of the assumed sensors types and their locations, $\underline{C}$, as outlined in reference [3.42]. In order to detect the optimal sensors locations, and subsequently the corresponding gain matrix $\underline{B}_c$, the following quantity is to be minimized:

$$J_2 = \text{Tr}[\underline{B}_c\underline{B}_c^T] \ . \tag{3.290}$$

It is obvious that the optimal gain matrix $\underline{B}_c$ is obtained when $\underline{C}$ has elements that are as large as possible. This requires that the sensors are placed at the most flexible locations of the structure since $\underline{C}$ depends on the deflection, velocity, or acceleration at certain locations. To show this, one may express $\underline{B}_c$ as

$$\underline{B}_c = \underline{G}_1\underline{C}^T[\underline{C}\underline{C}^T]^{-1} \ , \tag{3.291}$$

in which $\underline{G}_1$ is the optimal $[\underline{B}_c\underline{C}]$ matrix. It is obvious that the elements of $\underline{B}_c$ decrease as the elements of $[\underline{C}\underline{C}^T]$ increase.

It has been shown however, in reference [3.43], that the complete observability of the controlled structure is satisfied when only one measurement is taken. This is, in fact, recommended because of the minimum cost benefit and the direct determination of the gain matrix $\underline{B}_c$ [3.42]. In this case, equation (3.289) becomes

$$\left|1.0 + \underline{C}\phi(s)\underline{B}_c\right| = 0 \ . \tag{3.292}$$

The optimal control forces and their locations have been investigated in reference [3.44] using the modal control theory. Another easier and faster method which depends on the pole-assignment method is considered here. The feedback control forces $\underline{U}$ shall be expressed by

$$\underline{U} = -\underline{KX} \ , \tag{3.293}$$

in which K is the gain matrix of dimension r×n.

Substituting equation (3.293) into equation (3.212), the controlled system becomes

$$\dot{\underline{X}} = (\underline{A-BK})\underline{X} + \underline{d}_1 \;, \quad \underline{X}_0 \;. \tag{3.294}$$

The eigenvalues of the controlled structure are obtained from the following characteristic equation:

$$\Delta(s) = \left| s\underline{I}_n - \underline{A} + \underline{BK} \right| = 0 \;. \tag{3.295}$$

In a similar analysis as given previously, the poles of the controlled structure should satisfy

$$\Delta(s) = \left| \underline{I}_n + \underline{\phi}(s)\underline{BK} \right| = 0 \;, \tag{3.296}$$

which can be written as

$$\Delta(s) = \left| \underline{I}_r + \underline{K\phi}(s)\underline{B} \right| = 0 \;. \tag{3.297}$$

Specifying the structure poles for control, the optimal gain matrix $\underline{K}$ can be determined for each of the assumed control force locations $\underline{B}$. In order to determine the optimal control forces $(\underline{K})$, and subsequently their optimal locations $(\underline{B})$, the following quantity should be minimized:

$$J_3 = \int_0^\infty \underline{U}^T \underline{U} dt \;. \tag{3.298}$$

It can also be replaced by

$$J_3 = \mathrm{Tr}[\underline{K}^T \underline{K}] \;. \tag{3.299}$$

It is obvious that the optimal gain matrix $\underline{K}$ is obtained when $\underline{B}$ has the largest elements possible. This is accomplished when the control forces are placed at the most flexible locations of the structure. To illustrate this, the gain matrix $\underline{K}$ may be expressed as

$$\underline{K}^T = \underline{G}_2^T \underline{B} (\underline{B}^T \underline{B})^{-1} \;, \tag{3.300}$$

in which $\underline{G}_2$ is the optimal $[\underline{BK}]$ matrix. It is obvious that the elements of $\underline{K}$ decrease when the elements of $[\underline{B}^T\underline{B}]$ increase.

When a single control force is used in controlling a structure, equation (3.297) shall be

$$\left|1.0 + \underline{K}\phi(s)\underline{B}\right| = 0 \ , \tag{3.301}$$

from which the gain matrix $\underline{K}$ is directly obtained when prescribing the control force location.

3.7.3 *Numerical Examples*

3.7.3.1 Example 1

The first example represents a tall building of height L, idealized by a cantilever as shown in Figure 3.110. The building is controlled by a single control force u(t). The optimal observer which depends on the optimal control force, its location and the type of location of the sensor shall be designed.

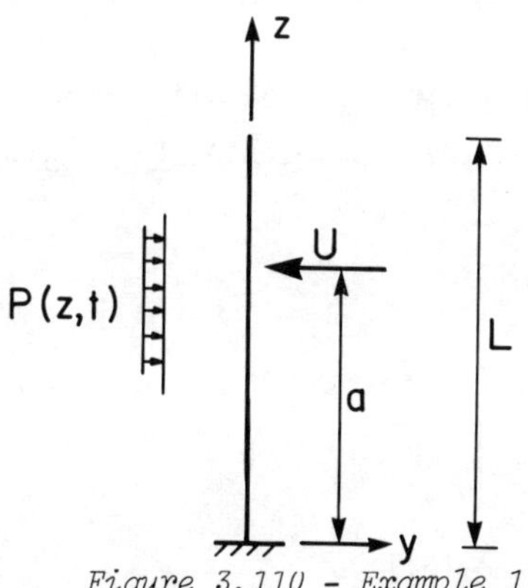

Figure 3.110 - Example 1

The equation of motion of the first two modes, Figure 3.111, are given by

$$(\ddot{h}_1+\omega_1^2 h_1)mL = \int_o^L P(x,t)\phi_1(z)dz+u(t)\phi_1(a) \ , \tag{3.302}$$

$$(\ddot{h}_2 + \omega_2^2 h_2)mL = \int_o^L P(z,t)\phi_2(z)dz + u(t)\phi_2(a) \ , \qquad (3.303)$$

in which $\phi_i(z)$ is the ith mode shape, a is the location of the control force, ω_j the natural frequency of the ith mode, m is the mass of the building per unit length, h_j is the generalized coordinate of the mode j, and $P(z,t)$ is an applied disturbance.

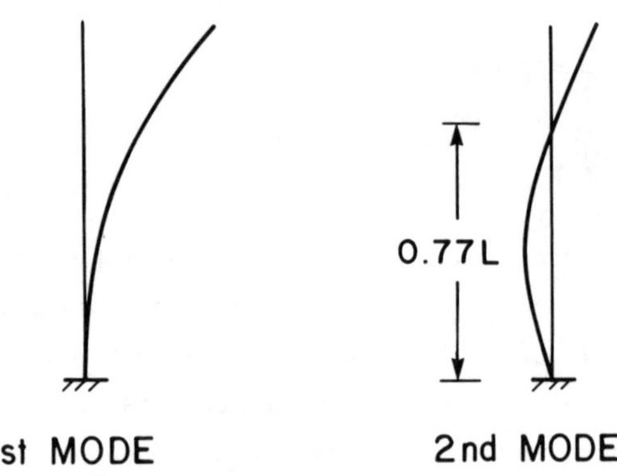

Figure 3.111 - Mode Shapes of Example 1

Equations (3.302) and (3.303) can be written in the state space form as equation (3.212), in which $\underline{A}$, $\underline{B}$ and $\underline{X}$ are given by

$$\underline{A} = \begin{bmatrix} 0 & 1 & 0 & 0 \\ -\omega_1^2 & 0 & 0 & 0 \\ 0 & 0 & 0 & 1 \\ 0 & 0 & -\omega_2^2 & 0 \end{bmatrix} \ , \qquad (3.304)$$

$$\underline{B}^T = \frac{1}{mL} [0 \quad \phi_1(a) \quad 0 \quad \phi_2(a)] \ , \qquad (3.305)$$

$$\underline{x}^T = [h_1(t) \quad \dot{h}_1(t) \quad h_2(t) \quad \dot{h}_2(t)] \ . \qquad (3.306)$$

A sensor shall be placed at the distance $z = b$ from the building base. The deflection sensor, velocity sensor, or or acceleration sensor shall provide a signal which can respectively, be expressed by

$$y(b,t) = [\phi_1(b) \quad 0 \quad \phi_2(b) \quad 0]\underline{X} , \tag{3.307}$$

$$\dot{y}(b,t) = [0 \quad \phi_1(b) \quad 0 \quad \phi_2(b)]\underline{X} , \tag{3.308}$$

$$\ddot{y}(b,t) = [-\omega_1^2\phi_1(b) \quad 0 \quad -\omega_2^2\phi_2(b) \quad 0]\underline{X} . \tag{3.309}$$

Assuming that it is possible to apply the control force alternatively at L/4, L/2, 3L/4, or L, the optimal location of the control force and its magnitude can be determined when the objective functions of equation (3.299), for the considered locations, are compared with each other. Table 3.7 shows this comparison for the specified poles and the numerical data given below. From Table 3.7 one reads off that the optimal control force location is as expected, at a = L.

Table 3.7 - Optimal Control and Locations

S	a	$(mL)\times\underline{B}^T$				$K\times10^{-4}$				J_3
0.5±J 1.794	$\frac{L}{4}$	[0	0.265	0	0.8193]	[0.4348	2.195	2.146	0.398]	11.574×10^8
&	$\frac{L}{2}$	[0	0.9249	0	1.4014]	[0.1245	0.639	1.2251	0.8178]	2.655×10^8
	$\frac{3L}{4}$	[0	1.7917	0	0.265]	[0.643	0.3247	6.63	4.32]	62.85×10^8
-1.0±J 12.365	L	[0	2.724	0	-1.9639]	[0.0423	0.2135	-0.895	-0.583]	1.19×10^8

Considering that the sensor can be placed at b = L/4, L/2, 3L/4, or L, the optimal gain matrix $\underline{B}_c$ can be determined from the comparison of the objective functions of equation (3.290). Table 3.8 shows this comparison for the three types of sensors and the specified observer poles. From Table 3.8 one concludes that the optimal sensor location is at b = L, for the three sensors, and that the smallest value for J_2 is the one corresponding to the case

218   *H.H.E. Leipholz and M. Abdel-Rohman*

when an acceleration sensor is being used. One notices that the $\underline{B}_c$ matrix, in this case, has the smallest elements.

Table 3.8 - Optimal Observers using Different Sensors

S	Type	b	$\underline{C}$				$\underline{B}_c^T$				J_2
$-5\pm J\ 12.5$ & $4\pm J\ 2$	Acceleration	L	[-10.59	0	299.7	0]	[-.99	-1.61	.024	.357]	3.7
		$\frac{3L}{4}$	[-6.973	0	-40.5	0]	[-1.51	-2.44	-.18	-2.65]	15.26
		$\frac{L}{2}$	[-3.6	0	-214	0]	[-2.94	-4.73	-.034	-.5]	31.26
	Velocity	L	[0	2.72	0	-1.96]	[-1.6	3.89	.357	-3.76]	31.95
		$\frac{3L}{4}$	[0	1.79	0	.265]	[-2.44	5.91	-2.65	27.9]	826.3
		$\frac{L}{2}$	[0	.925	0	1.4]	[-4.74	11.46	-.5	5.27]	181.83
	Deflection	L	[2.72	0	-1.96	0]	[3.89	6.26	-3.76	-54.67]	3057.26
		$\frac{3L}{4}$	[1.79	0	0.265	0]	[5.91	9.53	27.9	405]	164929.1
		$\frac{L}{2}$	[.925	0	1.4	0]	[11.46	18.46	5.27	76.6]	6367.4

The numerical data of this example are: L = 480 ft, m = 18 lb.s^2/in^2, ω_1^2 = 3.896 rad^2/s^2, ω_2^2 = 152.89 rad^2/s^2, and the first two mode shapes are respectively, given by

$$\phi_1(z) = \sin\alpha_1 z - \sinh\alpha_1 z + 1.3622(\cosh\alpha_1 z - \cos\alpha_1 z) \ ,$$

$$\phi_2(z) = \sin\alpha_2 z - \sinh\alpha_2 z + 0.9819(\cosh\alpha_2 z - \cos\alpha_2 z) \ ,$$

in which $\alpha_1 = (1.875/L)$, and $\alpha_2 = (4.694/L)$.

The final design step involves a check of the time t_f, in equations (3.282) to (3.284), required for convergence. It could be necessary to compare the current terminal time t_f with those due to alternative observer's poles, in order to find the optimal observer which reconciles the desire for high speed of convergence with the drive for optimal gain. This point shall be dealt with in the next example.

3.7.3.2 Example 2

The second example represents a single span bridge of length L, flexural rigidity EI, and mass m per unit length. The bridge is considered to be fixed at the left support and hinged at the right support as shown in Figure 3.112. For a moving load P(z,t), the equation of motion of the nth mode is given by

$$(\ddot{h}_n + \omega_n^2 h_n)mL = \int_o^L P(z,t)\phi_n(z)dz + U(t)\phi_n(a) \;, \qquad (3.310)$$

in which ω_n is the natural frequency of the nth mode, $h_n(t)$ is the generalized coordinate of the nth mode, $\phi_n(z)$ is the nth mode shape, and a is the location of the concentrated control force.

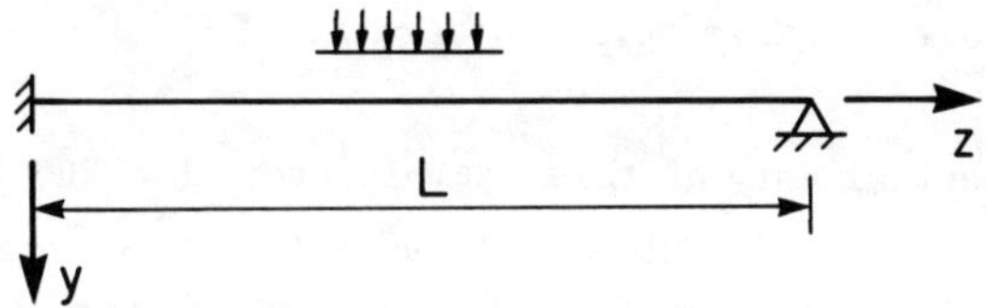

Figure 3.112 - Example 2

Considering the numerical data given below, the equations of motion of the first three modes, Figure 3.113, can be expressed in state form of equation (3.212), in which $\underline{A}$, $\underline{B}$ and $\underline{X}$ are given by

$$\underline{A} = \begin{bmatrix} 0 & 1 & 0 & 0 & 0 & 0 \\ -0.215 & 0 & 0 & 0 & 0 & 0 \\ 0 & 0 & 0 & 1 & 0 & 0 \\ 0 & 0 & -2.257 & 0 & 0 & 0 \\ 0 & 0 & 0 & 0 & 0 & 1 \\ 0 & 0 & 0 & 0 & -9.826 & 0 \end{bmatrix}, \qquad (3.311)$$

$$\underline{B}^T = \frac{1}{mL} \, [0 \quad \phi_1(a) \quad 0 \quad \phi_2(a) \quad 0 \quad \phi_3(a)] \; , \tag{3.312}$$

$$\underline{X}^T = [h_1(t) \quad \dot{h}_1(t) \quad h_2(t) \quad \dot{h}_2(t) \quad h_3(t) \quad \dot{h}_3(t)] \; . \tag{3.313}$$

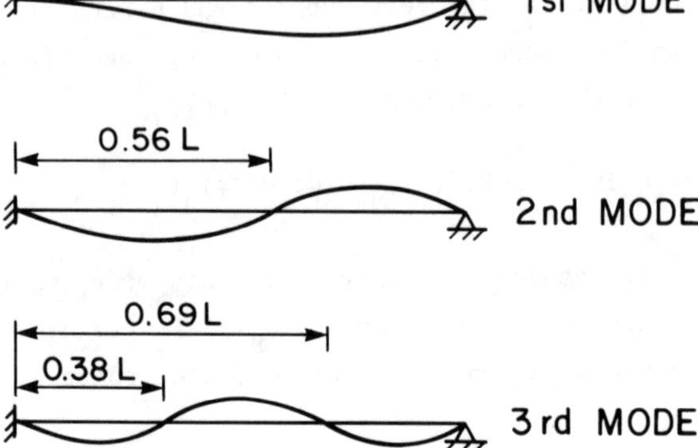

Figure 3.113 - Mode Shapes of Example 2

The numerical data of this example are: $L = 200$ ft, $EI = 30 \times 10^{11}$ lb.in^2, $m = 100$ lb.s^2/in^2, $\omega_1^2 = 0.215$ rad^2/sec^2, $\omega_2^2 = 2.2575$ rad^2/s^2, $\omega_3^2 = 9.8263$ rad^2/s^2, and the first three mode shapes are, respectively,

$$\phi_1(z) = -1.0007(\sinh\alpha_1 z - \sin\alpha_1 z) + \cosh\alpha_1 z - \cos\alpha_1 z \; ,$$

$$\phi_2(z) = -(\sinh\alpha_2 z - \sin\alpha_2 z) + \cosh\alpha_2 z - \cos\alpha_2 z \; ,$$

$$\phi_3(z) = -(\sinh\alpha_3 z - \sin\alpha_3 z) + \cosh\alpha_3 z - \cos\alpha_3 z \; .$$

where $\alpha_1 = 3.927/L$, $\alpha_2 = 7.068/L$, and $\alpha_3 = 10.21/L$.

The placement of a sensor at $z = b$ shall provide for deflection, velocity, and acceleration sensor, respectively,

$$\underline{C} = [\phi_1(b) \quad 0 \quad \phi_2(b) \quad 0 \quad \phi_3(b) \quad 0] \; , \tag{3.314}$$

$$\underline{C} = [0 \quad \phi_1(b) \quad 0 \quad \phi_2(b) \quad 0 \quad \phi_3(b)] \; , \tag{3.315}$$

$$\underline{C} = [-\omega_1^2\phi_1(b) \quad 0 \quad -\omega_2^2\phi_2(b) \quad 0 \quad -\omega_3^2\phi_3(b) \quad 0] \; . \tag{3.316}$$

Trying all feasible sensor locations, it was found that a deflection sensor should be placed at b = 0.75L, a velocity sensor at b = 0.65L, and the acceleration sensor at b = 0.62L. Table 3.9 shows the values of J_2 when the three sensors are placed at the most flexible locations.

Table 3.9 - Design of Optimal Observers for Different Sensors

Case	1			2			3			Sensor Type
Eigen-values	$-3 \pm J$ 0.464 $-4 \pm J$ 1.502 $-2 \pm J$ 3.135			$-2.5 \pm J$ 0.464 $-3.5 \pm J$ 1.502 $-1.5 \pm J$ 3.135			$-2 \pm J$ 0.464 $-3 \pm J$ 1.502 $-1 \pm J$ 3.135			
b	0.44L	0.66L	0.75L	0.44L	0.66L	0.75L	0.44L	0.66L	0.75L	Deflection
$J_2 \times 10^{-3}$	42.79	52.28	38.31	10.86	12.62	9.599	2.65	3.15	2.339	Deflection
t_f	3.145 sec			3.210 sec.			3.50 sec.			Deflection
J_1	0.001256			0.000593			0.000344			Deflection
e_{max}	0.052			0.0296			0.0216			Deflection
b	0.47L	0.65L	0.72L	0.47L	0.65L	0.72L	0.47L	0.65L	0.72L	Velocity
$J_2 \times 10^{-3}$	140.1	128.9	145.8	31.7	29.4	33.4	6.19	5.75	6.52	Velocity
t_f	3.34 sec.			3.42 sec.			3.59 sec.			Velocity
J_1	0.004539			0.001946			0.00103			Velocity
e_{max}	0.171			0.100			0.058			Velocity
b	0.51L		0.62L	0.51L		0.62L	0.51L		0.62L	Acceleration
$J_2 \times 10^{-3}$	370.5		338.85	94.8		88.7	22.23		20.74	Acceleration
t_f	3.01 sec.			3.275 sec.			3.55 sec.			Acceleration
J_1	0.000507			0.000211			0.000103			Acceleration
e_{max}	0.052			0.029			0.014			Acceleration

One notices that, for this specific example, the optimal locations of the three sensors are different. Also, the gain provided by the deflection sensor is smaller than that provided by the velocity and acceleration sensors. This is mainly attributed to the value of ω_1 which was less than 1.0 rad/s, such that the contribution of the first mode in the velocity and acceleration measurements is trivial.

The values of t_f and J_1 corresponding to initial conditions $h_1 = \dot{h}_1 = 0.01$ are also shown in Table 3.9 for the assumed observer poles. A comparison of J_2 and the time t_f, which is required for convergence towards the estimation, leads to the optimal observer. For this example, a deflection sensor placed at $b = 0.75L$ with $J_2 = 9.59 \times 10^3$ and $t_f = 3.21$ sec will provide an optimal observer.

The optimal location of a concentrated control force and the feedback gain of this force can be determined from Table 3.10, which presents the values of J_3 for the most flexible locations. In considering three modes in the design, the control force should be placed at $a = 0.24L$.

Table 3.10 - Optimal Control and Location

S	a	$K \times 10^{-4}$						$J_3 \times 10^{-14}$
-0.1±J 0.464	0.24L	[-3.28	102.35	11.97	141.04	235.71	182.32]	0.1193
	0.485L	[-1.41	44.02	21.97	258.8	-293.1	-226.7]	0.2067
-0.3±J 1.502	0.615L	[-1.33	41.73	-27.6	-325.31	-351.49	-271.87]	0.3058
-0.5±J 3.135	0.84L	[-2.33	72.8	-12.19	-143.67	249.51	192.99]	0.1256

3.7.4 *Conclusions*

It has been shown that the optimal observer is not only the one which is characterized by an optimal gain matrix $\underline{B}_c$, but the location of sensors and their types, $\underline{C}$, the locations of the control forces and their optimal design, $\underline{B}\ \underline{U}$, are also decisive factors which should be considered in the design of observers.

In order to design an optimal observer, [3.45] the sensors should be placed at the most flexible locations on the structure, according to the sensor types. This shall provide the minimum gain

$\underline{B}_c$ for the assumed observer poles. If the time required for convergence, t_f, is not short enough, one can try other observer poles
with negative real parts greater than the previously assumed. A
comparison of the reduction in t_f and the increase in $\underline{B}_c$ indicates
the observer which compromises between the gain and the time for
convergence. To minimize the input control energy to the observer,
the control forces must be placed at the most flexible locations in
the structure, and they are designed to provide the desirable
eigenvalues for the controlled structure.

It has also been shown that the optimal sensor type and
its location depends on the frequencies and mode shapes of the
structure. The acceleration sensor is preferable for structures
with frequencies greater than 1.0 rad/s, whereas deflection or
velocity sensors are used for structures with frequencies less
than 1.0 rad/s.

The method presented here depended on trying all feasible
locations, desirable poles, and comparing results. Although such
procedure appears to be primitive, the calculations involved are
very much simpler than those involved in using the stochastic
optimal control method. It has been shown that the latter method
depends as well, on trials, guessing in each trial the variances
of observation noise and disturbances in order to solve for a large
order Riccati matrix equation.

3.8 REFERENCES

[3.1] MITCHELL, S. and BOORMANN, G.F., "Vehicle Loads and Highway
       Bridge Design", *Proceedings of the ASCE, J. of the Structural
       Division*, Vol. 83, No. ST4, July, 1957, Paper No. 1302.

[3.2] BIGGS, J.M., SUER, H.S. and LOUW, J.M., "The Vibration of
       Simple Span Highway Bridges", *Transactions of the ASCE*,
       Vol. 124, 1959.

[3.3] LINGER, D.A. and HULSTOS, C.L., "Dynamics of Highway Bridges",
       Bulletin No. 17, Iowa Highway Research Board, Iowa State
       Highway Commission, No. 2, 1960.

[3.4]   OEHLER, L.T., "Vibration Susceptibilities of Various
        Highway Bridge Types", *Proc. of the ASCE, J. of the Struc-
        tural Division*, Vol. 83, No. ST4, July, 1957, Paper No.
        1318.

[3.5]   BIGGS, J.M., *Introduction of Structural Dynamics*, McGraw-
        Hill Book Company, New York, New York, 1964.

[3.6]   ABDEL-ROHMAN, M. and LEIPHOLZ, H.H.E., "Active Control of
        Flexible Structures", *Proc. of the ASCE, J. of the Struc-
        tural Division*, Vol. 104, No. ST8, August, 1978, pp. 1251-1266.

[3.7]   DORF, R.C., *Modern Control Systems*, Addison-Wesley Pub-
        lishing Co., Ontario, 1967.

[3.8]   DI STEFANO, J.J., STUBBERUD, A.R. and WILLIAMS, I.J.,
        Feedback and Control Systems, Schaum's Outline Series,
        McGraw-Hill Book Company, New York, New York, 1967.

[3.9]   QUINTANA, V.H., "Conventional and Modern Control System
        Analysis and Design", *Internal Technical Report*, Depart-
        ment of Electrical Engineering, University of Waterloo,
        Ontario, Canada, March, 1976.

[3.10] ZUK, W. and CLARK, R.H., *Kinetic Architecture*, Van Nostrand
        Reinhold Company, New York, New York, 1970.

[3.11] ROORDA, J., "Active Damping in Structures", *Technical
        Report Aero No. 8*, Cranfield Institute of Technology,
        Bedford, England, July, 1971.

[3.12] BROGAN, W.L., *Modern Control Theory*, Q.P.I. Series, New York,
        New York, 1974.

[3.13] BRYSON, A.E. and HO, Y.C., *Applied Optimal Control*, Hemi-
        sphere Publishing Corp., Washington, D.C., 1975.

[3.14] ATHANS, M. and FALB, P.L., *Optimal Control*, McGraw-Hill
        Book Company, New York, 1966.

[3.15] PORTER, B. and CROSSLEY, T.R., *Modal Control: Theory and
        Applications*, Taylor and Francis, London, England, 1972.

[3.16] MARTIN, C.R. and SOONG, T.T., "Modal Control of Multistory
        Structures", *Proc. of the ASCE, Journal of the Engineering
        Mechanics Division*, Vol. 102, August, 1976, Paper No. 12321,
        pp. 613-623.

[3.17] DAVISON, E.J., "On Pole Assignment in Linear Systems with
        Incomplete State Feedback", *IEEE Trans. on Automatic
        Control*, Vol. AC-15, No. 13, June, 1970, pp. 348-351.

[3.18] HEYMAN, M., "Comments on Pole Assignment in Multi-Input
        Controllable Linear Systems", *IEEE Trans. on Automatic
        Control*, Vol. AC-13, No. 6, December, 1968, pp. 748-749.

[3.19] WONHAM, W.M., "On Pole Assignment in Multi-Input Controllable
        Linear Systems", *IEEE Trans. on Automatic Control,* Vol.
        AC-12, No. 6, December, 1967, pp. 660-665.

[3.20] LUENBERGER, D.G., "An Introduction to Observers", *IEEE
        Trans. on Automatic Control,* Vol. AC-16, No. 6, December,
        1971, pp. 596-602.

[3.21] DAVISON, E.J., "On Pole Assignment in Linear Systems with
        Incomplete State Feedback", *IEEE Trans. on Automatic Control,*
        Vol. AC-15, No. 3, June, 1970, pp. 348-351.

[3.22] CHEN, C.T., *Introduction to Linear System Theory,* Holt,
        Reinhart and Winston, New York, New York, 1970.

[3.23] LUENBERGER, D.G., *Optimization of Vector Space Methods,*
        John Wiley, New York, 1969.

[3.24] McCAUSLAND, I., *Introduction to Optimal Control,* John
        Wiley and Sons, Toronto, Ontario, Canada, 1969.

[3.25] QUINTANA, V.H., "Some Numerical Methods for Solving Optimal
        Control Problems", *Ph. D. Thesis,* University of Toronto,
        Ontario, Canada, May, 1970.

[3.26] ABDEL-ROHMAN, M. and LEIPHOLZ, H.H.E., "Structural Control
        by Pole Assignment Method", *Proc. of the ASCE, J. of the
        Engineering Mechanics Division,* Vol. 104, No. EM5,
        October, 1978, pp. 1159-1175.

[3.27] ANDERSON, B.D.O. and MOORE, J.B., *Linear Optimal Control,*
        Prentice-Hall Inc., Englewood Cliffs, New Jersey, 1971.

[3.28] DAVISON, E.J. and AMKI, M.C., "The Numerical Solution of
        the Matrix Riccati Differential Equation", *IEEE Trans. on
        Automatic Control,* Vol. AC-18, No. 1, February, 1973.

[3.29] SAVAS, E.S., *Computer Control of Industrial Processes,*
        McGraw-Hill Book Co., New York, New York, 1965.

[3.30] ABDEL-ROHMAN, M., QUINTANA, V.H. and LEIPHOLZ, H.H.E.,
        "Optimal Control of Civil Engineering Structures", *Proc.
        of the ASCE, J. of the Engineering Mechanics Division,*
        Vol. 106, No. EM1, February, 1980, pp. 57-73.

[3.31] SCHORN, G., "Feedback Control of Structures", *Ph. D. Thesis,*
        Department of Civil Engineering, University of Waterloo,
        Ontario, Canada, March, 1975.

[3.32] YANG, J-N. and YAO, J.T.P., "Formulation of Structural
        Control", *Technical Report No. CE-STR-74-2,* School of Civil
        Engineering, Purdue University, West Lafayette, Indiana,
        1974.

[3.33] YANG, J-N., "Application of Optimal Control Theory to Civil
        Engineering Structures", *Proc. of the ASCE, J. of the
        Engineering Mechanics Division,* Vol. 101, No. EM6, December,
        1975, pp. 819-838.

[3.34] FRÝBA, L., *Vibration of Solids and Structures under Moving Loads*, Academia Publishing House of the Czechoslovak Academy of Sciences, Prague, 1972.

[3.35] BOLOTIN, V.V., *The Dynamic Stability of Elastic Systems*, Holden-Day, Inc., San Francisco, California, 1964.

[3.36] KOLOUSEK, V., *Dynamics in Engineering Structures*, Butterworth and Co. Ltd., Toronto, 1973.

[3.37] ABDEL-ROHMAN, M. and LEIPHOLZ, H.H.E., "Automatic Active Control of Structures", *Proc. of the ASCE, J. of the Structural Division*, Vol. 106, No. ST3, March, 1980, pp. 663-677.

[3.38] KWAKERNAAK, H. and SIVAN, R., *Linear Optimal Control Systems*, Wiley-Interscience, New York, New York, 1972.

[3.39] LIN, Y.K., *Probabilistic Theory of Structural Dynamics*, Robert E. Krieger Publishing Co., Huntington, New York, 1976.

[3.40] TUNG, C.C., PENZIEN, J. and HORONJEFF, R., "The Effect of Runway Uneveness on the Dynamic Response of Supersonic Transports", *NASA CR-119*, University of California, Berkeley, California, October, 1964.

[3.41] ABDEL-ROHMAN, M. and LEIPHOLZ, H.H.E., "A General Approach to Active Structural Control", *Proc. of the ASCE, J. of the Engineering Mechanics Division*, Vol. 105, No. EM6, December, 1979, pp. 1007-1023.

[3.42] ABDEL-ROHMAN, M., "Active Control of Large Structures", *Proc. of the ASCE, J. of the Engineering Mechanics Division*, Vol. 108, No. EM5, October, 1982, pp. 719-730.

[3.43] JUANG, J.N., SAE-UNG, S. and YANG, J.N., "Active Control of Large Building Structures", *Proceedings of the IUTAM Symposium on Structural Control*, held at University of Waterloo, Ontario, Canada, North Holland Publishing Co., New York, 1980, pp. 663-676.

[3.44] SOONG, T.T. and CHANG, M.I.J., "On Optimal Configuration in Theory of Modal Control", *Proceedings of the IUTAM Symposium on Structural Control*, held at University of Waterloo, Ontario, Canada, North Holland Publishing Co., New York, 1980, pp. 723-738.

[3.45] ABDEL-ROHMAN, M., "Design of Optimal Observers for Structural Control Systems", *IEEE, Journal of Control and Applications*, Part D, Vol. 131, No. 4, July, 1984, pp. 158-163.

Chapter IV

Automatic Active Control of Tall Buildings

4.1 INTRODUCTION

In this chapter, the automatic control techniques, which are applied
to control tall buildings against wind forces, are presented. It
is intended to develop a practically useful control mechanism and
to consider the time delay effect on the effectiveness of the control. The tall buildings to be considered here are those which can
be modelled as cantilever tube type structures. The tall buildings'
response against wind shall be considered first, and various control
techniques shall be analyzed in detail.

4.2 BUILDING RESPONSE DUE TO WIND FORCES

4.2.1 *Wind Forces*

The tall building considered shall be idealized as a cantilever
tube type structure of height L and with square cross-section as
shown in Figure 4.1. The relative wind velocity at any location x
is expressed as [4.21,4.22]

$$U(x,t) = \{[\overline{U}(x)+v_1(x,t)-\dot{w}_1(x,t)]^2+[v_2(x,t)-\dot{w}_2(x,t)]^2+v_3^2(x,t)\}^{1/2} ,$$

$$(4.1)$$

in which $\overline{U}(x)$ is the mean wind speed; $v_1(x,t)$ is the fluctuating wind speed component in w_1-direction; $v_2(x,t)$ is the fluctuating wind speed component in w_2-direction; $w_1(x,t)$ is the building deflection in wind direction; $w_2(x,t)$ is the building deflection perpendicular to wind direction; and $v_3(x,t)$ is the fluctuating wind speed component in the vertical direction as shown in Figure 4.2.

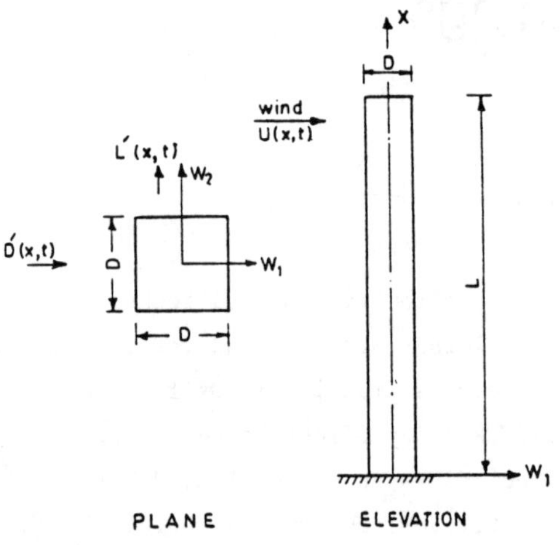

Figure 4.1 - Tall Building Geometry

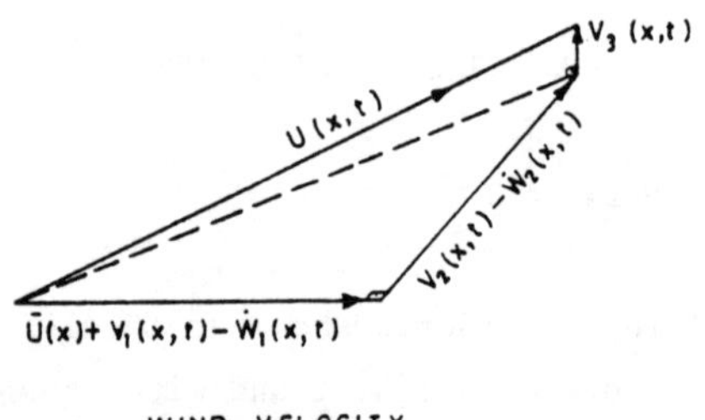

Figure 4.2 - Wind Velocity Resolution

The wind forces applied on the building are drag forces, $D(x,t)$, and lift forces, $L(x,t)$. These forces are calculated from

$$D(x,t) = g_1\overline{U}^2(x,t)+g_2\dot{\overline{U}}(x,t) \ , \qquad (4.2)$$

$$L(x,t) = g_3\overline{U}^2(x,t)f(t) \ , \qquad (4.3)$$

in which ρ is the air density, D is the building width, C_d is the drag coefficient, C_L is the lift coefficient, C_m is the coefficient of virtual mass, $f(t)$ is the function of the time history of oscillating vortices, and $g_1 = 0.5\rho C_d D$, $g_2 = (\pi/4)\rho C_m D^2$, $g_3 = 0.5\rho C_L D$.

By imposing the simplifying conditions that $|v_1-\dot{w}_1| > |v_2-\dot{w}_2| > |v_3|$, and corresponding conditions for the time derivatives, equations (4.2) and (4.3) become

$$D'(x,t) = g_1\{\overline{U}^2(x)+2\overline{U}(x)[v_1(x,t)-\dot{w}_1(x,t)]+[v_1(x,t)-\dot{w}_1(x,t)]^2\}$$

$$+g_2[\dot{v}_1(x,t)-\ddot{w}_1(x,t)] \ , \qquad (4.4)$$

$$L'(x,t) = g_3f(t)\{\overline{U}(x)^2+2\overline{U}(x)[v_1(x,t)-\dot{w}_1(x,t)]+[v_1(x,t)-\dot{w}_1(x,t)]^2$$

$$+1/2[v_2(x,t)-\dot{w}_2(x,t)]^2\} \ , \qquad (4.5)$$

where the structure has also been considered in terms of a simplified aeroelastic model.

Neglecting torsional motion, i.e., assuming the resultant of the wind forces to pass through the elastic centre of any cross section, the equations of motion in the w_1 and w_2 directions are, respectively,

$$(EI)_1 \frac{\partial^4 w_1}{\partial x^4} + C_1\frac{\partial w_1}{\partial t} + m_1 \frac{\partial^2 w_1}{\partial t^2} = D'(x,t) \ , \qquad (4.6)$$

$$(EI)_2 \frac{\partial^4 w_2}{\partial x^4} + C_2 \frac{\partial w_2}{\partial t} + m_2 \frac{\partial^2 w_2}{\partial t^2} = L'(x,t) \ , \qquad (4.7)$$

in which $(EI)_1$ is the flexural rigidity in the w_1 direction, C_1 is the damping in the w_1 direction; and m_1 is the mass per unit length in the w_1 direction.

Solution of equations (4.6) and (4.7) is usually obtained in normal mode form using the assumption that

$$w_1(x,t) = \sum_{i=1}^{\infty} \phi_i(x)b_i(t) \ , \tag{4.8}$$

$$w_2(x,t) = \sum_{i=1}^{\infty} \psi_i(x)a_i(t) \ , \tag{4.9}$$

in which $\phi_i(x)$ and $\psi_i(x)$ are mode shapes in the w_1 and w_2 directions, respectively; and $b_i(t)$ and $a_i(t)$ are the generalized coordinates in the w_1 and w_2 directions, respectively.

Applying integral transformation, and considering only the dominant mode in each direction, one obtains

$$(\ddot{b}_1 + 2\zeta_1\omega_1\dot{b}_1 + \omega_1^2 b_1)mL = F_1 + G_1 \ , \tag{4.10}$$

$$(\ddot{a}_1 + 2\zeta_1\omega_1\dot{a}_1 + \omega_1^2 a_1)mL = F_2 + G_2 \ , \tag{4.11}$$

in which ω_1 is the first mode's natural frequency; ζ_1 is the first mode's damping ratio, and m is the mass per unit length assumed constant in both w_1 and w_2 directions.

In equations (4.10) and (4.11), the terms $(F_1 + G_1)$ and $(F_2 + G_2)$ are, respectively, obtained from

$$F_1 + G_1 = \int_0^L \phi_1(x)D'(x,t)dx \ , \tag{4.12}$$

$$F_2 + G_2 = \int_0^L \psi_1(x)L'(x,t)dx \ . \tag{4.13}$$

Keeping the first mode only, $D'(x,t)$ and $L'(x,t)$ can be written as

$$D'(x,t) \simeq g_1\{\overline{U}^2(x) + 2\overline{U}(x)v_1(x,t) - 2\overline{U}(x)\phi_1(x)\dot{b}_1(t) + v_1^2(x,t)$$

$$-2v_1(x,t)\phi_1(x)\dot{b}_1(t) + \phi_1^2(x)\dot{b}_1^2(t)\} + g_2\dot{v}_1(x,t) - g_2\phi_1(x)\ddot{b}_1(t) \ ,$$

$$\tag{4.14}$$

$$L'(x,t) \simeq g_3 f(t)\{\overline{U}^2(x)+2\overline{U}(x)v_1(x,t)-2\overline{U}(x)\phi_1(x)\dot{b}_1(t)+v_1^2(x,t)$$

$$-2v_1(x,t)\phi_1(x)\dot{b}_1(t)+\phi_2^2(x)\dot{b}_1^2(t)+0.5v_2^2(x,t)-v_2(x,t)\psi_1(x)\dot{a}_1(t)$$

$$+0.5\psi_1^2(x)\dot{a}_1^2(t)\} \ . \tag{4.15}$$

The fluctuating components are usually small as compared with the mean components. Therefore, quadratic terms can be neglected to yield:

$$F_1+G_1 = g_1\!\int_o^L [\overline{U}^2(x)+2\overline{U}(x)v_1(x,t)-2\overline{U}(x)\phi_1(x)\dot{b}_1(t)-2v_1(x,t)\phi_1(x)\dot{b}_1(t)$$

$$+\phi_1^2(x)\dot{b}_1^2(t)]\phi_1(x)dx + g_2\int_o^L [\dot{v}_1(x,t)-\phi_1(x)\ddot{b}_1(t)]\phi_1(x)dx \ , \tag{4.16}$$

$$F_2+G_2 = g_3 f(t)\int_o^L [\overline{U}^2(x)+2\overline{U}(x)v_1(x,t)-2\overline{U}(x)\phi_1(x)\dot{b}_1(t)$$

$$-2v_1(x,t)\phi_1(x)\dot{b}_1(t)+\phi_1^2(x)\dot{b}_1^2(x)-v_2(x,t)\psi_1(x)\dot{a}_1(t)$$

$$+0.5\psi_1^2\dot{a}_1^2(t)]\psi_1(x)dx \ , \tag{4.17}$$

where

$$F_1 = g_1\int_o^L [\overline{U}^2(x)+2\overline{U}(x)v_1(x,t)]\phi_1(x)dx + g_2\int_o^L \dot{v}_1(x,t)\phi_1(x)dx \ , \tag{4.18}$$

$$G_1 = g_1\int_o^L [-2\overline{U}(x)\phi_1(x)\dot{b}_1-2v_1(x,t)\phi_1(x)\dot{b}_1(t)+\phi_1^2(x)\dot{b}_1^2]\phi_1(x)dx$$

$$- g_2\int_o^L \ddot{b}_1(t)\phi_1^2(x)dx \ , \tag{4.19}$$

$$F_2 = g_3 f(t)\int_o^L [\overline{U}^2(x)+2\overline{U}(x)v_1(x,t)]\psi_1(x)dx \ , \tag{4.20}$$

$$G_2 = g_3 f(t)\int_o^L [-2\overline{U}(x)\phi_1(x)\dot{b}_1-2v_1(x,t)\phi_1(x)\dot{b}_1+\phi_1^2(x)\dot{b}_1^2$$

$$-v_2(x,t)\psi_1(x)\dot{a}_1+0.5\psi_1^2(x)\dot{a}_1^2]\psi_1(x)dx \ , \tag{4.21}$$

in which F_1 and F_2 represent external actions due to wind, and
G_1, G_2 represent aerodynamic forces due to coupling with the
building. In many treatments of wind effects on tall buildings,
the aerodynamic forces are disregarded.

In the following, a summary for both stochastic and
deterministic solutions of the problem will be given.

4.2.2 *Stochastic Response*

Wind is a very complex phenomenon not easily representable by
mathematical models since there are so many random variables
involved controlling its behaviour. Among the spectra assumed to
represent the fluctuating components along the wind direction
[4.33],

$$S_v(x,n) = \frac{U_*^2}{n} \frac{200f}{(1+50f)^{5/3}} \tag{4.22}$$

is being proposed in which n is the frequency, U_* is the shear
velocity of the wind, x is the height from the ground, and
$f = (nx)/\overline{U}(x)$, where $\overline{U}(x)$ is the mean velocity at height x.

Firstly, the above spectrum is neither a rational nor an
even function and can therefore not be represented easily by a
stochastic process in the time domain. Secondly, the variable U_*
is a random variable since it depends on the roughness of the
terrain. It can be obtained from

$$\overline{U}(x) = 2.5U_* \ln \frac{x}{x_0} \tag{4.23}$$

in which x_0 is the roughness of the terrain under consideration.

The pressure acting on a body at any point of the eleva-
tion **x** from the ground in a steady flow of velocity $\overline{U}(x)$ is given by

$$P(x) = \frac{1}{2} \rho \overline{U}^2(x) C_p(x) \, , \tag{4.24}$$

in which ρ is the air density and C_p is the pressure coefficient
at height x.

In the case of unsteady flow, $U(x) = \overline{U}(x)+u(x,t)$, the wind pressure is approximately

$$P(x,t) = \overline{P}(x)+P'(x,t) \ , \tag{4.25}$$

where

$$\overline{P}(x) = \frac{1}{2} \rho C_p(x)\overline{U}^2(x) \ , \tag{4.26}$$

$$P'(x,t) = \rho C_p(x)\overline{U}(x)u(x,t) \ . \tag{4.27}$$

Equations (4.26) and (4.27) indicate also that the drag and lift coefficients are random variables and not constants. However, they are averaged at every height. Equation (4.26) resembles the term F_1 derived in the previous section except that now the effect of the lift forces (F_2) has been neglected for the sake of simplicity.

Therefore, if one is interested in the mean response along the wind, one uses equations (4.10) and (4.26) and, considering the first mode only, one gets

$$\overline{y}(x) = \frac{\displaystyle\int_o^L \overline{P}(x)\phi_1(x)dx}{\omega_1^2 mL} \tag{4.28}$$

$$= 1/2 \ \rho C_D D \ \frac{\displaystyle\int_o^L \overline{U}^2(x)\phi_1(x)dx}{\omega_1^2 mL} \ . \tag{4.29}$$

In order to determine the characteristics of the fluctuating response, the co-spectrum of the pressure at a point x must be used which is given by

$$S_p'(x,n) = \rho^2 C_p^2\overline{U}^2(x)S_u(x,n) \ , \tag{4.30}$$

in which S_u is the spectral density of the longitudinal velocity fluctuations at elevation x, e.g., equation (4.22).

The co-spectrum at two points of heights x_1 and x_2 with coordinates (z_1,x_1) and (z_2,x_2) is given by [4.33]

$$S^c_{p_1p_2}(x_1,x_2,n) = S^{1/2}_p(x_1,n)S^{1/2}_p(x_2,n)Coh(z_1,z_2,x_1,x_2,n)N(n) \quad , \quad (4.31)$$

in which Coh is the across-wind correlation coefficient, and $N(n)$ is the along-wind correlation coefficient

$$Coh(z_1,z_2,x_1,x_2,n) = \exp\left[-\frac{n[C_x^2(x_1-x_2)^2+C_z^2(z_1-z_2)^2]^{1/2}}{0.5[\overline{U}(x_1)+\overline{U}(x_2)]}\right] \quad (4.32)$$

$N(n) = 1.0$ if only one point was considered or two points are coinciding.

The spectrum of the along-wind response is obtained as

$$S_y(x,n) = |H(x,n)|^2 S_p(x,n) \quad , \quad (4.33)$$

in which $H(x,n)$ is the transfer function of the structure. Considering one mode only, $H(x,n)$ is given by

$$|H^2(x,\omega)|^2 = \left|\frac{1}{mL[-\omega^2+2\zeta_1\omega_1\omega+\omega_1^2]}\right|^2 = \frac{1}{m^2L^2[(\omega_1^2-\omega^2)^2+4\zeta_1^2\omega_1^2\omega^2]} \quad , \quad (4.34)$$

in which $\omega = 2\pi n$.

The mean square value of the fluctuating along-wind deflection is obtained as

$$\sigma_y^2(x) = \int_{-\infty}^{\infty} S_y(x,n)dn \quad . \quad (4.35)$$

The mean square value of the fluctuating along wind acceleration is obtained as

$$\sigma_y^2(x) = \int_o^{\infty} 16\pi^4 n^4 S_y(x,n)dn \quad . \quad (4.36)$$

The expected value of the largest peak occurring in an NT-year period is

$$y_{max}(x) = K_y(x)\sigma_y(x) \ , \tag{4.37}$$

in which $K(x)$ is the peak factor given by

$$K_y(x) = [2 \ln V_y(x)T]^{1/2} + \frac{0.577}{[2 \ln V_y(x)T]^{1/2}} \ , \tag{4.38}$$

where $V_y(x)$ is given as

$$V_y(x) = \left[\frac{\int_0^\infty n^2 S_y(x,n)\,dn}{\int_0^\infty S_y(x,n)\,dn} \right]^{1/2} . \tag{4.39}$$

The magnitude of $K_y(x)$ is usually about 3.0 to 4.0.

Similarly, the largest peak along-wind acceleration is

$$\ddot{y}_{max}(x) = K_{\ddot{y}}(x)\sigma_{\ddot{y}}(x) \ , \tag{4.40}$$

in which $K_{\ddot{y}}(x) \simeq 4.0$.

In all previous equations, the effect of lift forces has been neglected.

Thus, the spectrum of along-wind fluctuating wind forces is

$$S_F(x,n) = \int_0^L \int_0^L \rho^2 C_D^2(x)\overline{U}^2(x)S_u(x,n)\phi_1^2(x)\,d^2x \ . \tag{4.41}$$

The mean square response is obtained from

$$\sigma_y^2(x) = \int_0^\infty |H(x,n)|^2 S_F(x,n)\,dn = \int_0^\infty \int_0^L \int_0^L \frac{\rho^2 C_D^2(x)\overline{U}^2(x)\phi_1^2 S_u(x,n)\,dxdxdn}{16\pi^4 m^2 L^2 n_1^4 [(n_1^2-n^2)^2+4\zeta_1^2 n^2 n_1^2]} \ . \tag{4.42}$$

The difficulty of carrying out this integration is obvious. In many research papers, graphical charts are used to evaluate this integral [4.33].

                                    H.H.E. Leipholz and M. Abdel-Rohman

The mean square response of along wind acceleration is

$$\sigma_{\ddot{y}}^2(x) = 16\pi^4 \int_0^\infty n^4 |H(x,n)|^2 S_F(x,n)\,dn \ . \tag{4.43}$$

Equations (4.42) and (4.43) can be simplified and put in a form similar to

$$\sigma_y^2 = \text{Constant} \int_0^\infty S_u(x,n) |H(n,x)|^2 dn \ . \tag{4.44}$$

If $S_u(x,n)$ represents a white noise of intensity S_o as shown in Figure 4.3, then the integral of equation (4.44) can be evaluated to yield

$$\sigma_y^2(x) = \text{Constant} \times \frac{\pi n_1}{4\zeta_1} S_0 \ . \tag{4.45}$$

If the damping in the structure is small, i.e., uncontrolled, then the response of $|H(n)|^2$ is as shown in Figure 4.4, indicating that the integration in equation (4.42) is basically over $n_1 - \varepsilon < n < n_1 + \varepsilon$, where ε is very small.

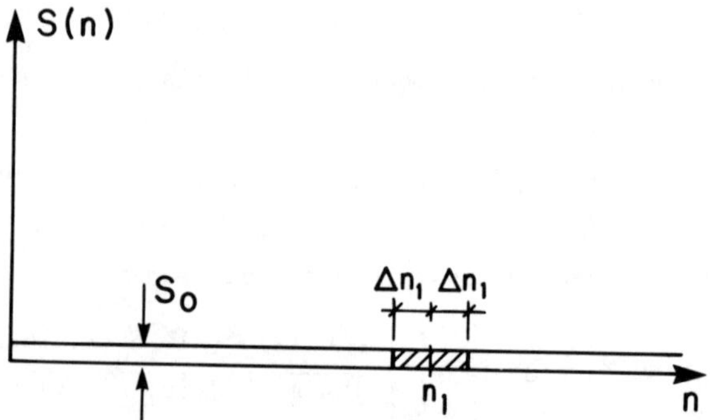

Figure 4.3 - White Noise

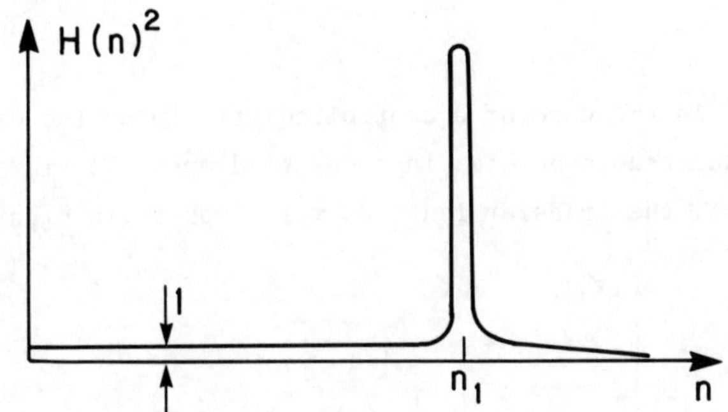

Figure 4.4 - Transfer Function of ODOF System

In reality, wind cannot be idealized as white noise. The spectrum of $S_u(x,n)$ is, in general, as shown in Figure 4.5. Therefore, in case of uncontrolled structures one has the following formula:

$$\sigma_y^2(x) = \text{Constant}\left[\int_0^{n_1-\varepsilon} |H(n)|^2 S_u(x,n)dn + \int_{n_1-\varepsilon}^{n_1-\varepsilon} |H(n)|^2 S_u(x,n)dn \right.$$

$$\left. + \int_{n_1+\varepsilon}^{\infty} |H(n)|^2 S_u dn \right]. \tag{4.46}$$

If $\zeta_1 \ll 1.0$, i.e., small, then

$$\sigma_y^2(x) \cong \left[\int_0^{\infty} S_u(x,n)dn + \frac{\pi n_1}{4\zeta_1} S_u(x,n_1)\right] \times \text{Constant} . \tag{4.47}$$

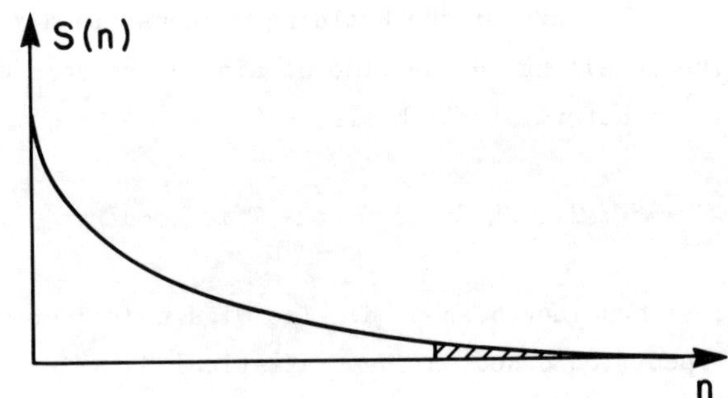

Figure 4.5 - General Noise

In the case of a controlled structure, the damping is
usually desired to be high in order to dampen the vibration. In
this event, the transfer function will look as in Figure 4.6.

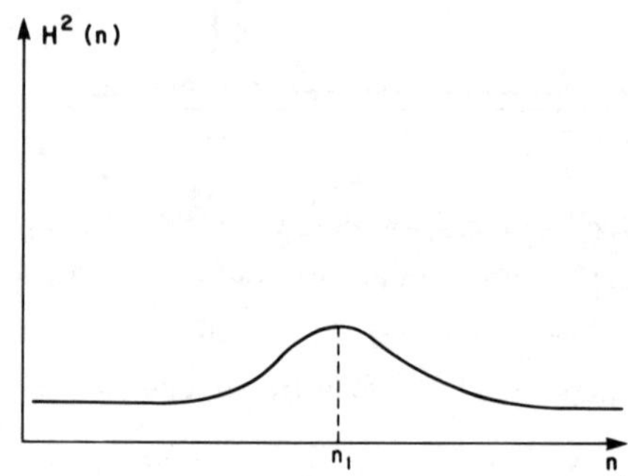

Figure 4.6 - Transfer Function of Dampened ODOF System

The mean square response is then obtained by the general
formula

$$\sigma_y^2(x) = const.x \int_o^\infty |H(x,n)|^2 S_u(x,n)dx \ . \tag{4.48}$$

The above approach depends completely on using the
frequency domain to evaluate the response characteristics. However,
if one can model the wind forces in the time domain, one easily
gets the time history of the building response in a stochastic
sense. The modelling in the time domain can be done on both the
stochastic or deterministic basis.

4.2.3 *Deterministic Analysis in the Time Domain*

This process has long been used. The wind characteristics are
uniquely specified based on the statistical data, i.e., the mean
wind speed, the various fluctuating components of wind speeds, the

drag, and lift coefficients, etc. With these parameters, one can solve equations (4.10) and (4.11) easily in the usual fashion.

One may write equation (4.10) in state space form as

$$\dot{\underline{X}} = \underline{A}\,\underline{X} + \underline{d} \; , \tag{4.49}$$

in which $\underline{X} = [b_1 \; \dot{b}_1]^T$, and

$$\underline{A} = \begin{bmatrix} 0 & 1 \\ -\omega_1^2 & -2\zeta_1\omega_1 \end{bmatrix} , \tag{4.50}$$

$$\underline{d} = \frac{1}{mL} \begin{bmatrix} 0 \\ \underline{F}_1 + \underline{G}_1 \end{bmatrix} . \tag{4.51}$$

Then, the solution of equation (4.49) is given by

$$\underline{X}(t) = \underline{\phi}(t,t_0)\underline{X}(t_0) + \int_{t_0}^{t} \underline{\phi}(t,\tau)\underline{d}(\tau)d\tau \; , \tag{4.52}$$

in which $\underline{\phi}(.,.)$ is the transition matrix of the system obtained from the solution of

$$\dot{\underline{\phi}}(.,.) = \underline{A}\,\underline{\phi}(.,.) \; , \quad \underline{\phi}(t_0,t_0) = \underline{I} \; , \tag{4.53}$$

in which I is the identity matrix.

For example, one can use the wind speed spectrum in evaluating the deterministic fluctuating components. This can be done in the form [4,11]

$$V(t) = \sqrt{2} \sum_{i=1}^{n} |2S_v(\omega_i)\Delta w|^{1/2} \cos(\omega_i t + \phi_i) \; , \tag{4.54}$$

in which $S_v(\omega_i)$ is the wind speed spectrum, ϕ_i is the random phase angle uniformly distributed in the interval $(0-2\pi)$, n is the number of frequency components, and ω_i and Δw are given by

$$\omega_i = (i-0.5)\Delta w \; , \quad i = 1,2,\ldots,n \; , \tag{4.55}$$

$$\Delta w = \frac{\omega_u}{n} \; , \tag{4.56}$$

where w_u is the upper cutoff frequency.

Equation (4.52) usually is solved by numerical integration to get the deflection and acceleration of the building at any height.

4.2.4 *Stochastic Analysis*

Here, the wind is modelled by a stochastic process of known properties and similar to the one which is expected to prevail. For example, if the wind spectrum is stationary, as usually assumed, and uncorrelated with time, one may assume it as a white noise of certain intensity to be designed according to the data collected. In many cases, however, wind cannot be modelled as stationary white noise. Here, the technique of stochastic modelling can be used in which one has to find a dynamic system driven by a white noise such that the characteristics of the output are very close to the actual output data collected.

Usually, a dynamic system with a set of first order differential equations driven by the white noise fulfils the objective. One only needs to determine the parameters of the model and the intensity of the white noise.

Suppose such a model can be represented by

$$\dot{X}_f = A_f X_f + G(t)W(t) \ , \tag{4.57}$$

in which A_f is a constant and $w(t)$ is a white noise of intensity $V(t)$.

Augmenting equation (4.57) with equation (4.49) one gets

$$\begin{bmatrix} \dot{X} \\ \dot{X}_f \end{bmatrix} = \begin{bmatrix} A & C \\ 0 & A_f \end{bmatrix} \begin{bmatrix} X \\ X_f \end{bmatrix} + \begin{bmatrix} 0 \\ G(t)W(t) \end{bmatrix} \tag{4.58}$$

in which $\underline{d} = \underline{C}X_f$.

The combined equations shall be expressed as

$$\dot{\underline{X}}_c = \underline{A}_c \underline{X}_c + \underline{G}_c(t)\underline{W}_c(t) \ ; \quad \underline{X}_c = \underline{0} \ . \tag{4.59}$$

Solution of equation (4.59) is similar to the expression in equation (4.52). The mean response of the state $\underline{X}_c(t)$ is obtained from

$$\underline{m}_{X_c}(t) = \underline{\phi}_c(t,t_0)\underline{m}_{X_c}(t_0) , \qquad (4.60)$$

in which $\underline{\phi}_c$ is the transition matrix of $\underline{A}_c$. The mean square response of the process is

$$E[\underline{X}_c\underline{X}_c^T(t)] = \underline{P}_{X_cX_c}(t) + \underline{m}_{X_c}(t)\underline{m}_{X_c}^T(t) , \qquad (4.61)$$

$$E[\underline{X}_c\underline{X}_c^T(t_0)] = \underline{P}_{X_cX_c}(t_0) + \underline{m}_{X_c}(t_0)\underline{m}_{X_c}^T(t_0) , \qquad (4.62)$$

where $\underline{P}_{X_cX_c}(t)$ is the covariance matrix obtained from

$$\underline{\dot{P}}_{X_cX_c}(t) = \underline{A}_c\underline{P}_{X_cX_c} + \underline{P}_{X_cX_c}\underline{A}_c^T + \underline{G}_c\underline{V}\,\underline{G}_c^T ; \quad \underline{P}_{X_cX_c}(t_0) , \qquad (4.63)$$

and where $\underline{V} = E[\underline{W}_c\underline{W}_c^T(t)]$ is the intensity of white noise.

4.3 ACTIVE CONTROL OF TALL BUILDINGS

Figure 4.7 describes the basic loop of active control of any structure. One or various sensors are placed at certain locations in order to collect information about the behaviour of the structure. These data are fed into a data processor system (called filter or observer) which estimates, on the basis of the measurements, the actual state of the structure. The designed control forces are, in general, proportional to the current structure states, depending on the objective of the control. The controlled response is again observed by the sensors and the control intends to regulate the structures' motion if it is subjected to steady state forces, or to dampen the motion if the structure is subjected to residual forces.

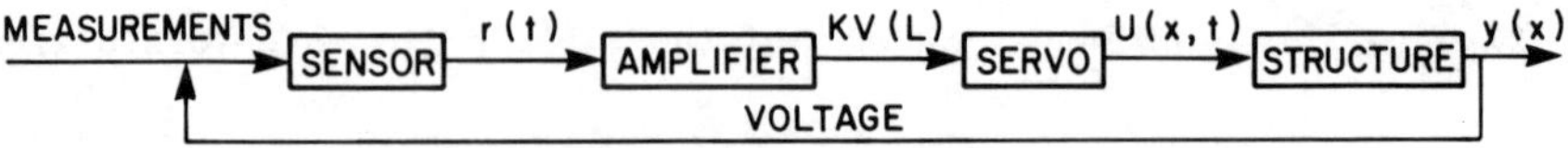

Figure 4.7 - Basic Control Loop

The questions of what types of sensors to choose, where
to place them, the design of the data processor system, the design
of the optimal control law, and the location of the control forces
have been considered in Section 3.7. That section gave indeed
important answers to those questions. The sensor and the control
force should be placed at the most flexible location on the structure. In a tall building this location is at the top of the building.
It provides optimal control and optimal filtering. The acceleration sensor provides minimum gain for the filter.

The most difficult problem encountered in tall buildings'
control is the design of the practical control mechanism needed to
generate the desired control forces. We shall now describe
possible control mechanisms which can be implemented in practice.

4.3.1 *Movable Appendages*

The proposal of using movable appendages was made by Klein et al
in 1972 [4.10]. The appendage which is placed at the top of the
building is either deployed or folded depending on the building
response to the wind. In their design [4.10], they assumed that
in order to repulse some of the wind forces, the appendage is
fully deployed when the velocity of the building is opposite to the

mean wind direction, and the appendage is fully folded if the
velocity of the building is in the same wind direction.

In a recent investigation by Soong [4.11], where the
appendage movement was designed via optimal control theory, the
controlled response and the consumed energy were better than in
Klein's design [4.10].

A description for the control mechanism is given in
Figure 4.8, where a plate hinged at one edge of the building, and
several links hinged at certain locations on the plate and rolling on
the roof of the building at the other ends, are shown. The control
system shall regulate the motion of the roller supports via actuators
in order to defold or deploy the plate according to properly
designed movements.

4.3.2 *Tuned Mass Dampers*

Tuned mass dampers have long been used in the passive control of
tall buildings. The coupling between the tuned mass and the building
introduces damping to the structure. This damping contributes to
the dissipation of the oscillations in a shorter time.

Since active control techniques have now become popular
among structural engineers, the idea of using active control for
tuned mass dampers has been taken up. It became thus possible to
control the coupling between the tuned mass damper and the building by
actuators. In this way, higher damping and fast tuning with the
building motion could be provided. A practical application of an
active tuned mass damper was implemented by the MTS Systems Corpor-
ation [4.13,4.14] in the Citicorp Center, Manhattan, New York.
A large scale tuned mass damper was designed to reduce the first
mode building sway. A diagram of that system is shown in
Figures 4.9 and 4.10.

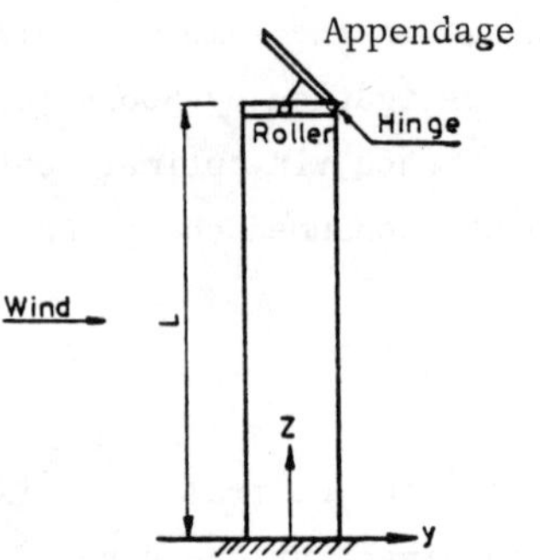

Figure 4.8 - Appendage Mechanism

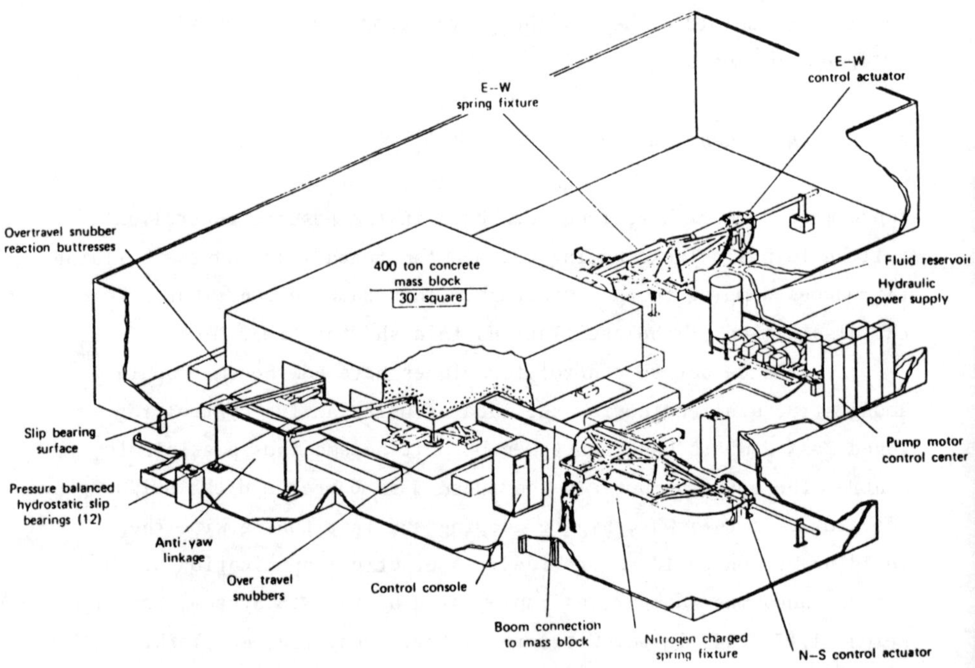

Figure 4.9 - MTS Tuned Mass Damper System, Citicorp Center,
New York City. Taken from Reference [4.33].

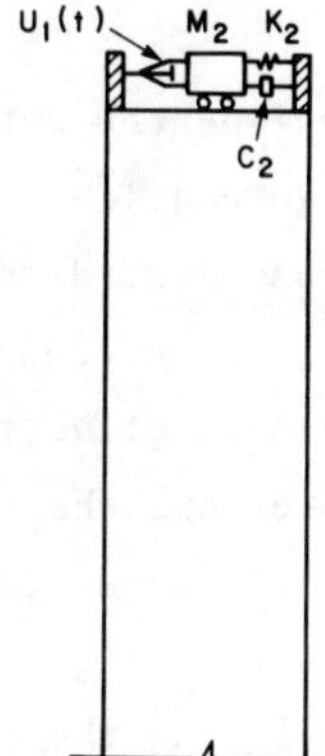

Figure 4.10 - Active Tuned Mass Damper

One expects that many difficulties will be encountered
with respect to the installation of the mass and the providing of
a "frictionless" motion according to the control law. Problems
like time delay effect, effectiveness of the damper in the presence
of random wind shall be considered specifically in this chapter.

4.3.3 *Active Tendons*

In his early proposals, Zuk [4.3] has shown that there is the possibility of controlling most of the types of structures by means of
active tendons. Tendons have long been used in the static control
of bridges, suspension bridges, prestressing and wind bracings.
They provide passive stiffness to the structure and therefore
reduce the amplitude of oscillations. However, the frequency of
the oscillations increases due to tendons. This could cause excessive vibrations and discomfort for occupants of the building. For
these reasons, researchers have proposed the idea of
actively operating the tendons using actuators. By releasing and
tensioning the tendons one can add both active stiffness and damping

to the structure. This idea has been used in a number of applications [4.4-4.6].

In a recent proposal by the authors [4.20], the tendons
were arranged as shown in Figure 4.11. Through this arrangement,
one can control the sway in any wind direction as well as the torsion
of the building. The tendons can be connected between the top of the
building and the base to yield an effective and optimal control
force, as explained in the previous chapter. However, if this
should not be practical, one can connect the other end of the tendons at a location which is not moving very much, as shown in
Figure 4.11. Problems of designing the control law, including time
delays, and assuming "stochastic" wind shall be considered in the
next sections.

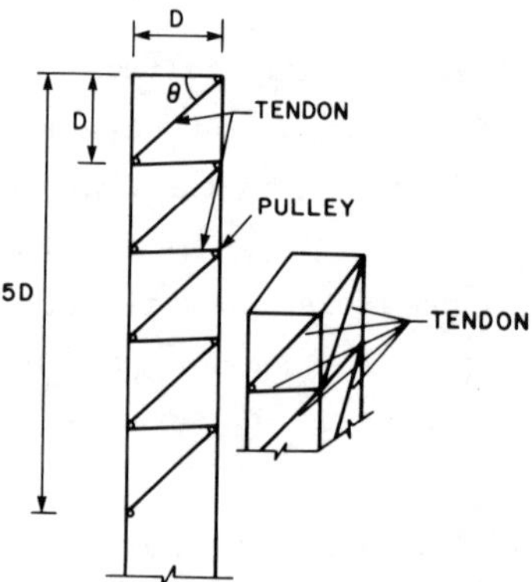

Figure 4.11 - Control Mechanism

4.3.4 *Practical Control Mechanisms*

By analyzing the previously mentioned three mechanisms, one will
be able to show the advantages and disadvantages of each mechanism.
The ideal mechanism to be used in practice can then be recommended.

4.4 CONTROL USING ACTIVE TENDONS

In this section, an example for controlling a tall building using
active tendons is given and the advantages and disadvantages of
this control mechanism will be discussed. It has been shown in
Section 3.4 that designing the optimal control forces by the optimal
control method is a time consuming and difficult process. In contrast, the proposed method [4.24] of using pole assignments provides
a fast approach for optimal design. It has also been shown in
Section 4.1 that if the aerodynamic forces G_1 and G_2 are taken into
account, the building equations become nonlinear. It is very difficult to design the optimal control forces for nonlinear systems.
In references [4.7,4.8], it has been concluded that designing the
control forces for a simplified linear system and applying these
forces to the nonlinear system could provide acceptable results
for the nonlinear case.

4.4.1 *Design of the Control Force*

The simplified linear system's equation of motion in the w_1-direction,
considering one mode only, is expressed as

$$\ddot{b}_1 + 2\zeta_1\omega_1\dot{b}_1 + \omega_1^2 b_1 = \frac{F_1}{mL} \, . \tag{4.61}$$

The mechanism shown in Figure 4.12 is used to control
the vibration of the building. It consists of four prestressed
tendons which are fixed at the top corners of the building, one on
each facade, and pass down through systems of pulleys until they
reach a certain building level. At this level, there are four
actuators installed with control systems and connected with the
tendons. A system of passive control forces is generated once the
building is subjected to a deflection. These passive forces are
shown in Figures 4.12 and 4.13. The equation of motion in

w_1-direction is

$$\ddot{b}_1 + 2\zeta_1\omega_1\dot{b}_1 + \omega_1^2 b_1 = \frac{F_1}{mL} - \frac{1}{mL}\frac{2EA}{D}\cos^3\theta\left[\sum_{i=1}^{5}\Delta_i\right]b_1 \, , \quad (4.62)$$

in which E is Young's modulus, A is the cross section of one tendon, θ is the angle between a tendon and the horizontal, and $\Delta_i = \phi(L-iD+D) - \phi(L-iD)$, $i = 1,2,\ldots,5$.

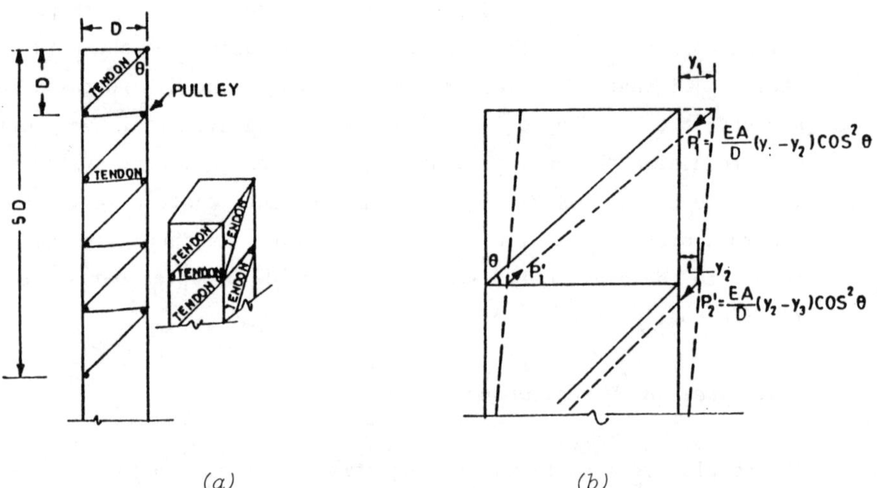

Figure 4.12 - (a) Control Mechanism
(b) Passive Control Forces

The increase in stiffness in the first mode due to the passive control force is apparent. That means, equation (4.62) can be expressed as

$$\ddot{b}_1 + 2\zeta_1\omega_1\dot{b}_1 + \omega_1'^2 b_1 = \frac{F_1}{mL} \, , \quad\quad\quad (4.63)$$

in which ω_1' is the frequency of the building including the pre-stressed tendons' effect.

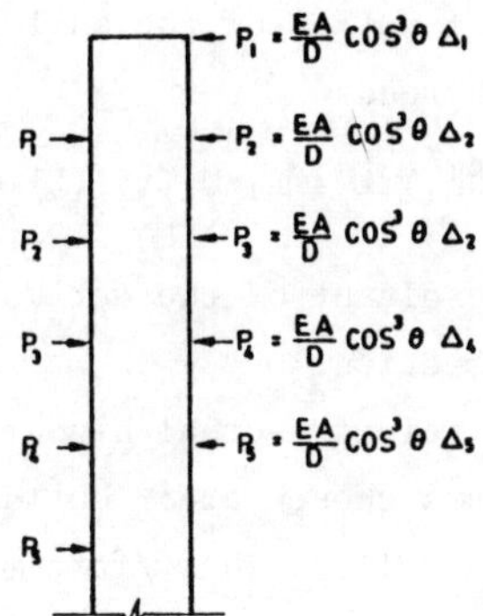

Figure 4.13 - Passive Control Forces

The active control forces, for the mechanism shown in
Figure 4.11, shall be calculated and applied as shown in Figure
4.14. The distance 5D can be alternatively increased until it becomes L. In that case, only the control force at the top of the
building would be apparent. However, for practical reasons and to
get rid of some problems concerning connections, the distance 5D
was chosen to cover the most flexible region of the building as
has been recommended in Section 3.7.

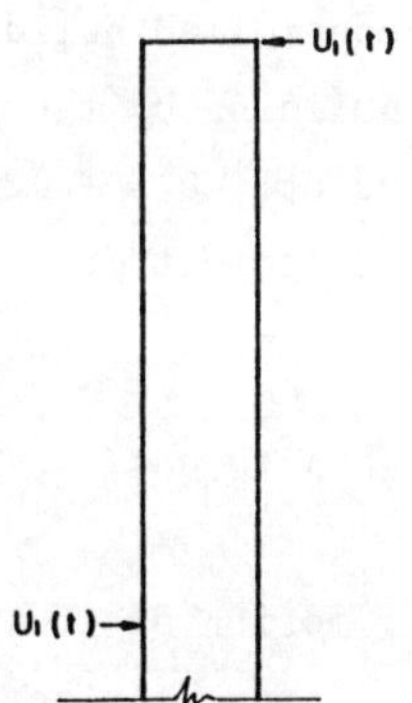

Figure 4.14 - Active Control Forces

The equation of motion of the building under the effect of active control forces reads

$$(\ddot{b}_1 + 2\zeta\omega_1\dot{b}_1 + \omega_1'^2 b_1)mL = F_1 - U_1(t)\phi(L) + U_1(t)\phi(L-5D) \quad , \quad (4.64)$$

in which $U_1(t)$ is the resultant of the active control force for two cables in the w_1 direction.

An attempt to select factual high rise building parameters was made. The data chosen are: building height L = 1000 ft, width D = 100 ft, mass m = 11000 lb.s^2/ft^2, ω_1 = 2 rad/sec, and damping ratio ζ_1 = 0.01.

Equation (4.64) can be written in the state space form as

$$\begin{bmatrix} \dot{b}_1 \\ \ddot{b}_1 \end{bmatrix} = \begin{bmatrix} 0 & 1 \\ -4.971 & -.04 \end{bmatrix} \begin{bmatrix} b_1 \\ \dot{b}_1 \end{bmatrix} + \begin{bmatrix} 0 \\ \dfrac{-1.32}{mL} \end{bmatrix} U_1(t) + \begin{bmatrix} 0 \\ \dfrac{F_1}{mL} \end{bmatrix} \quad ,$$

$$(4.65)$$

in which $[\phi(L)-\phi(L-5D)]$ = 1.32, $\omega_1'^2$ = 4.971; A = 1.0 ft^2, θ = 45°, Δ_1 = 0.2752, Δ_2 = 0.2738, Δ_3 = 0.2692, Δ_4 = 0.2595, and Δ_5 = 0.2432.

It is obvious from equation (4.65) that the passive controlled building has poles at (-0.2 ± J2.2295) as compared with (-0.02 ± J2.0) for the uncontrolled building. In order to introduce more damping to the building by the active control technique, the eigenvalues are assumed at (-0.5 ± J2.22).

Using the pole assignment method, the control force U_1 is found to be

$$U_1 = \frac{11\times10^6}{1.32} [0.2074 \quad 0.96][b_1 \quad \dot{b}_1]^T \quad . \tag{4.66}$$

The equations of motion of the building in the w_1-direction and w_2-direction are, respectively, after including the aerodynamic wind forces,

$$(\ddot{b}_1+2\zeta_1\omega_1\dot{b}_1+\omega_1'^2 b_1)mL = F_1+G_1-1.32U_1 \ , \tag{4.67}$$

$$(\ddot{a}_1+2\zeta_1\omega_1\dot{a}_1+\omega_1'^2 a_1)mL = F_2+G_2-1.32U_2 \ , \tag{4.68}$$

in which U_2 is given by Eq.(4.66), but replacing b_1 and $\dot{b}_1$ by a_1 and $\dot{a}_1$, respectively.

Let the wind first be treated by a deterministic analysis. Stochastic control shall be considered in a later section.

4.4.2 *Response Analysis*

For simplicity, the wind shall be considered not to be a function of x. This means that the height does not affect the wind intensity. Although the assumption is not always realistic, it is nevertheless assumed to prevail for the sake of simplicity, as the major concern is to obtain a feeling for the effect of the active control forces on the controlled building response.

Assuming $\bar{U}$ = 70 fps; the fluctuating wind components are $v_1 = 5\cos(t)$ fps and $v_2 = 2\sin(t)$ fps. The function $f(t)$ was assumed as

$$f(t) = \sin\left(\frac{2\pi S\bar{U}}{D}\right)t \ , \tag{4.69}$$

in which S is the Strouhal number, which was taken to be 0.19. For a drag coefficient of C_d = 1.2, a lift coefficient of C_L = 1.2, mass coefficient of C_m = 0.5, and an air density of ρ = 0.0763 pcf, the wind was assumed to be acting for 20 seconds only, after which time the wind gust was assumed to disappear.

The building is assumed to have been at rest before the wind started to blow. Under this condition, the controlled and the uncontrolled responses of the building in both w_1 and w_2 directions are determined. In this application, it is also assumed that the accelerometer is placed at the top of the building, and that it supplied continuous measurements to the filter which subsequently processed the measurements to yield the estimated states for b_1 and $\dot{b}_1$. Figures 4.15 to 4.18 provide descriptions for both controlled and uncontrolled responses of the building [4.20].

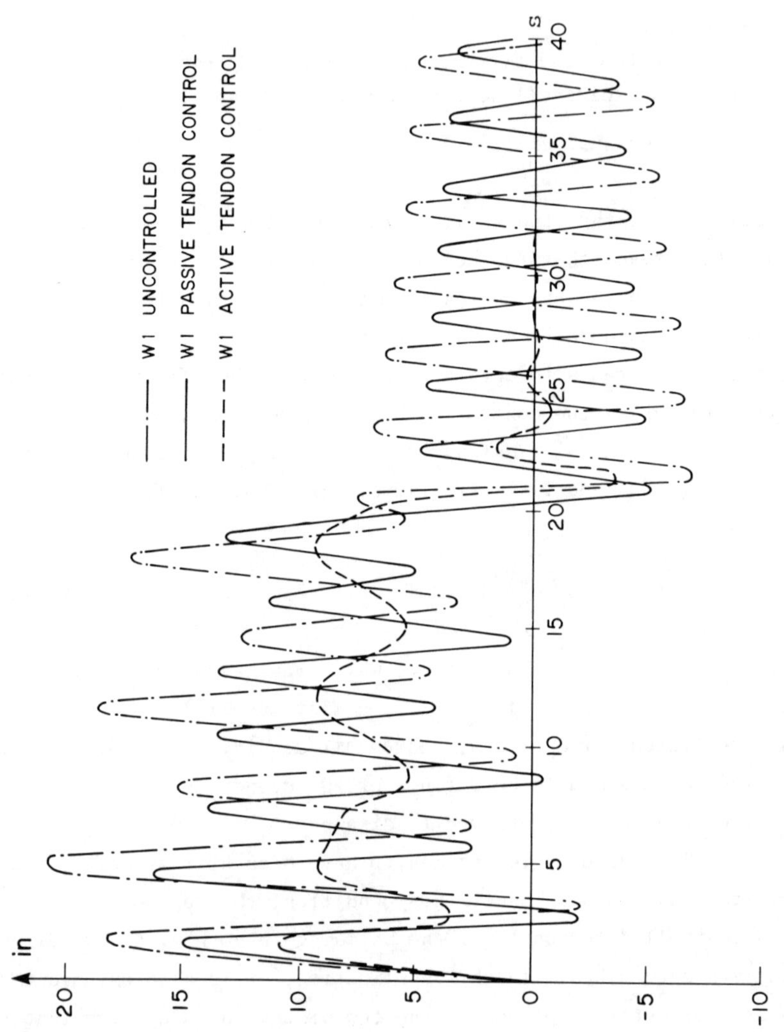

Figure 4.15 – Along-Wind Deflection Response Using Tendons

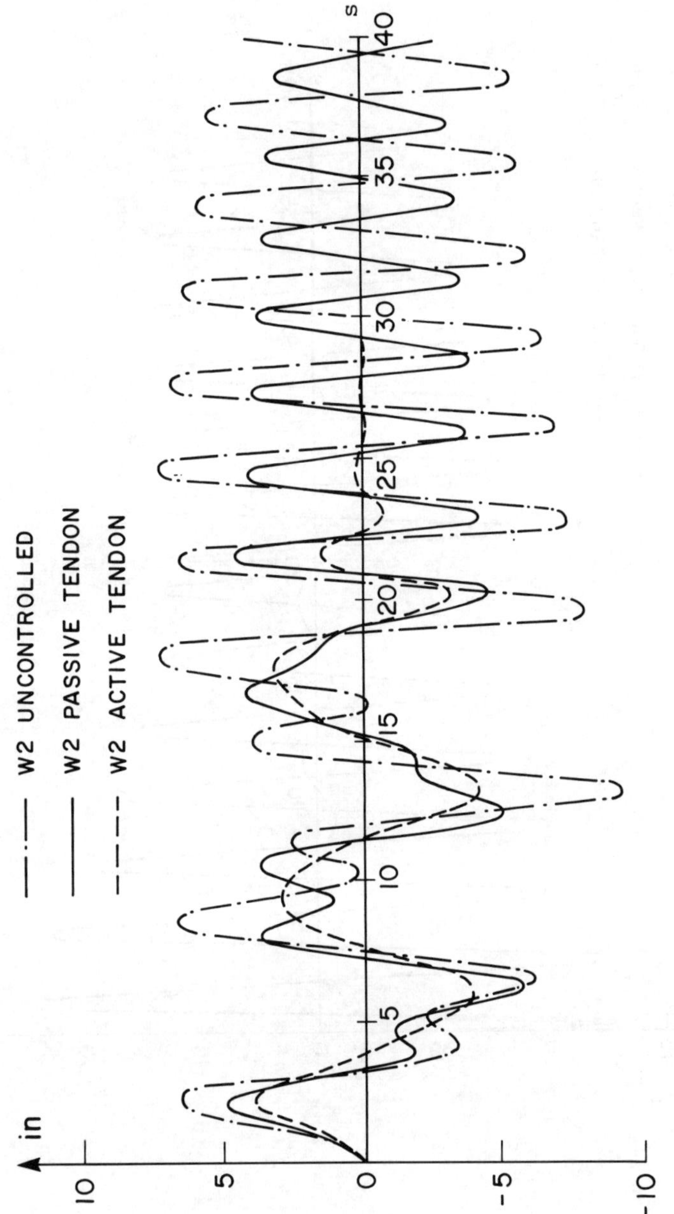

Figure 4.16 - Across-Wind Deflection Response Using Tendons

                    H.H.E. Leipholz and M. Abdel-Rohman

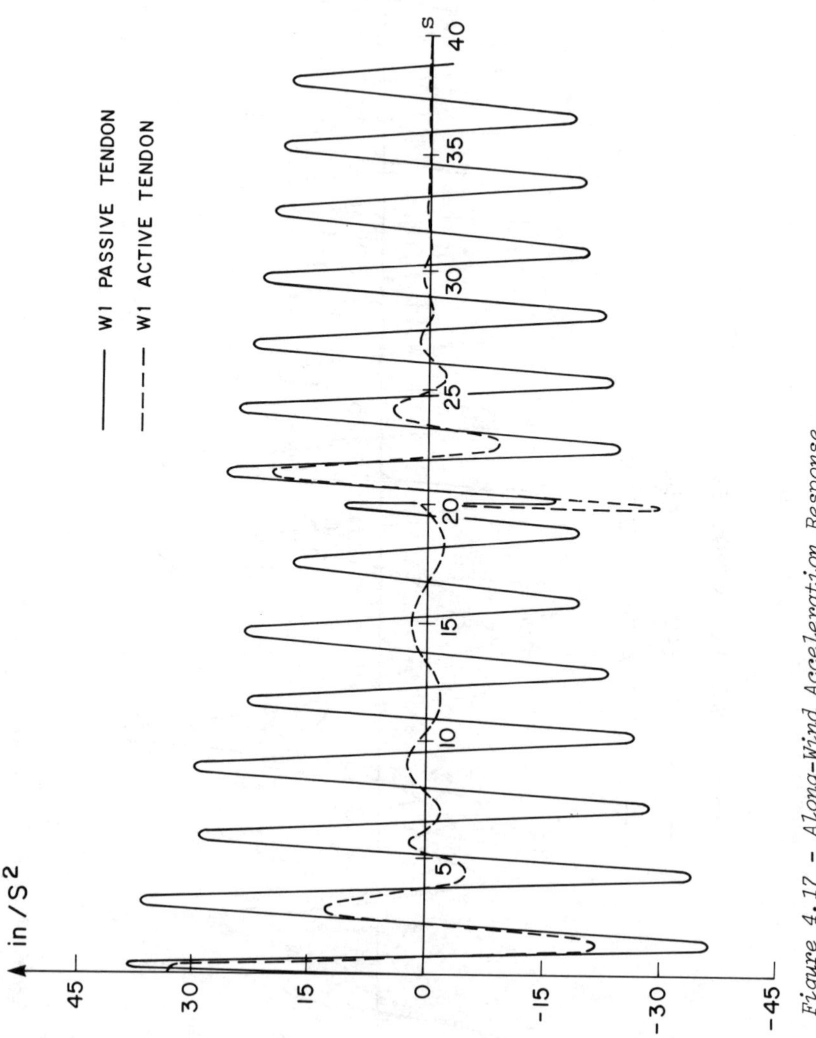

Figure 4.17 - Along-Wind Acceleration Response

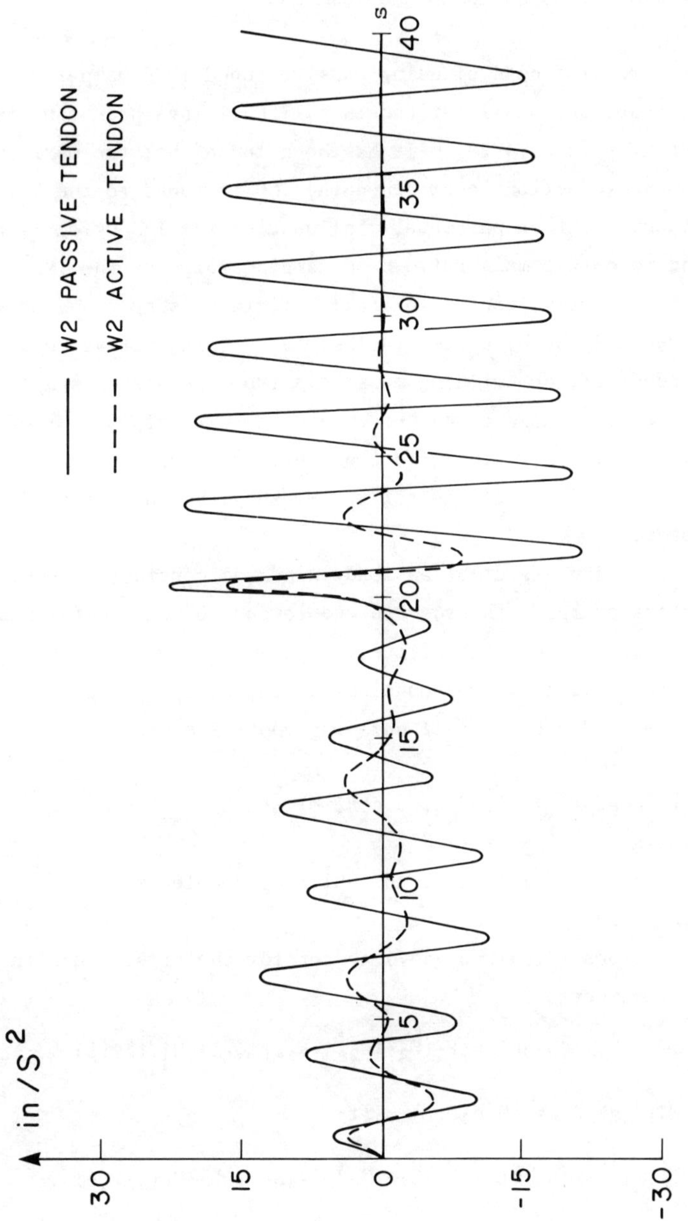

Figure 4.18 - Across-Wind Acceleration Response

4.5   CONTROL USING TUNED MASS DAMPERS

The appropriateness of using passive tuned mass dampers to effectively dampen the vibrations in tall buildings has been studied in references [4.27,4.28]. It has been found that the maximum building response reduction occurs when the TMD is tuned to the building frequency. Other parameters influencing the TMD effectiveness are found to be the mass ratio, and damping ratio of the TMD.

Active control of tall buildings using tuned mass dampers has recently been applied [4.13-4.15]. Using active control and 2 percent of the building mass as a TMD, the steady sway could for example be reduced by 40 percent in the Citicorp Center of New York City [4.13]. The effect of unsteady wind and time delay on the effectiveness of the active TMD shall be investigated in this chapter.

The effect of unsteady wind was considered to be as in the previous example, in order to enable one to compare this control with that one in which active tendons were used, and to enable one to come up with clearer conclusions on how effective the active tuned mass damper in tall building control may be.

4.5.1 *Design of the Active Control Law*

The equation of motion of the building coupled with the active TMD, Figure 4.19, can be derived by applying an integral transformation to equations (4.6) and (4.7) to get for the first modes in the w_1 and w_2 directions

$$(\ddot{b}_1+2\zeta_1\omega_1\dot{b}_1+\omega_1^2 b_1)mL = F_1+G_1+(C_2\dot{Z}_1+K_2 Z_1)\phi(L)-U_1(t)\phi(L) \ , \qquad (4.70)$$

$$M_2\ddot{Z}_1+C_2\dot{Z}_1+K_2 Z_1 = -M_2\ddot{b}_1\phi(L)+U_1(t) \ , \qquad (4.71)$$

$$(\ddot{a}_1+2\zeta_1\omega_1\dot{a}_1+\omega_1^2 a_1)mL = F_2+G_2+(C_2\dot{Z}_2+K_2 Z_2)\psi(L)-U_2(t)\psi(L) \ , \qquad (4.72)$$

$$M_2\ddot{Z}_2+C_2\dot{Z}_2+K_2 Z_2+K_2 Z_2 = -M_2\ddot{a}_1\psi(L)+U_2(t) \ . \qquad (4.73)$$

In these equations $Z_1 = wT_1-w_1$, $Z_2 = wT_2-w_2$, wT_1 is the motion of
the TMD in w_1 direction, wT_2 is the motion of TMD in w_2 direction,
M_2 is the mass of TMD, C_2 is the damping of TMD, K_2 is the stiffness
of TMD, $U_1(t)$ is the active control force in w_1-direction, and
$U_2(t)$ is the active control force in w_2-direction.

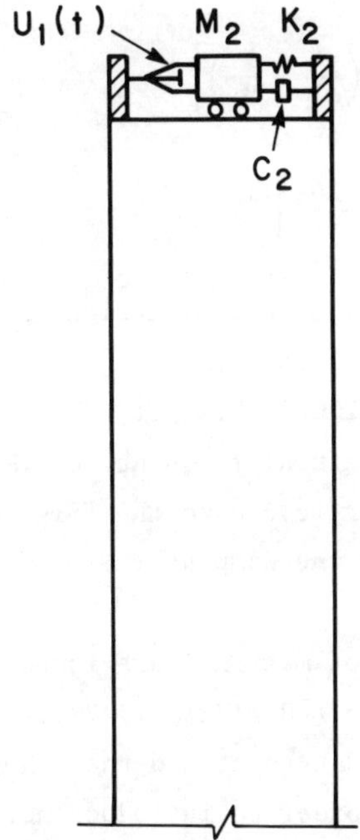

Figure 4.19 - Active Tuned Mass Damper

Writing equations (4.70) and (4.71) in state space form
as

$$\dot{\underline{X}} = \underline{A}\,\underline{X} + \underline{B}\,\underline{U} + \underline{d} \ , \quad \underline{U} = U_1, \text{ and} \qquad (4.74)$$

where $\underline{X} = [b_1\ \dot{b}_1\ Z_1\ \dot{Z}_1]^T$,

$$\underline{A} = \begin{bmatrix} 0 & 1 & 0 & 0 \\ -\omega_1^2 & -2\zeta_1\omega_1 & \dfrac{2K_2}{mL} & \dfrac{2C_2}{mL} \\ 0 & 0 & 0 & 1 \\ 2\omega_1^2 & 4\zeta_1\omega_1 & -\left(\dfrac{4K_2}{mL} + \dfrac{K_2}{M_2}\right) & -\left(\dfrac{4C_2}{mL} + \dfrac{C_2}{M_2}\right) \end{bmatrix} , \quad (4.75)$$

$$\underline{B}^T = \begin{bmatrix} 0 & \dfrac{-2}{mL} & 0 & \left(\dfrac{4}{mL} + \dfrac{1}{M_2}\right) \end{bmatrix} , \quad (4.76)$$

$$\underline{d}^T = \begin{bmatrix} 0 & \dfrac{F}{mL} & 0 & \dfrac{-2F_1}{mL} \end{bmatrix} . \quad (4.77)$$

G_1 has been neglected for the sake of designing the optimal control force U_1 for a simplified linear system.

Selecting M_2 = 220000 lb.s^2/ft, K_2 = 839960 lb/ft, C_2 = 128960 lb.s/ft, the natural frequency of TMD is then 1.953 rad/sec, which is very close to the first mode building frequency, 2 rad/sec; also the damping ratio of TMD was assumed to be 0.15.

The eigenvalues of matrix $\underline{A}$ are found to be $\lambda_{1,2}$ = (-0.223±J2.2198) and $\lambda_{3,4}$ = (-0.1132±J1.7479). The first two poles provide damping of 10 percent and the other two poles provide damping of 6 percent. In order to introduce more damping using active control forces, the poles are assumed at $\lambda_{1,2}$ = -0.5±J2.22 and $\lambda_{3,4}$ = -0.4±J1.7479. In this case, all poles provide a damping of 22 percent. This enables one to compare the active TMD control with the active tendon control.

Using the pole assignment method, the control force was found to be

$$U_1 = 10^6 [3.487 \quad -1.025 \quad -.0767 \quad -.2695]\underline{X} . \quad (4.78)$$

4.5.2 *Controlled Response*

Substituting equation (4.78) into equation (4.74), one can obtain the active controlled response of the building and the TMD. It is shown in Figures 4.20 - 4.25.

A comparison between the along-wind response when using either active tendons or active TMD control is given in Figures 4.26 and 4.27. Table 4.1 presents also a comparison between tendon control and TMD control in terms of the minimum overall reduction which has been achieved for the maximum peaks of w_1, w_2, $\ddot{w}_1$, and $\ddot{w}_2$ with respect to the uncontrolled response. One observes that active tendon control is more effective than active TMD which proves that it does not pay off the energy supplied to it.

Table 4.1 - Reduction in Tall Building Response Using Tendon and TMD Systems with Respect to Uncontrolled Responses

Kind of Response	Active TMD	Active Tendons	Passive TMD	Passive Tendon
w_1	20.2%	47.6%	16.7%	21.4%
w_2	34.4%	54.8%	32.2%	36.5%
$\ddot{w}_1$ (first peak)	0.4%	15.4%	0.2%	0.1%
$\ddot{w}_2$ (first peak)	51.3%	71.8%	48.7%	30.7%

On the other hand, active tendons have required larger control energy for an efficient control. One of the objectives of this chapter is to investigate the effectiveness of active TMD in greater detail and to come up with recommendations on the parameters which improve efficiency. This shall be done in the next section.

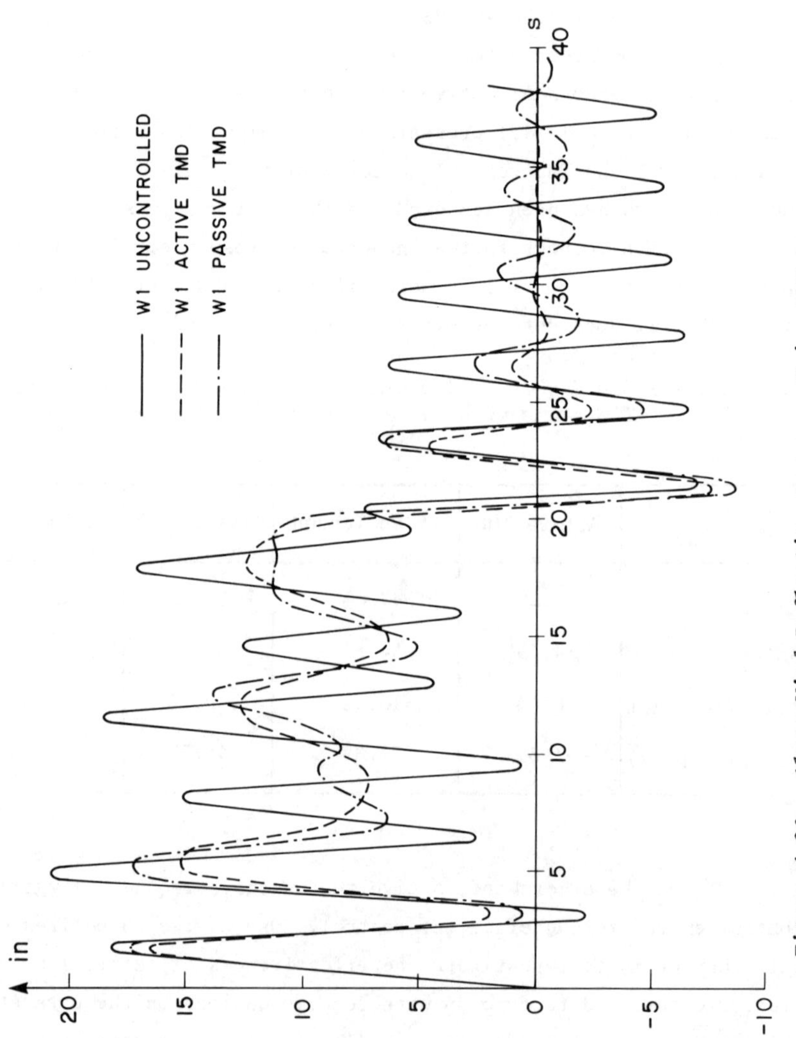

Figure 4.20 – Along-Wind Deflection Response Using TMD

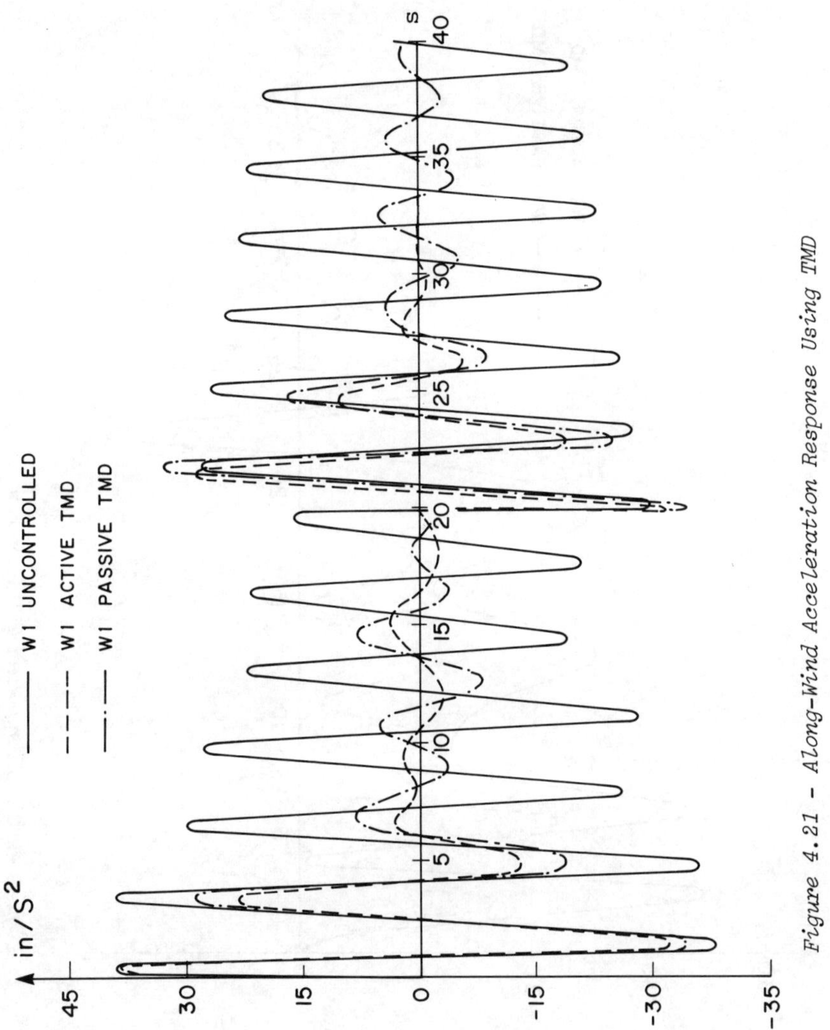

Figure 4.21 - Along-Wind Acceleration Response Using TMD

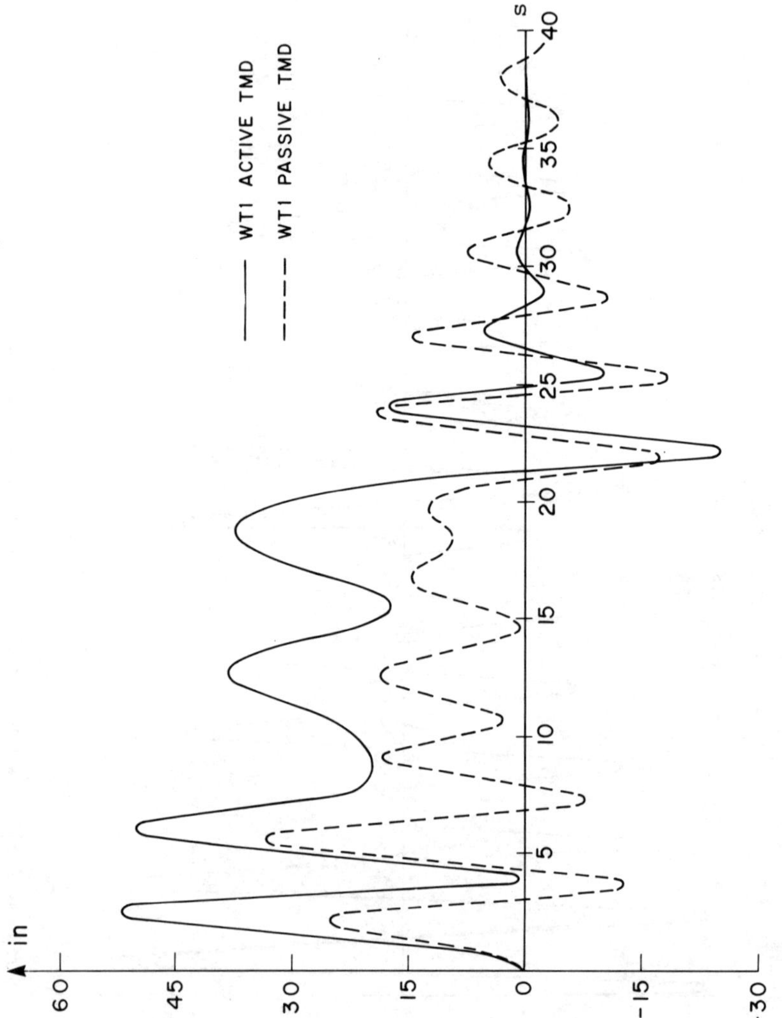

Figure 4.22 - Along-Wind TMD Response

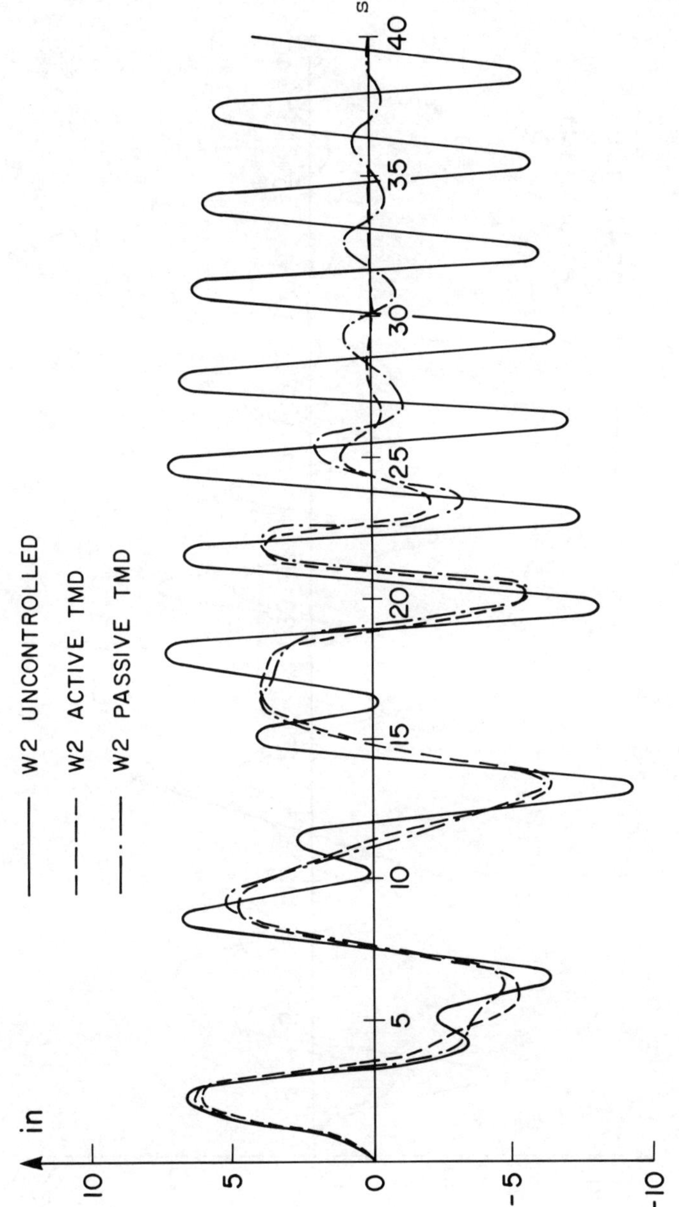

Figure 4.23 - Across-Wind Deflection Response Using TMD

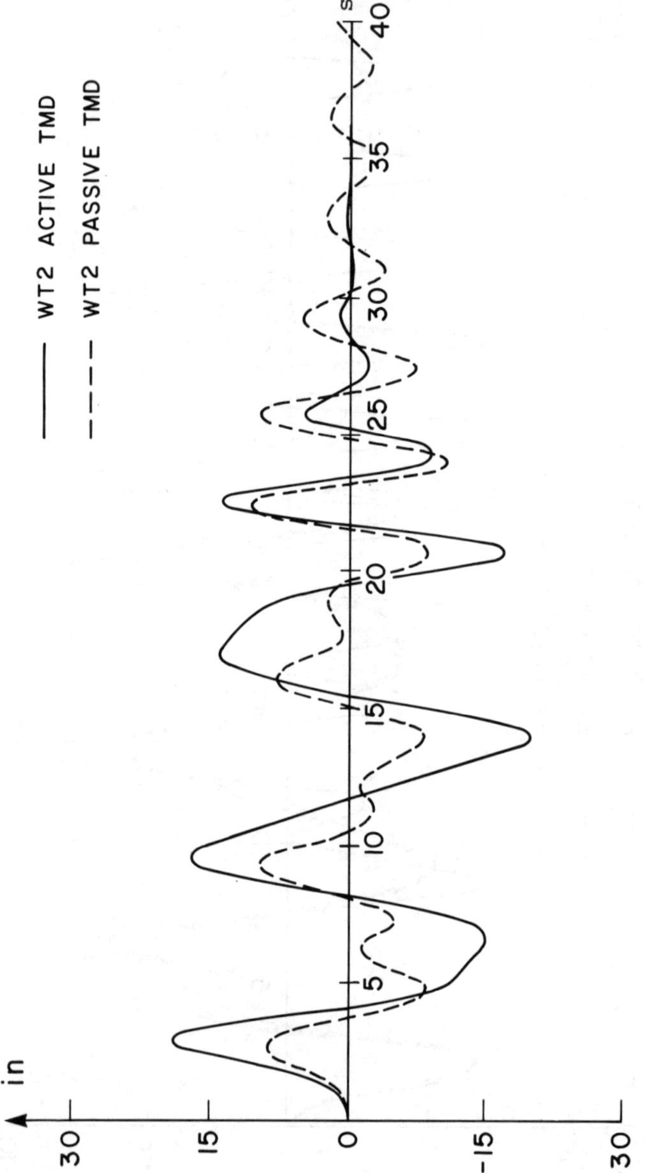

Figure 4.24 - Across-Wind TMD Response

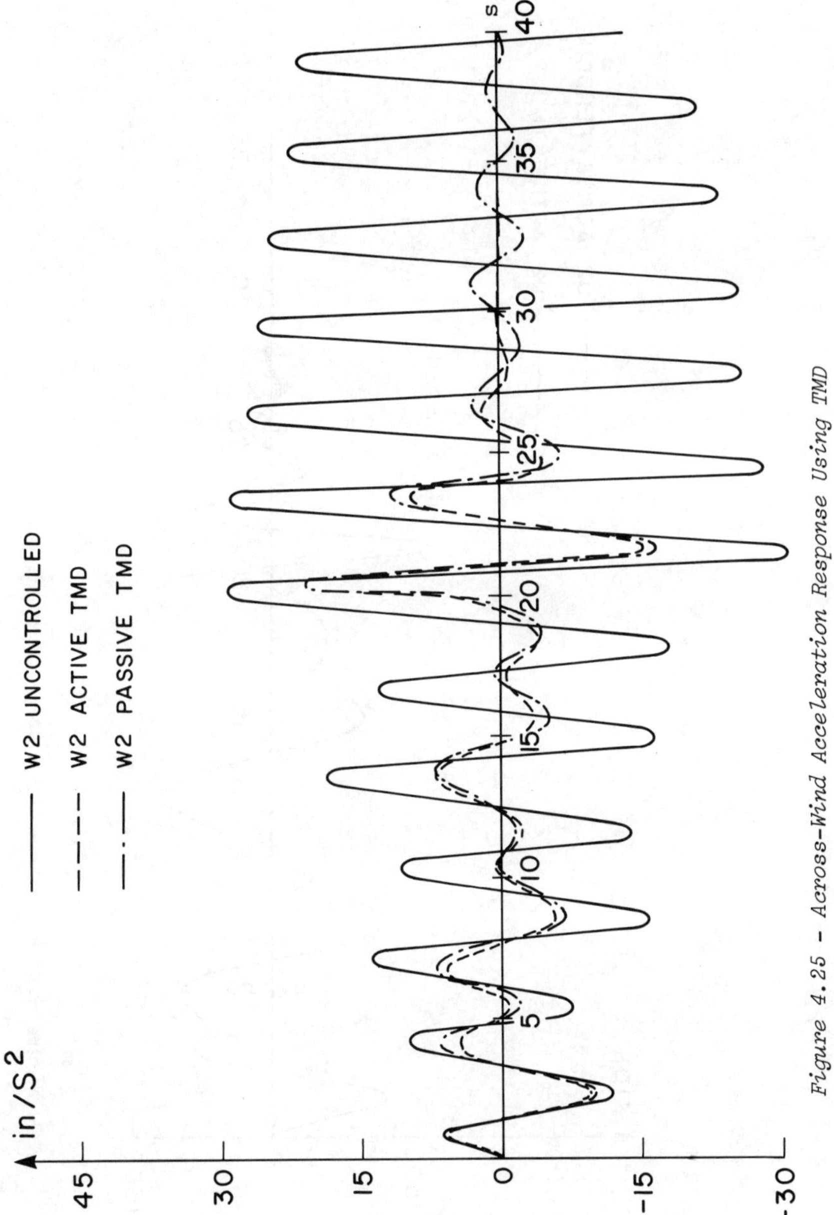

Figure 4.25 — Across-Wind Acceleration Response Using TMD

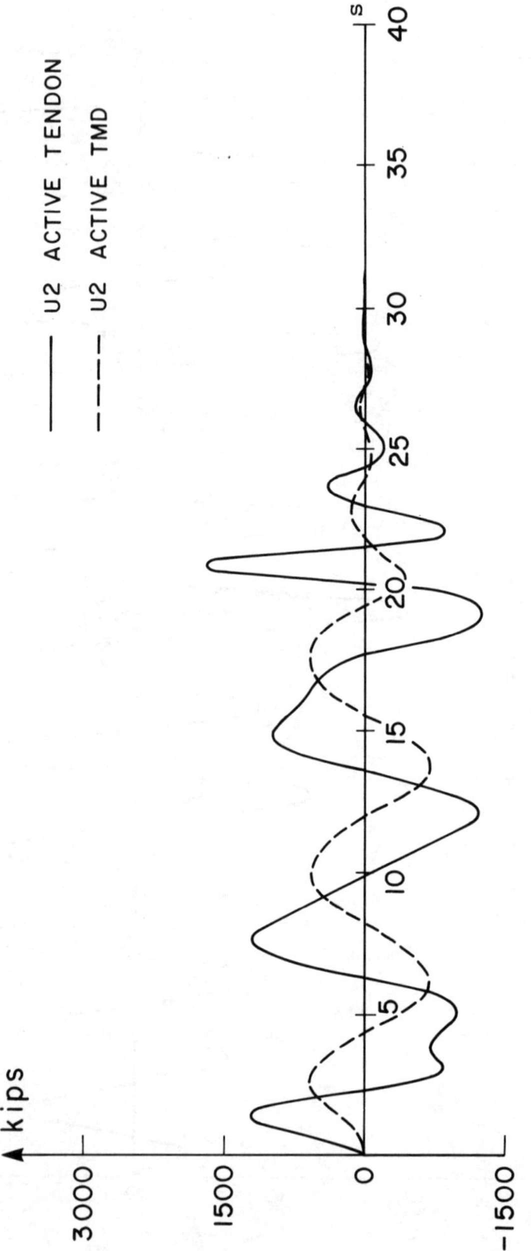

Figure 4.26 - Across-Wind Control Force Response

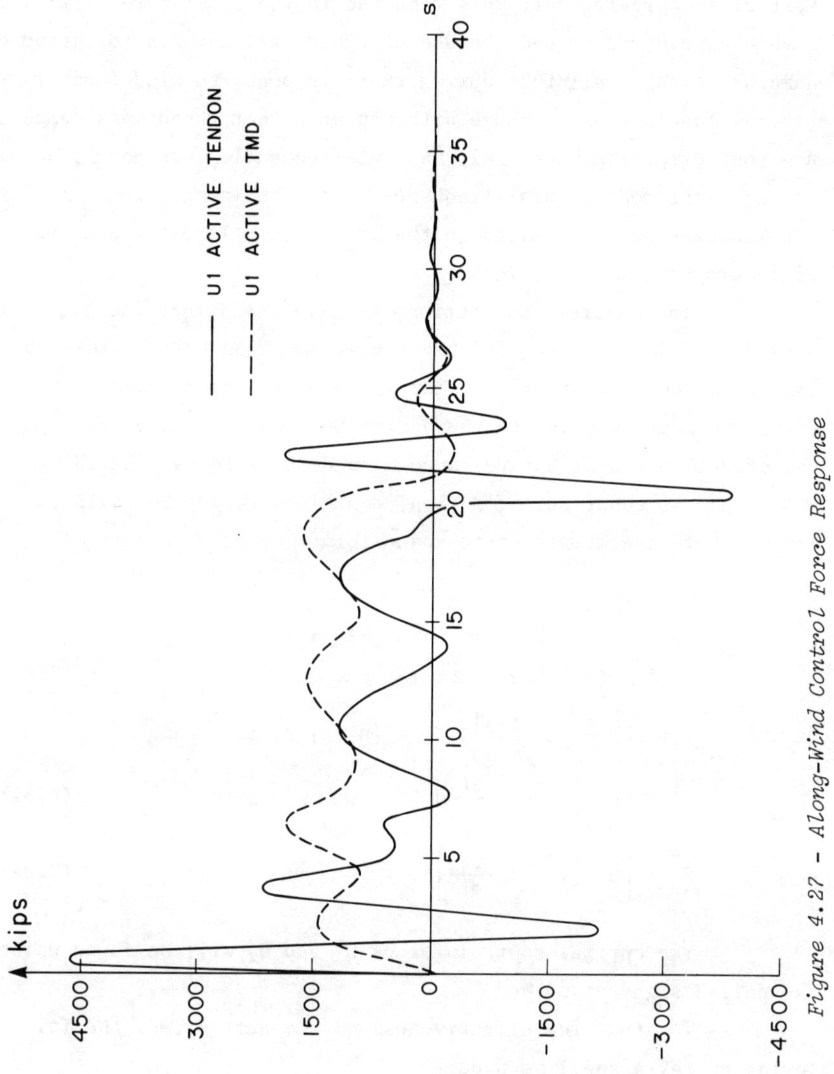

Figure 4.27 - Along-Wind Control Force Response

4.5.3 *Effectiveness of Active TMD*

Most of the previous attempts reported in the literature [4.13-4.15]
have been centred around the use of tuned mass dampers to introduce
damping to tall buildings during their response to wind forces. The
general conclusions of these attempts were that tuned mass dampers
are most effective when their natural frequencies are nearly equal
to the first mode natural frequency of the building. Tuned mass
dampers have been installed in the CN Tower in Toronto, and the
Citi-Corp Center in New York.

In the previous section, we have shown that the active tuned
mass damper does not pay off the energy supplied to it. This conclusion which is contrary to what is commonly assumed among structural engineers has led to the desire to study, in more detail,
the effectiveness of active tuned mass dampers in tall buildings.

The equations of motion, equations (4.70) to (4.73),
are put into the matrix state space form to read

$$\dot{\underline{X}}_1 = \underline{A}\,\underline{X}_1 + \underline{B}\,\underline{U}_1 + \underline{F}_1 \ , \tag{4.79}$$

$$\dot{\underline{X}}_2 = \underline{A}\,\underline{X}_2 + \underline{B}\,\underline{U}_2 + \underline{F}_2 \ , \tag{4.80}$$

in which $\underline{X}_1 = [b_1 \ \dot{b}_1 \ z_1 \ \dot{z}_1]^T$, $\underline{X}_2 = [a_1 \ \dot{a}_1 \ z_2 \ \dot{z}_2]^T$, and

$$\underline{F}_1^T = \left[0 \quad \frac{F_1}{mL} \quad 0 \quad \frac{-2F_1}{mL} \right] \ , \tag{4.81}$$

$$\underline{F}_2^T = \left[0 \quad \frac{F_2}{mL} \quad 0 \quad \frac{-2F_2}{mL} \right] \ . \tag{4.82}$$

The optimal control forces U_1 and U_2 will be found using
the pole-assignment method.

To study the effectiveness of the active TMD, the following criteria shall be used:

$$J_1 = \int_0^T (b_1^2 + \dot{b}_1^2)\,dt \ , \tag{4.83}$$

$$J_2 = \int_0^T (a_1^2 + \dot{a}_1^2)\,dt \ , \qquad\qquad (4.84)$$

$$J_3 = \int_0^T (wT_1^2 + \dot{w}T_1^2)\,dt \ , \qquad\qquad (4.85)$$

$$J_4 = \int_0^T (wT_2^2 + \dot{w}T_2^2)\,dt \ , \qquad\qquad (4.86)$$

$$J_5 = \int_0^T 10^{-10}(U_1^2 + U_2^2)\,dt \ . \qquad\qquad (4.87)$$

Equations (4.83) and (4.84) represent the total response of the structure. Equations (4.85) and (4.86) represent the total response of the TMD. Equation (4.87) evaluates the cost of the control energy.

For the same building parameters as considered in the previous section, and for wind acting during 20 seconds only, various eigenvalues have been assumed as desirable poles for the controlled building. The results of equations (4.83) to (4.87) were evaluated and then tabulated in Table 4.2. In this way, a comparison is provided between the response of one passive and nine active controlled TMD, due to both the wind forces and the self-excited wind forces, (F_1, G_1, F_2, G_2).

Studying Table 4.2, one finds that the reduction in the total building response, over $T = 40$ seconds, due to an application of active tuned mass dampers is trivial as compared with the reduction in building response due to passive tuned mass damper, except in cases 8, 9 and 10 in which reductions of 17, 20 and 30 percent, respectively, occurred, but at the expense of a large control energy consumption. Moreover, the responses of all tuned mass dampers have been increased, except in case 7 in which a negligible reduction in the building response occurred. One concludes that any noticeable reduction in the tall building response, using active tuned mass dampers, is at the expense of a greater energy consumption and an increase in the dampers' response.

Table 4.2 - Response Criteria for T = 40 Seconds

	Case	Pole	J_1	J_2	J_3	J_4	J_5
Passive TMD	1	-0.223 ±J 2.2198 -0.1132±J 1.7479	6.556	1.2279	104.82	23.169	0.0
Active TMD	2	-0.223 ±J 2.2198 -0.5 ±J 1.7479	6.1334	1.2276	101.67	23.566	36,035
Active TMD	3	-0.223 ±J 2.2198 -0.5 ±J 1.6786	5.986	1.1955	125.03	28.859	64,679
Active TMD	4	-0.223 ±J 2.2198 -1.0 ±J 1.7479	6.24	1.217	122.84	27.099	68,867
Active TMD	5	-0.223 ±J 2.2198 -1.0 ±J 1.518	6.016	1.1787	193.98	40.29	207,661
Active TMD	6	-0.223 ±J 2.2198 -2.0 ±J 1.7579	6.299	1.1905	102.91	24.03	48,423
Active TMD	7	-0.223 ±J 2.2198 -2.0 ±J 2.2279	6.417	1.2013	89.66	21.118	64,591
Active TMD	8	-0.5 ±J 2.22 -0.405 ±J 1.7479	5.477	1.11	208.89	47.368	193,329
Active TMD	9	-0.5 ±J 2.1742 -1.0 ±J 1.7479	5.214	1.035	393.5	77.45	659,163
Active TMD	10	-3.386 ±J 2.22 -3.0 ±J 1.7479	4.499	0.7488	501.81	128.24	919,940

If the same cases of Table 4.2 are studied with regard to the maximum peak responses, as shown in Table 4.3, one observes that in cases 2 to 7 no remarkable reduction of the maximum building response can be observed. In the other cases, reductions up to 20 percent for along wind response are reported, and only in case 10 a reduction of 15 percent for across wind response has been noticed. However, those cases required the application of huge control forces.

In order to study the effect of active tuned mass dampers on the free vibration response, the differences between the response criteria of equations (4.83) t0 (4.87) for T = 40 seconds and for T = 20 seconds are reported in Table 4.4. The ratios between the values following from these criteria versus those due to passive TMD of Table 4.4 are given in Table 4.5. The last column in

Table 4.5 represents the percentage of the control energy response with respect to the total control energy response of Table 4.2. One notices the effectiveness of active tuned mass dampers during the free vibration response, especially for cases 8 and 9 in which only 3.2 and 5 percent, respectively, of the total energy response were required to provide a reduction of more than 40 percent in the building free vibration response.

In order to be sure that self-excited wind forces do not change the results of the previous analysis and the conclusions based on it, Table 4.3 has been redone neglecting the self-excited wind forces (G_1 and G_2). The results tabulated in Table 4.6 show that the previous conclusions remain valid [4.30].

Table 4.3 - Maximum Peak Responses

	Case	w_1 inch	w_2 inch	wT_1 inch	wT_2 inch	U_1 kips	U_2 kips
Passive TMD	1	17.51	6.35	33.72	11.49	0.0	0.0
Active TMD	2	17.45	6.27	35.2	12.83	964.16	405.98
Active TMD	3	17.32	6.24	39.83	14.78	1081.7	491.5
Active TMD	4	17.55	6.14	39.45	14.09	1722.2	598.2
Active TMD	5	17.29	6.12	49.13	18.46	2194.0	819.0
Active TMD	6	17.62	6.12	33.52	13.91	2071.5	631.0
Active TMD	7	17.75	6.14	27.2	13.35	2278.0	617.0
Active TMD	8	16.77	6.27	52.17	19.84	1834.0	692.6
Active TMD	9	16.51	6.18	68.84	27.61	3196.6	1232.4
Active TMD	10	13.99	5.42	103.25	33.06	12813.0	4181.0

Table 4.4 - Response Criteria for Free Vibration

	Case	J_1	J_2	J_3	J_4	J_5
Passive TMD	1	1.356	0.564	40.5	13.919	0.0
Active TMD	2	1.133	0.347	26.5	9.66	5,271
Active TMD	3	1.066	0.32	29.33	10.41	6,266
Active TMD	4	1.20	0.352	34.06	10.86	12,669
Active TMD	5	1.10	0.317	40.46	11.59	20,576
Active TMD	6	1.211	0.35	37.6	11.75	14,510
Active TMD	7	1.26	0.364	37.28	12.0	10,678
Active TMD	8	0.817	0.24	38.26	13.4	6,274
Active TMD	9	0.73	0.20	59.1	14.78	33,558
Active TMD	10	0.47	0.07	167.9	30.84	298,169

Table 4.5 - Response Criteria Ratios for Free Vibration

	Case	J_1 ratio	J_2 ratio	J_3 ratio	J_4 ratio	Energy Ratio
Passive TMD	1	1.0	1.0	1.0	1.0	.....
Active TMD	2	0.835	0.615	0.654	0.694	14.6%
Active TMD	3	0.786	0.567	0.724	0.748	9.7%
Active TMD	4	0.885	0.624	0.84	0.78	18.4%
Active TMD	5	0.811	0.562	0.999	0.832	9.9%
Active TMD	6	0.893	0.62	0.928	0.844	29.9%
Active TMD	7	0.929	0.645	0.92	0.862	16.5%
Active TMD	8	0.60	0.425	0.945	0.963	3.2%
Active TMD	9	0.538	0.355	1.459	1.062	5.0%
Active TMD	10	0.347	0.124	4.145	2.214	32.4%

Table 4.6 – Maximum Peak Response Neglecting Self-Excited Wind

	Case	w_1 inch	w_2 inch	wT_1 inch	wT_2 inch	U_1 kips	U_2 kips
Passive TMD	1	16.84	6.36	32.94	11.43	0.0	0.0
Active TMD	2	16.91	6.23	33.94	12.75	898.7	404.2
Active TMD	3	16.78	6.21	38.17	14.7	1035.0	490.3
Active TMD	4	16.98	6.18	37.93	14.13	1605.0	597.8
Active TMD	5	16.72	6.12	46.68	18.37	2046.6	817.10
Active TMD	6	16.98	6.14	32.77	13.84	1936.4	629.7
Active TMD	7	17.09	6.16	27.08	13.28	2312.0	619.8
Active TMD	8	16.21	6.25	49.85	19.77	1708.0	691.2
Active TMD	9	15.90	6.16	65.35	27.52	2981.6	1229.0
Active TMD	10	13.08	5.4	96.38	32.98	13573.0	4161.0

4.5.4 *Concluding Remarks*

The following general conclusions can now be drawn:

(1) Active tuned mass dampers are not as effective as expected
in the case of a typical wind excitation which causes both forced
and free vibrations. The passive TMD is in this respect more effective when taking into account the control energy cost and the
total building response.

(2) Active tuned mass dampers are only effective with respect
to controlling the free and steady state response. They require in
this case small amounts of control energy and provide high damping
to the building.

The above conclusions follow as a result of the minor role that damping plays in suppressing the first peak of the forced vibration and of the absence of any active stiffness caused by the coupling between TMD and the tall building. Damping is very powerful only for free vibrations. Therefore, in our search for an active TMD of practical value, we have come to the point of realizing that it could only be used during the free or steady state vibration. Designing active TMD's should therefore be investigated in more detail in the following.

4.5.5 *Design of an Active TMD*

In the previous section, the active tuned mass damper (TMD) was designed with a natural frequency approximately equal to the first mode natural frequency of the building. The optimal design of the active TMD was restricted to finding the optimal control law [4.15,4.20]. However, one should realize that the active control technique introduces to both active TMD and the building active stiffness and active damping which change the structural parameters of the controlled structure. Therefore, to design an optimal active TMD, optimization should also be concerned with the TMD parameters.

The building is assumed to be subjected to stationary white noise wind forces and the effectiveness of the active TMD is to be measured in terms of the mean square controlled response, the effective damping, and the consumed control energy.

If the along-wind response only is taken into consideration, then one deals with equations (4.74) to (4.78).

When one specifies the TMD parameters in equations (4.75) to (4.78) and when the desired eigenvalues are specified for the controlled structure, then the optimal control law can easily be found. Let the optimal control law be expressed as

$$U(t) = [G_1 \quad G_2 \quad G_3 \quad G_4]\underline{X} \ , \tag{4.88}$$

in which G_1, G_2, G_3, and G_4 are elements of the designed gain matrix.

Substituting equation (4.88) into equations (4.74) to (4.78), one obtains

$$\ddot{b}_1 + 2\zeta_1\omega_1\dot{b}_1 + \omega_1^2 b_1 = f(t) + \frac{2C_2}{M_1}\dot{Z}_1 + \frac{2K_2}{M_1}Z_1 - \frac{2}{M_1}(G_1 b_1 + G_2\dot{b}_1 + G_3 Z_1 + G_4\dot{Z}_1) \ , \tag{4.89}$$

$$\ddot{Z}_1 + 2\zeta_2\omega_2\dot{Z}_1 + \omega_2^2 Z_1 = -2\ddot{b}_1 + \frac{1}{M_2}(G_1 b_1 + G_2\dot{b}_1 + G_3 Z_1 + G_4\dot{Z}_1) \ . \tag{4.90}$$

4.5.5.1 Stochastic Analysis

Equations (4.89) and (4.90) are transformed into the frequency domain using Laplace or Fourier transformation, since the equations' coefficients are time invariant. This results in the following transfer matrix relation:

$$\begin{bmatrix} H_{1,1}(\omega) & H_{1,2}(\omega) \\ H_{2,1}(\omega) & H_{2,2}(\omega) \end{bmatrix} \begin{bmatrix} H_b(\omega) \\ H_z(\omega) \end{bmatrix} = \begin{bmatrix} 1.0 \\ 0.0 \end{bmatrix} \ , \tag{4.91}$$

in which $H_b(\omega)$ is the complex frequency response function of $b_1(t)$, $H_z(\omega)$ is the complex frequency response function of $Z_1(t)$, $f(t)$ was assumed equal to $\exp(i\omega t)$, and $H_{1,1}(\omega)$, $H_{1,2}(\omega)$, $H_{2,1}(\omega)$ and $H_{2,2}(\omega)$ are respectively, given by

$$H_{1,1}(\omega) = -\omega^2 + 2\left(\zeta_1\omega_1 + \frac{G_2}{M}\right)i\omega + \left(\omega_2^2 + \frac{2G_1}{M_1}\right), \qquad (4.92)$$

$$H_{1,2}(\omega) = \frac{2}{M_1}(G_3-K_2) + \frac{2}{M_1}(G_4-C_2)i\omega, \qquad (4.93)$$

$$H_{2,1}(\omega) = -\left(2\omega^2 + \frac{G_1}{M_2}\right) - \frac{G_2}{M_2}i\omega, \qquad (4.94)$$

$$H_{2,2}(\omega) = -\omega^2 + \left(2\zeta_2\omega_2 - \frac{G_4}{M_2}\right)i\omega + \left(\omega_2^2 - \frac{G_3}{M_2}\right). \qquad (4.95)$$

Solving equation (4.91) for $H_b(\omega)$ and $H_z(\omega)$, one obtains

$$H_b(\omega) = \left[-\omega^2 + \left(2\zeta_2\omega_2 - \frac{G_4}{M_2}\right)i\omega + \left(\omega^2 - \frac{G_3}{M_2}\right)\right]\bigg/ \Delta, \quad (4.96)$$

$$H_z(\omega) = \left[2\omega^2 + \frac{G_1}{M_1} + \frac{G_2}{M_2}i\omega\right]\bigg/ \Delta, \qquad (4.97)$$

in which

$$\Delta = H_{1,1}H_{2,2} - H_{1,2}H_{2,1}. \qquad (4.98)$$

The spectral density function of the building and the TMD response due to the spectrum of the wind forces, $S_F(\omega)$, is given respectively, as

$$S_b(\omega) = |H_b(\omega)|^2 S_F(\omega), \qquad (4.99)$$

$$S_z(\omega) = |H_z(\omega)|^2 S_F(\omega). \qquad (4.100)$$

The mean square responses of the building and the TMD are respectively,

$$E[b_1^2] = \int_{-\infty}^{\infty} S_b(\omega)\,d\omega = \int_{-\infty}^{\infty} |H_b(\omega)|^2 S_F(\omega)\,d\omega, \qquad (4.101)$$

$$E[z^2] = \int_{-\infty}^{\infty} |H_z(\omega)|^2 S_F(\omega)\,d\omega. \qquad (4.102)$$

For simplicity, the wind forces are assumed as stationary white noise forces with intensity S, i.e., $S_F(\omega) = S$. The mean square response of the building, $E[b_1^2]$, can then be found according to references [4.27,4.31], as

$$E[b_1^2] = S\pi \frac{\left[\frac{B_0^2}{A_0}(A_2A_3-A_1A_4)+A_3(B_1^2-2B_0B_2)+A_1B_2^2\right]}{A_1(A_2A_3-A_1A_4)-A_0A_3^2} , \qquad (4.103)$$

in which A_0, A_1, ..., A_4 are the coefficients of the denominator of the transfer function $H_b(\omega)$, and B_0, B_1, ..., B_3 are the coefficients of the numerator of the transfer function $H_b(\omega)$, arranged as

$$H_b(\omega) = \frac{-i\omega^3 B_3-\omega^2 B_2+i\omega B_1+B_0}{\omega^4 A_4-i\omega^3 A_3-\omega^2 A_2+i\omega A_1+A_0} . \qquad (4.104)$$

The mean square response of the TMD is given by an expression similar to equation (4.103), but with $B_0 = G_1/M_2$, $B_1 = G_2/M_2$ and $B_2 = -2.0$.

4.5.5.2 The Design Process for an Active TMD

For a tall building, parameters ω_1, ξ_1, and M_1 are assumed to be given, and it is required to design the TMD parameters M_2, ξ_2, and ω_2 in order to provide a certain degree of control for the building under steady state or free vibration. If this objective could not be achieved by using a passive TMD, one should introduce active control. Yet, the control energy should also be minimized.

The desirable degree of control can be prescribed by assigning new eigenvalues for the structure, or by requiring that an objective function in terms of the controlled response and the control energy be satisfied. Since the optimal control method depends, to a great extent, on the proper choice of a certain objective function, the pole assignment method shall be used instead.

The design process consists in assuming desirable eigenvalues for the controlled building, and in determining the gain matrix of equation (4.88) using the pole assignment method. This

requires knowledge of the numerical values of M_2, ζ_2, and ω_2.
Assuming values for these parameters and evaluating the consumed
control energy, the mean square building response, and TMD response,
one can deduce the optimal parameters of the TMD. An index for the
control energy is for example

$$G = G_1^2 + G_2^2 + G_3^2 + G_4^2 \; . \tag{4.105}$$

4.5.5.3 Numerical Investigation

In order to evaluate the effective damping introduced by means of
active control, equation (4.103) is equated to the mean square
response of the uncontrolled building with effective damping ζ_e.
In this case, the effective damping is obtained from

$$\zeta_e = \frac{\pi S}{E[b_1^2] 2\omega_1^3} \; , \tag{4.106}$$

in which ω_1 is the uncontrolled natural frequency of the building.

To evaluate $E[b_1^2]$, one has to specify the white noise
intensity S. In order to omit the dependence of $E[b_1^2]$ on the
value of S, equation (4.103) is normalized with respect to a single
degree of freedom system with parameters ζ_1 and ω_1 driven by the
same white noise S. Equation (4.103) in the normalized form
becomes

$$E[b_1^2]_n = \frac{E[b_1^2] 2\omega_1^2 \zeta_1}{\pi S} \; . \tag{4.107}$$

The building considered in the previous sections is used
as an example here again. The mass ratios between TMD mass and
building mass are, respectively, $\mu = 0.005$, 0.01, 0.015 and 0.02.
The damping ratio $\zeta_2 = 0.09$, 0.12 and 0.15 were assumed, respectively, for the TMD. The frequency ω_2 was considered as a ratio
of ω_1, i.e., ω_2/ω_1 was a variable parameter.

Assuming first $\lambda_{1,2} = -0.5 \pm J2.22$ and $\lambda_{3,4} = -0.4 \pm J1.748$,
and finding the relation between G of equation (4.105) and (ω_2/ω_1)

for all possible μ and ξ_2, one comes up with Figure 4.28. One deduces that there is only a little difference between (ω_2/ω_1) for optimal active and optimal passive TMD. However, when ω_1 was changed through changing the stiffness of the building, great differences were found between the optimal passive and optimal active TMD frequencies, as shown in Figures 4.29 - 4.38. One concludes that the optimal TMD parameters do not necessarily coincide with the optimal passive TMD parameters [4.29].

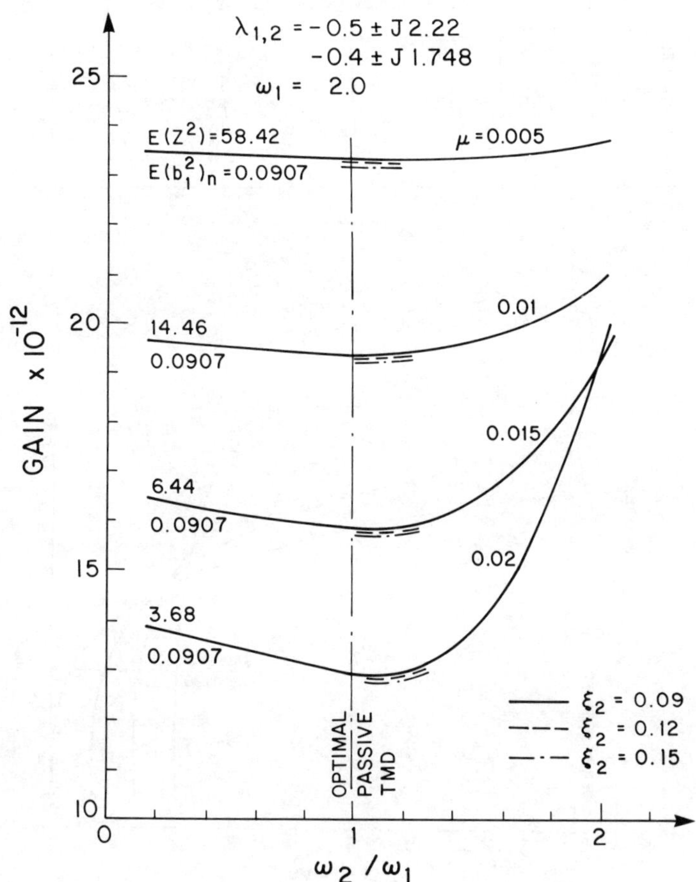

Figure 4.28 - *Optimal ATMD Parameters*

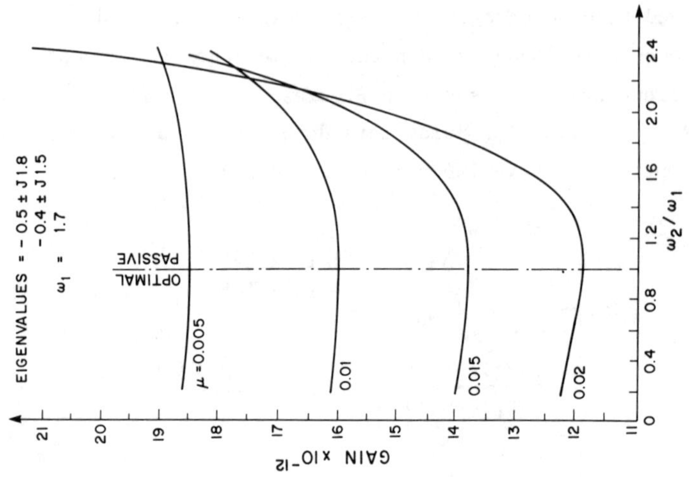

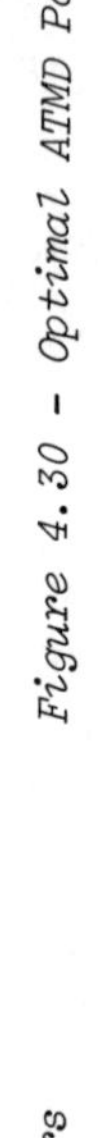

Figure 4.30 − Optimal ATMD Parameters

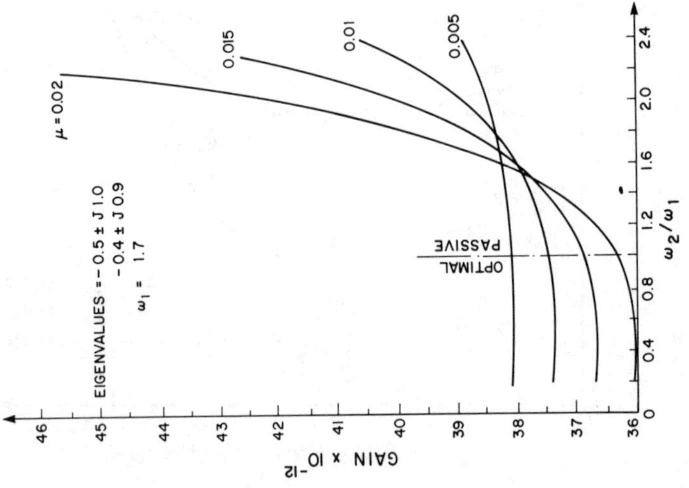

Figure 4.29 − Optimal ATMD Parameters

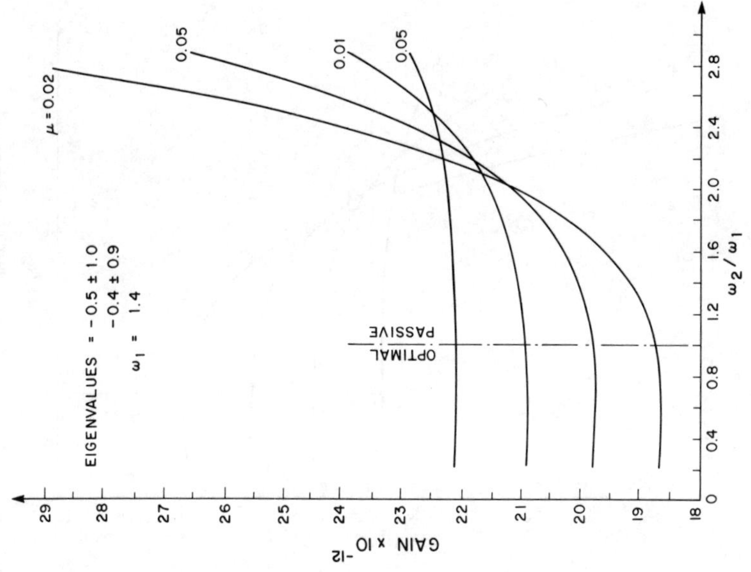

Figure 4.32 − Optimal ATMD Parameters

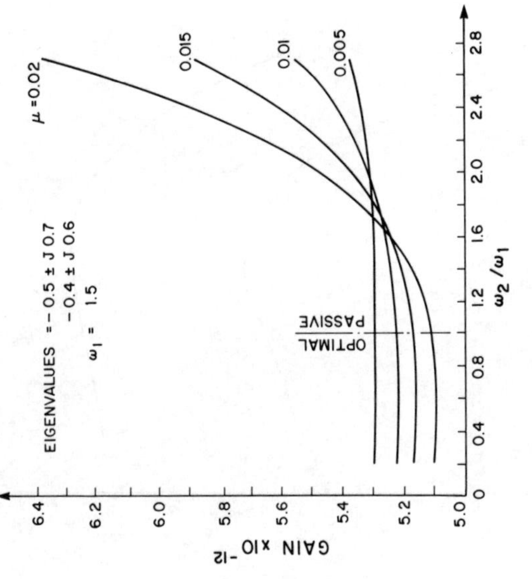

Figure 4.31 − Optimal ATMD Parameters

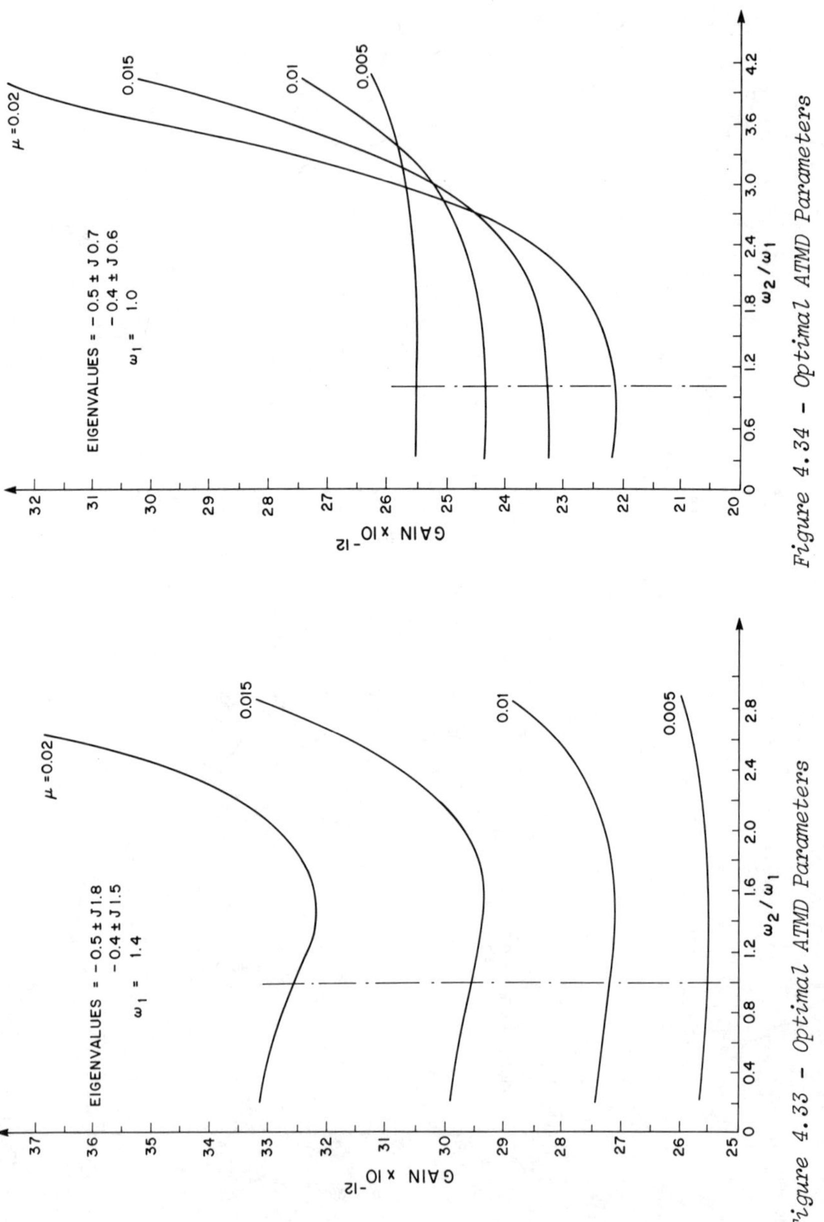

Figure 4.34 – Optimal ATMD Parameters

Figure 4.33 – Optimal ATMD Parameters

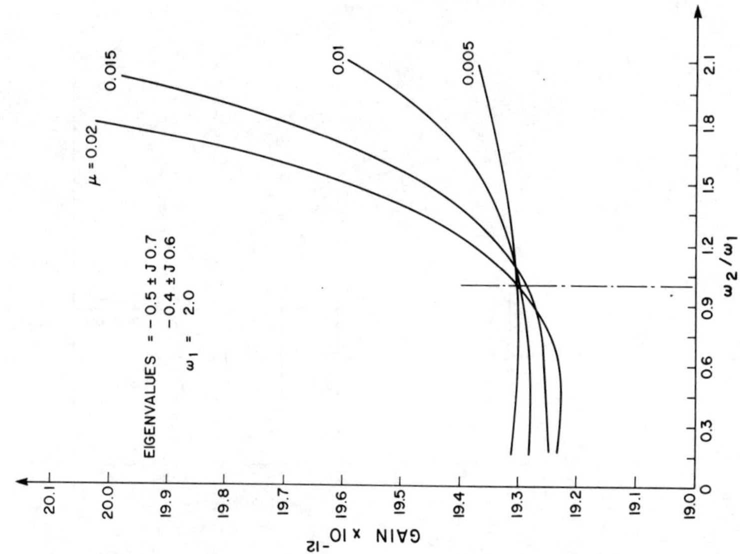

Figure 4.36 - Optimal ATMD Parameters

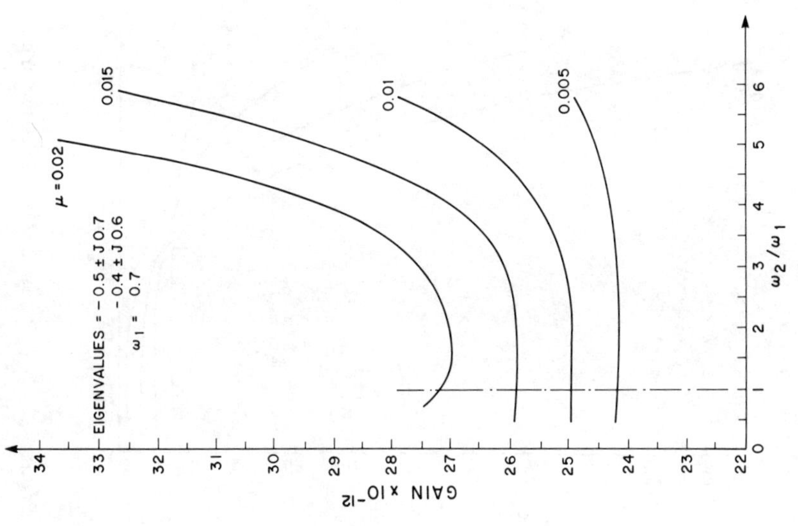

Figure 4.35 - Optimal ATMD Parameters

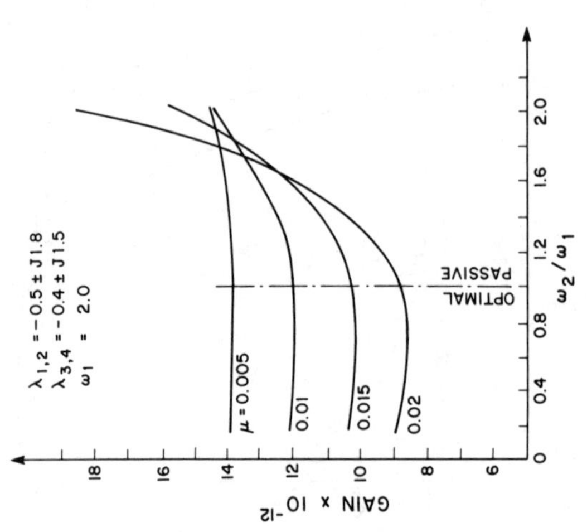

Figure 4.38 – Optimal ATMD Parameters

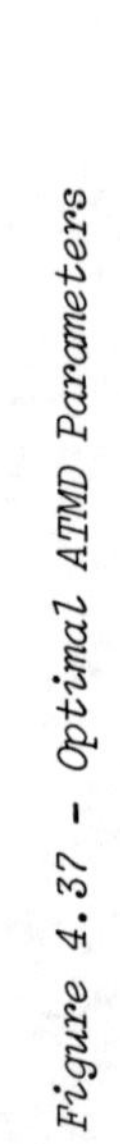

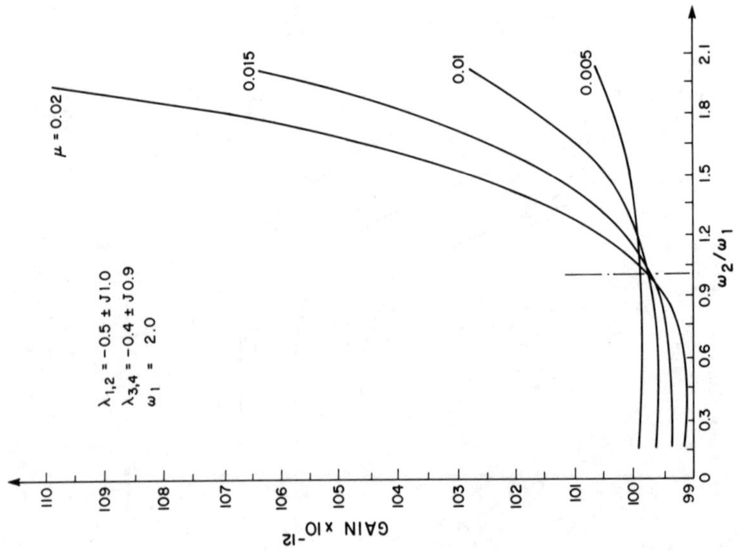

Figure 4.37 – Optimal ATMD Parameters

Example

A tall building has the natural frequency ω_1 = 1.228 rps, the damping coefficient ζ_1 = 0.01, and the mass M_1 = 13.42 × 10^6 lb.sec^2/ft.
For M_2 = 0.02 M_1, the optimal PTMD was determined with ω_2 = 1.204
rps, and ζ = 0.07. The eigenvalues of the passive controlled
system are (-0.068 ± J1.392) and (-0.035 ± J1.06). The mean square
building response $E[b_1^2]_n$ = 0.1557, the effective damping ζ_e = 0.064,
and the mean square PTMD response $E[Z^2]$ = 6.14. To increase the
effective damping, new eigenvalues are specified to be (-0.35 ±
J1.392), and (-0.35 ± J1.06). Figure 4.39 indicates that the minimum energy index, G, lies at ω_2 = 1.375 rps. This design provides
an effective damping ζ_e = 0.1255, mean square building response
$E[b_1^2]_n$ = 0.0796, and a mean square ATMD response $E[Z_1^2]_n$ = 4.786 by
using the gain matrix

$$\underline{K} = 10^6[2.769 \quad -1.69 \quad 0.05 \quad -0.352] .$$

The mean square response $E[b_1^2]_n$ and the effective damping ζ_e for
both ATMD and PTMD are shown in Figures 4.40 and 4.41, respectively.

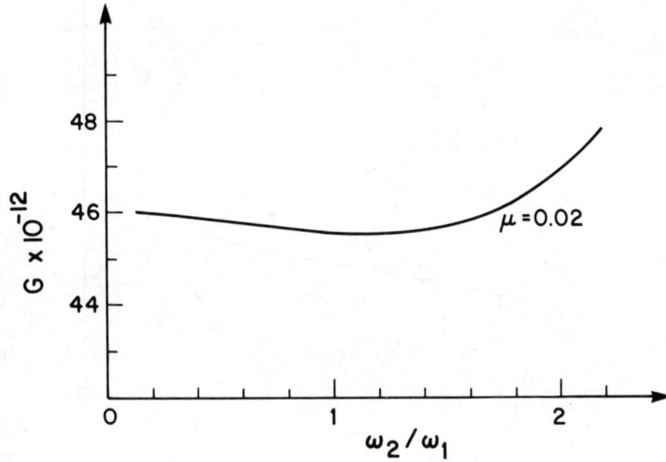

Figure 4.39 - Optimal ATMD Parameters

     H.H.E. Leipholz and M. Abdel-Rohman

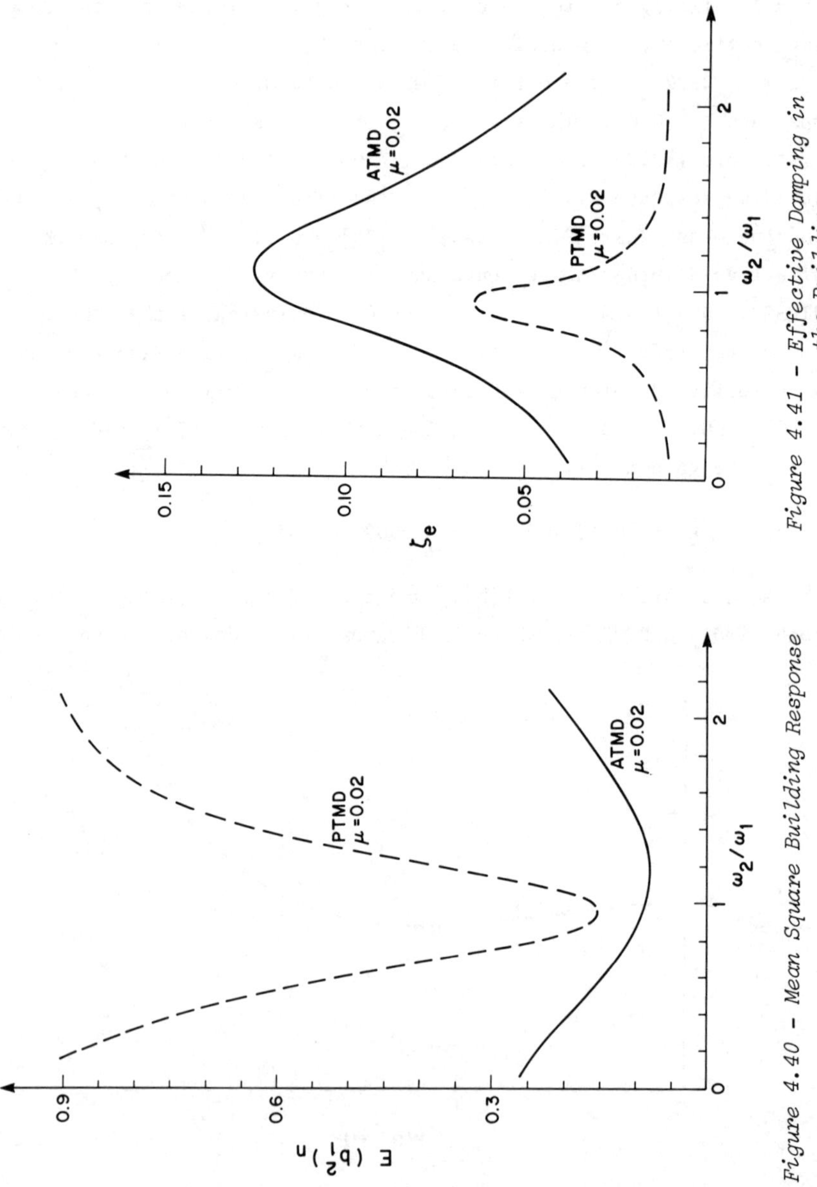

Figure 4.41 - Effective Damping in the Building

Figure 4.40 - Mean Square Building Response

4.5.5.4 Conclusions

The use of an active TMD in tall buildings is only effective when
the building is responding to steady state or free vibrations.
The optimal TMD parameters are not necessarily the same as those
of the optimal passive TMD. One has to find the proper values for
these parameters through numerical experimentation.

4.6 ACTIVE CONTROL BY MEANS OF APPENDAGES

The active control of tall buildings response due to wind using
aerodynamic appendage was first proposed by Klein et al [4.10].
This control mechanism was an outcome of the need to save control
energy and the desire to use the wind power in generating control
forces. The control scheme was a simple on-off system. The appendage is fully deployed when the velocity at the top of the building
is in opposite direction to the main wind direction. The appendage
is fully folded when the velocity at the top is along the main
wind direction (see Figures 4.42 to 4.44).

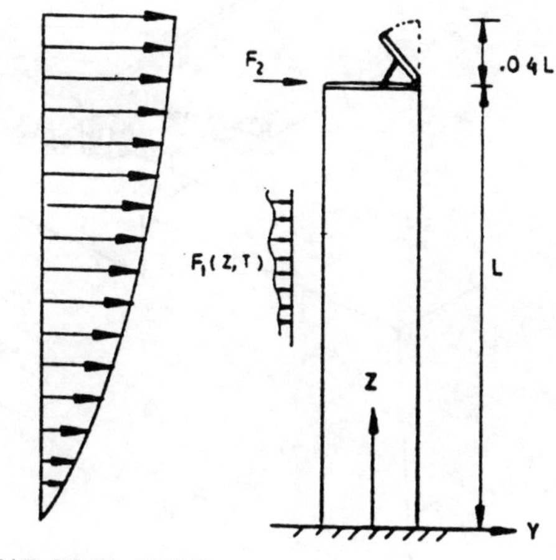

Figure 4.42 - Wind and Control Forces on a Tall Building

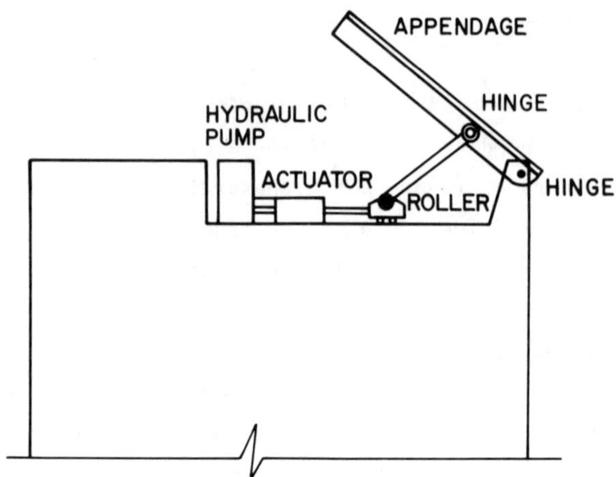

Figure 4.43 - The Appendage Mechanism

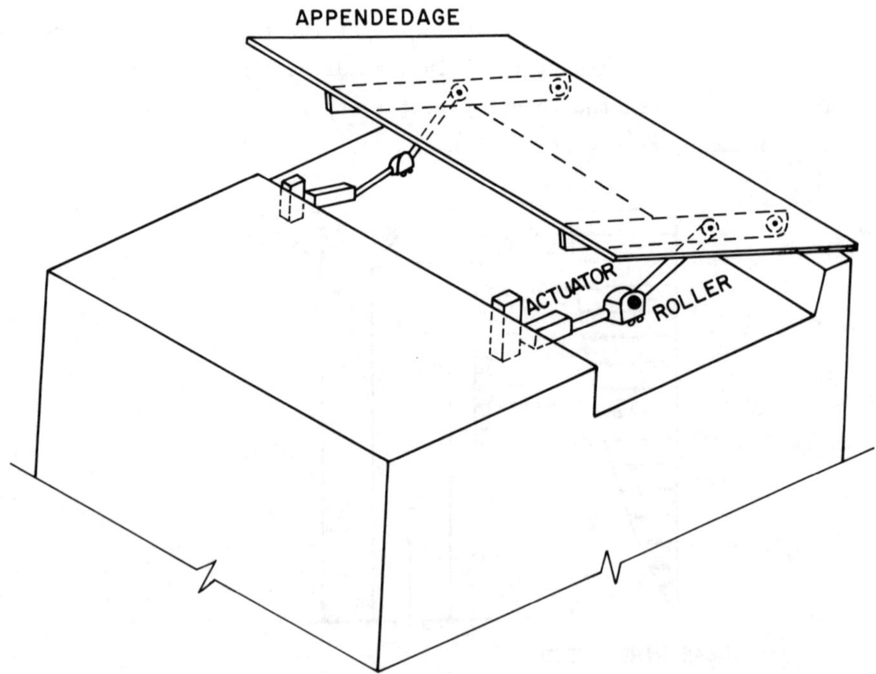

Figure 4.44 - The Appendage Mechanism

Soong has considered this control mechanism and investigated the possibility of getting a better design for the appendage movement. He has used optimal control theory and has obtained better results than Klein [4.11]. Moreover, Soong has investigated, experimentally, this solution in suppressing the vibration of a tall building model in a wind-tunnel [4.12].

4.6.1 *Design of Optimal Appendage Movement*

Considering the building motion along wind direction, the equation of motion is given by

$$EI \frac{\partial^4 w_1}{\partial x^4} + C \frac{\partial w_1}{\partial t} + m \frac{\partial^2 w_2}{\partial t^2} = F_1(x,t) + F_2(x,t)U(t)\delta(x-L) \, , \qquad (4.108)$$

in which $F_1(x,t)$ are the drag wind forces on the building, $F_2(x,t)$ are the drag wind forces on the appendage, $\delta(x-L)$ is the Dirac delta function, and $U(t)$ is the control variable for the appendage movement which equals 1.0 when the appendage is fully deployed. The drag wind force on the building, neglecting the self-excited wind forces, is given by

$$F_1(x,t) \frac{A_s}{2} \frac{\rho}{L} C_D(x,t)U^2(x,t) \, , \qquad (4.109)$$

in which A_s is the windward area of the building, L is the height of the building, $C_D(x,t)$ is the drag coefficient of the building, and $U(x,t)$ is the along-wind speed at height x.

The drag wind force on the appendage is given by

$$F_2(x,t) = \frac{A_p}{2} \rho C_D'(x,t)U^2(L,t) \, , \qquad (4.110)$$

in which A_p is the windward area of the appendage, C_D' is the drag coefficient for the appendage, and $U(L,t)$ is the wind speed at height L.

The wind speed is expressed as the sum of a mean component and a fluctuating component. For simplicity, the fluctuating component is assumed to be variable only with time, i.e.,

$$U(x,t) = \overline{U}(x) + u(t) \ . \tag{4.111}$$

Whatever the way is of simulating the fluctuating component u(t), which is responsible for the oscillations, our concern is to develop correspondingly the best way for the appendage movement. Herein, the simulation is not that important because the control force is also a function of the fluctuating component. To show this, equation (4.111) is substituted in equations (4.109) and (4.110) to obtain, respectively,

$$F_1(x,t) = \frac{A_s}{2} \frac{\rho}{L} C_D[\overline{U}^2(x) + 2\overline{U}(x)u(t) + u^2(t)] \ , \tag{4.112}$$

$$F_2(x,t) = \frac{A_p}{2} \rho C_D'[\overline{U}^2(L) + 2\overline{U}(L)u(t) + u^2(t)] \ . \tag{4.113}$$

Applying an integral transformation, and considering only the first mode, one obtains

$$\ddot{b}_1 + 2\zeta_1\omega_1\dot{b}_1 + \omega_1^2 b_1 = \frac{1}{mL} \int_0^L F_1(x,t)\phi_1(x)dx + \frac{1}{mL} \int_0^L F_2(x,t)\phi_1(x)U(t)\delta(x-L)dx. \tag{4.114}$$

The first term on the right hand side can be written, in more detail as

$$\frac{C_1}{mL}\left[\int_0^L \overline{U}^2(x)\phi_1(x)dx + \int_0^L 2\overline{U}(x)u(t)\phi_1(x)dx + u^2(t) \int_0^L \phi_1(x)dx \right] \ , \tag{4.115}$$

in which $C_1 = A_s\rho C_D/(2L)$, and C_D was considered constant. The second term on the right hand side of equation (4.114) can be written as

$$\frac{C_2}{mL} [\overline{U}^2(L) + 2\overline{U}(L)u(t) + u^2(t)]U(t)\phi_1(L) \ , \tag{4.116}$$

in which $C_2 = A_p\rho C_D'/2$, and C_D' is assumed to be constant.

For the sake of simplicity, equations (4.115) and (4.116) are expressed, respectively, by $d_1(t)$, and $b_1(t)U(t)$. The functions $d_1(t)$ and $b_1(t)$ are deterministic if the fluctuating component is deterministic, or stochastic if the wind speed is treated as a stochastic variable.

Equation (4.114) is expressed in state space form as

$$\dot{\underline{X}} = \underline{A}\,\underline{X} + \underline{B}\,U(t) + \underline{d} \,, \tag{4.117}$$

in which $\underline{X} = [b_1 \ \dot{b}_1]^T$, and $\underline{A}$, $\underline{B}$, $\underline{d}$ are given by

$$\underline{A} = \begin{bmatrix} 0 & 1 \\ -\omega_1^2 & -2\zeta_1\omega_1 \end{bmatrix} \,, \tag{4.118}$$

$$\underline{B}^T = [0 \ \ b_1(t)] \,, \tag{4.119}$$

$$\underline{d}^T = [0 \ \ d_1(t)] \,. \tag{4.120}$$

The control matrix $\underline{B}$ is time varying if only the control variable $U(t)$ is being considered. However, equation (4.117) can also be written as

$$\dot{\underline{X}} = \underline{A}\,\underline{X} + \underline{B}[b_1(t)U(t)] + \underline{d} \,, \tag{4.121}$$

in which $\underline{B}^T = [0 \ \ 1]$.

In this case, $\underline{B}$ is time-invariant and the complete system, equation (4.141), is considered as a time-invariant system because of $\underline{A}$ and $\underline{B}$. Equation (4.121) was considered by Soong [4.11] in his investigation, where he has taken the control as the full term $b_1(t)U(t)$. It is obvious that $b_1(t)$ is related to the wind forces, and $U(t)$ is the control variable which governs the motion of the appendage to initiate the control force.

Soong [4.11] has used a quadratic objective function in the form of

$$J = \frac{1}{2} \int_0^T [\underline{X}^T\underline{Q}\underline{X} + b_1^2(t)U^2(t)]dt \,, \tag{4.122}$$

without any weighing coefficient added to $U^2(t)$ in order to get
as much benefit of the control from wind as possible.

In this case, the Hamiltonian is given by [4.16,4.25]

$$H = \frac{1}{2} \underline{X}^T \underline{Q} \underline{X} + \frac{1}{2} b_1^2 U^2 + \underline{\lambda}^T (\underline{A}\underline{X} + \underline{B}b_1 U + \underline{d}) \ , \tag{4.123}$$

which leads to the following optimality conditions

$$H_u = b_1^2 U + \underline{B}^T \underline{\lambda} b_1 = 0 \ , \tag{4.124a}$$

$$H_x = \underline{Q}\underline{X} + \underline{A}^T \underline{\lambda} = -\underline{\dot{\lambda}} \ , \quad \underline{\lambda}(T) = \underline{0} \ . \tag{4.124b}$$

From equation (4.124a), the control variable follows as

$$U = - \frac{\underline{B}^T \underline{\lambda}}{b_1} \ . \tag{4.125}$$

In order to let the appendage motion depend on the building response, the Lagrange variable, $\underline{\lambda}$, is expressed

$$\underline{\lambda} = \underline{P}\underline{X} \ , \tag{4.126}$$

in which $\underline{P}$ is an 2×2 Riccati matrix.

Substituting equation (4.126) into equation (4.124a) and
disregarding the wind forces $\underline{d}$, one obtains the Riccati equation

$$-\underline{\dot{P}} = \underline{Q} + \underline{A}^T \underline{P} + \underline{P}\underline{A} - \underline{P}\underline{B}\underline{B}^T \underline{P}, \quad \underline{P}(T) = \underline{0} \ , \tag{4.127}$$

in which $\underline{Q}$ is the weighting matrix, positive semi-definite.

The solution to equation (4.127) provides the Riccati
matrix $\underline{P}(t)$. The control variable is then

$$U(t) = \frac{-\underline{B}^T \underline{P}\underline{X}}{b_1} \ , \quad 0.0 \le U \le 1.0 \ . \tag{4.128}$$

The controlled building motion can subsequently be
expressed as

$$\underline{\dot{X}} = (\underline{A} - \underline{B}\underline{B}^T \underline{P})\underline{X} + \underline{d} \ . \tag{4.129}$$

Observing equation (4.127), one notices that the matrices $\underline{Q}$, $\underline{A}$ and $\underline{B}$ are time-invariant. Therefore, $\underline{P}(t)$ tends to become constant as T increases. Usually, wind gusts durate for many seconds and that is why Soong has considered the case of a steady state $\underline{P}$. For this special case, one also can determine U(t) using the pole assignment method, because in this case the gain matrix equals $\underline{B}^T\underline{P}$. This means that equation (4.130) can also be written as

$$\dot{\underline{X}} = (\underline{A} - \underline{B}\ \underline{K})\underline{X} + \underline{d} \tag{4.130}$$

which is the general form of a closed-loop controlled system with a gain matrix $\underline{K}$. Equation (4.127) indicates that the optimal matrix $\underline{P}$ will be obtained using the sensitivity analysis on $\underline{Q}$ only. Values of $\underline{Q}$ should be as large as possible in order to provide an effective control. However, from the computational point of view, this causes overflow during the execution. That is why the authors prefer to use a weighing factor on U(t) as well. In this case, equation (4.122) can be written as

$$J = \frac{1}{2} \int_0^T \left[\underline{X}^T\underline{Q}\underline{X} + b_1^2\ \frac{U^2}{\eta} \right] dt\ . \tag{4.131}$$

Equation (4.127) becomes, accordingly,

$$-\dot{\underline{P}} = \underline{Q} + \underline{A}^T\underline{P} + \underline{P}\underline{A} - \underline{P}\underline{B}\eta\underline{B}^T\underline{P}\ , \quad \underline{P}(T) = 0\ , \tag{4.132}$$

in which η is a weighing coefficient for control.

4.6.2 *Contributions*

Soong in his investigation obtained the control law for U(t) depending on the knowledge of $b_1(t)$ or $d_1(t)$. The objective function was formulated in such a way that the control variable is altogether $b_1(t)U(t)$. However, one can obtain a better control when the characteristics of the wind are taken into account.

For the nonhomogeneous state equation, equation (4.121), and the Hamiltonian equation (4.123), the Lagrange variable is expressed as

$$\underline{\lambda} = \underline{P}\underline{X} + \underline{q} \, , \tag{4.133}$$

in which $\underline{q}$ was introduced to take care of the nonhomogeneous part in the state equation.

Substituting equation (4.133) into equations (4.124), one obtains

$$-\underline{\dot{P}} = \underline{Q} + \underline{P}\underline{A} + \underline{A}^T\underline{P} - \underline{P}\underline{B}\eta\underline{B}^T\underline{P} \, , \quad \underline{P}(T) = \underline{0} \, , \tag{4.134}$$

$$-\underline{\dot{q}} = \underline{P} \, \underline{d} - \underline{P}\underline{B}\eta\underline{B}^T\underline{q} + \underline{A}^T\underline{q} \, , \quad \underline{q}(T) = \underline{0} \, . \tag{4.135}$$

The control variable, $U(t)$, shall also be given by

$$U(t) = - \frac{\underline{B}^T\eta}{b_1} \, (\underline{P}\underline{X}+\underline{q}) \, , \quad 0 \leq U(t) \leq 1.0 \, . \tag{4.136}$$

The controlled building state equation is

$$\underline{\dot{X}} = \underline{A}\underline{X} - \underline{B}\eta\underline{B}^T\underline{P}\underline{X} - \underline{B}\eta\underline{B}^T\underline{q} + \underline{d} = (\underline{A}-\underline{B}\eta\underline{B}^T\underline{P})\underline{X} - \underline{B}\eta\underline{B}^T\underline{q} + \underline{d} \, . \tag{4.137}$$

It is obvious from equation (4.137) that the designed control does not only control the parameters of the building, but alleviates the magnitude of the wind forces $\underline{d}$, if the vector $\underline{q}(t)$ is determined properly. The only way in which to carry out such design is to try different values for $\underline{Q}$ and η and check the controlled response from case to case.

The controlled response shall be investigated using the following criteria

$$J_1 = \int_o^T b_1^2 dt \, , \tag{4.138}$$

$$J_2 = \int_o^T \ddot{b}_1^2 dt \, , \tag{4.139}$$

$$J_3 = \int_o^T U^2(t)dt \ . \tag{4.140}$$

Values of J_1 and J_2 shall be compared with those of the uncontrolled building, whereas the value of J_3 will be compared with the solutions given by Soong and Klein [4.10,4.11].

4.6.3 *Example 1*

The tall building considered in the previous sections is considered here again. That means one has the following data: $L = 1000$ ft, $m = 11000$ lb.sec^2/ft^2, $\omega_1 = 2$ rad/sec, $\zeta = 0.01$.

The wind forces are modelled as in [4.11,4.32]:
$\overline{U}(X) = \overline{U} = 150$ ft/sec, $C_D = C_D' = 1.2$, $A_S = 150$ L, $A_p = 0.04A_s$, $\rho = 0.0763$ pcf. The fluctuating component $u(t)$ has been modelled as

$$u(t) = 2 \sum_{i=1}^{25} [2S_u(\omega_i)\Delta\omega]^{1/2} \cos(\omega_i t + \phi_i) \ , \tag{4.141}$$

$$\omega_i = (i-0.5)\Delta\omega \ , \tag{4.142}$$

in which ϕ_i is the random phase angle, uniformly distributed in the interval $(0,2)$, $\Delta\omega = \omega_u/25$, ω_u is the upper cutoff frequency.

The spectrum S_u was assumed to read

$$S_u(\omega_i) = \frac{2K\Phi^2 |\omega_i|}{\pi^2 \left[1 + \left(\dfrac{\phi\omega_i}{\pi U_0}\right)^2\right]^{4/3}} \ , \tag{4.143}$$

in which K is the surface drag coefficient, ϕ is the scale turbulence, and U_0 is the mean wind velocity at $X = 33$ ft.

The following numerical values were assumed to hold:
$U_0 = 150$ fps, $\overline{U}(X) = \overline{U} = 150$ fps, $\phi = 2000$ ft, $K = 0.04$, $C_D = C_D' = 1.2$, $\omega_u = 40$ rad/sec.

The three possible solutions from Klein, Soong and the authors are summarized here as follows:

4.6.3.1 Klein's Method

$$U(t) = \begin{cases} 1.0 \, , & \dot{b}_1 \leq 0.0 \, , \\[2mm] 0.0 \, , & \dot{b}_1 > 0.0 \, . \end{cases} \tag{4.144}$$

4.6.3.2 Soong's Method

$$U(t) = \begin{cases} 1.0 \, , & U \geq 1.0 \, . \\[3mm] -\dfrac{B^T \eta}{b_1} \, \underline{PX} \, , & 0 < U < 1.0 \, , \\[3mm] 0.0 \, , & U < 0.0 \, . \end{cases} \tag{4.145}$$

4.6.3.3 The Authors' Method

$$U(t) = \begin{cases} 1.0 \, , & U > 1.0 \, , \\[3mm] -\dfrac{B^T \eta}{b_1} \, (\underline{PX} + \underline{q}) \, , & 0 \leq U \leq 1.0 \, , \\[3mm] 0.0 \, , & U < 0.0 \, . \end{cases} \tag{4.146}$$

Figures 4.45 to 4.47 show the respective responses of
the three objective functions of equations (4.138) to (4.140),
using T = 40 seconds for all methods. The numerical values are
summarized in Table 4.7. The results indicate that the authors'
method provides a better controlled response and a better appendage
movement.

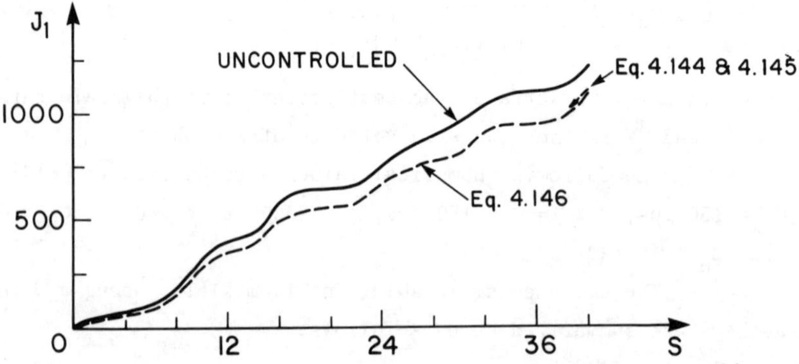

Figure 4.45 - Deflection Response $Q_{11} = Q_{22} = 22000, \; \eta = 10^{12}$

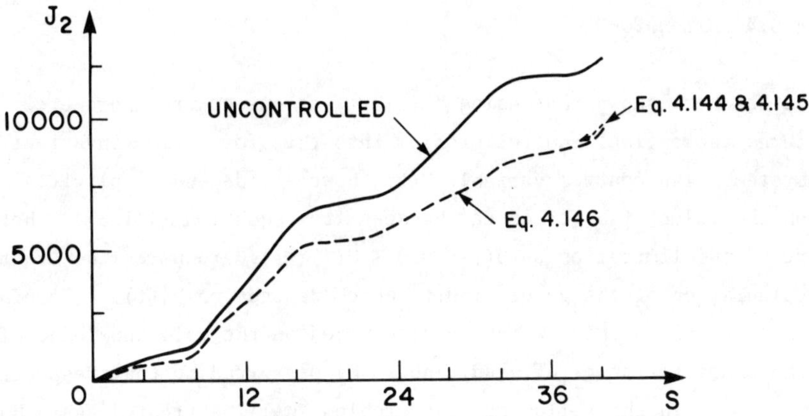

Figure 4.46 - Acceleration Response $Q_{11} = Q_{22} = 12000$, $\eta = 10^{12}$

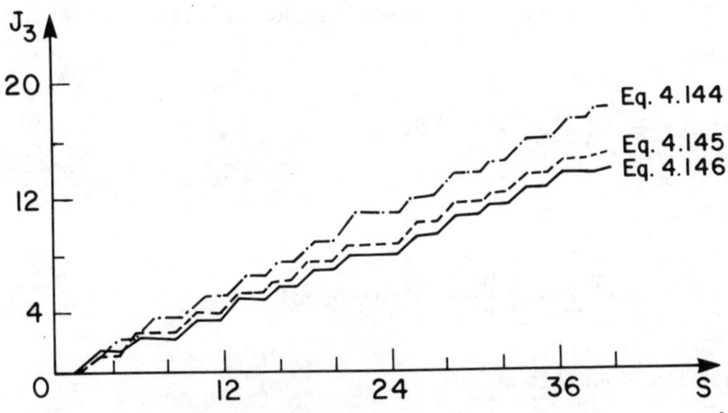

Figure 4.47 - Control Response $Q_{11} = Q_{22} = 12000$, $\eta = 10^{12}$

Table 4.7 - Comparison of Optimal Control Laws for $t_f = 40$ seconds

Case	J_1	J_2	J_3
Eq. (4.146)	1069.87	9431.5	13.79
Eq. (4.145)	1071.74	9688.4	15.24
Eq. (4.144)	1071.24	9478.4	18.35
Uncontrolled	1231.32	12200.87	-

4.6.4 *Extension*

It has been shown that using $b_1(t)U(t)$ as one control parameter
turns the optimal control problem into that for a time invariant
system. The control variable $U(t)$, however, depends explicitly
on the value of $b_1(t)$ as can be seen from equation (4.125). Therefore, the limitation on $U(t)$, $0.0 < U(t) < 1.0$ requires the exact
calculation of the value of $U(t)$ which depends on $b_1(t)$. Therefore,
the authors' method, which is also based on the full knowledge of
the characteristics of wind, should be preferred in that respect.

In the following, the problem shall be treated again but
considering $U(t)$ only as a control variable. Using equation (4.118),
in which $\underline{B}^T = [0 \quad b_1(t)]$, and assuming the objective function to
be given by

$$J = \frac{1}{2} \int_o^T \left(\underline{X}^T \underline{Q} \underline{X} + \frac{U^2}{\eta} \right) dt \ . \tag{4.147}$$

The Hamiltonian is then

$$H = \frac{1}{2} \underline{X}^T \underline{Q} \underline{X} + \frac{1}{2} \frac{U^2}{\eta} + \underline{\lambda}^T (\underline{A}\underline{X} + \underline{B}U + \underline{d}) \ . \tag{4.148}$$

The optimality conditions are

$$H_u = \frac{U}{\eta} + \underline{B}^T \underline{\lambda} = 0 \ , \tag{4.149a}$$

$$H_x = \underline{Q}\underline{X} + \underline{A}^T = -\dot{\underline{\lambda}} \ , \qquad \underline{\lambda}(T) = 0 \ . \tag{4.149b}$$

Using $\underline{\lambda} = \underline{P}\underline{X}$, one obtains

$$-\dot{\underline{P}} = \underline{Q} + \underline{P}\underline{A} + \underline{A}^T \underline{P} - \underline{P}\underline{B}\eta\underline{B}^T \underline{P} \ , \qquad \underline{P}(T) = \underline{0} \ , \tag{4.150}$$

$$U = -\eta \underline{B}^T \underline{P} \underline{X} \ . \tag{4.151}$$

Here, the Riccati matrix is time varying because of $\underline{B}(t)$, and $U(t)$
has a different value than in equation (4.125)

Similarly, using $\underline{\lambda} = \underline{P}\underline{X} + \underline{q}$, one has

$$-\underline{\dot{P}} = \underline{Q} + \underline{P}\underline{A} + \underline{A}^T\underline{P} - \underline{P}\underline{B}\eta\underline{B}^T p \ , \qquad \underline{P}(T) = \underline{0} \ , \tag{4.152}$$

$$-\underline{\dot{q}} = \underline{p}\underline{d} - \underline{P}\underline{B}\eta\underline{B}^T\underline{q} + \underline{A}^T\underline{q} \ , \tag{4.153}$$

$$U = \eta\underline{B}^T(\underline{P}\underline{X}+\underline{q}) \ , \tag{4.154}$$

which again indicates the dependence of $\underline{q}$ on $\underline{d}$.

One concludes then that considering $b_1(t)U(t)$ as one control variable provides a solution easier and faster because of the possibility of $\underline{P}$ to be a constant matrix.

4.6.5 *Example 2*

The example presented in Section 4.6.3 did not emphasize fully the advantage of the method developed by the authors as compared with the other methods. This is mainly due to the assumption that the mean wind component $U(x)$ is constant and equals $\overline{U}_0$. This assumption was introduced to simplify the presentation, however, the results are misleading. In the following, another example will be treated considering the variation of the mean wind component with height [4.41].

The tall building parameters are m = 25000 lb mass/ft, ω = 1.4 rad/sec, damping ratio ξ = 0.01. For simplicity, the mode shape $\phi_1(x)$ is assumed to be linear and to be given by

$$\phi_1(X) = \frac{2x}{L} \ . \tag{4.155}$$

The wind characteristics are the same as before, except that $\overline{U}_0$ = 66 fps. According to the logarithmic law, the mean wind velocity at any height x is given by [4.33]

$$\overline{U}(x) = 66 \ \frac{\ln\left(\dfrac{x}{x_0}\right)}{\ln\left(\dfrac{33}{x_0}\right)} \ , \tag{4.156}$$

in which x_0 is the roughness of the terrain.

From equation (4.155), one has the following identities:

$$\int_0^L \phi_1(x)dx = L , \tag{4.157}$$

$$\int_0^L \phi_1^2(x)dx = \frac{4L}{3} . \tag{4.158}$$

Equation (4.114) can then be written as

$$\frac{3C_1}{4mL}\left[\int_0^L \left(\frac{2X}{L}\right)66^2 \left(\ln \frac{x}{x_0}\middle/\ln \frac{33}{x_0}\right)^2 dx + u(t)\int_0^L 2\left(\frac{2X}{L}\right)66\left(\ln \frac{x}{x_0}\middle/\ln \frac{33}{x_0}\right)\right]dx$$

$$= + u^2(t)L . \tag{4.159}$$

Similarly, equation (4.116) becomes

$$\frac{3C_2}{4mL} 2U(t)\left[\overline{U}^2(L)+2\overline{U}(L)u(t)+u^2(t)\right] , \tag{4.160}$$

in which $C_1 = B\rho C_D/2L$, $C_2 = A_p\rho C_D'/2$. Considering the roughness of the terrain to be $x_0 = 0.8$ ft, equations (4.159) and (4.160), respectively, become

$$\frac{3C_1}{4mL} [210847.16 \times 66 + 2 \times 117660.67u(t) + 1000u^2(t)] , \tag{4.159'}$$

$$\frac{6C_2}{4mL} [126.5^2 + 253u(t) + u^2(t)]U(t) , \tag{4.160'}$$

where the following identities have been used

$$\int x \ln xdx = \frac{x^2}{2} (\ln x - 0.5) , \tag{4.161}$$

$$\int x \ln^2 dx = \frac{x^2\ln^2 x}{2} - \int x \ln xdx = \frac{x^2}{2} [\ln^2 x-(\ln x+0.5)] . \tag{4.162}$$

With these assumptions, the aerodynamic forces on the building and on the appendage can be calculated. The results for T = 25 secsonds are displayed in Figures 4.48 and 4.49.

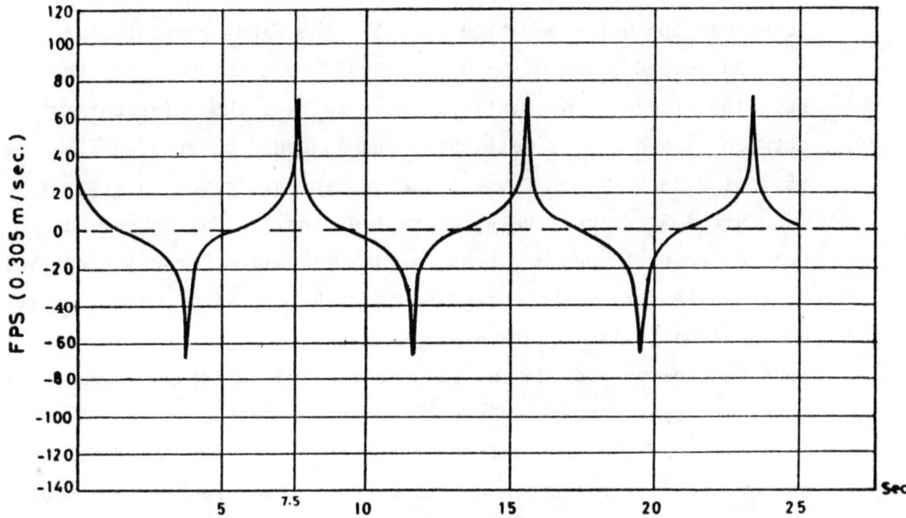

Figure 4.48 - *Velocity of Fluctuating Wind Component*

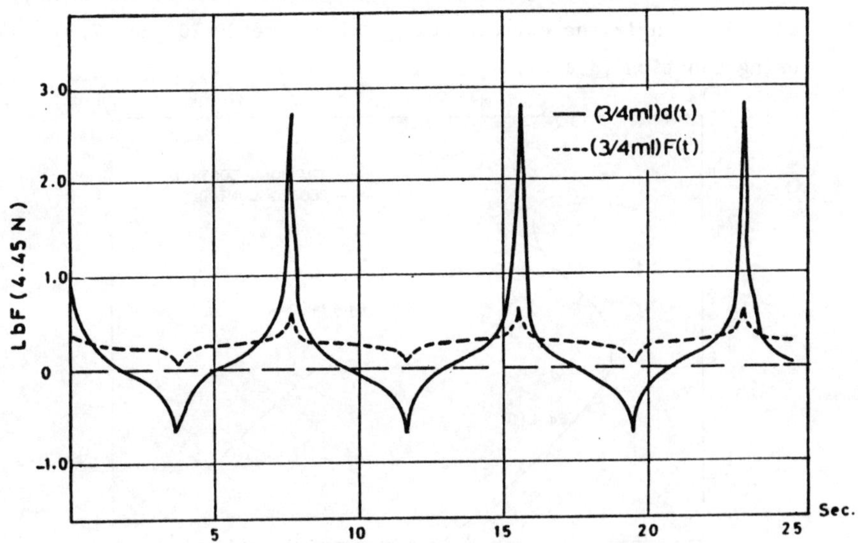

Figure 4.49 - *Generalized Wind Forces*

In order to compare the control using equation (4.146) with that one using equation (4.145), the final control time, t_f = 25 seconds, and Q_{11} = Q_{22} = 22,000 were assumed, but the weighing factor η was left variable so that the relationships between J_1 and J_3, as well as J_2 and J_3 could be studied for various values of η. These relationships are given in Figures 4.50 and 4.51 respectively, where equation (4.145) is called "closed control" and equation (4.146) is called "tracking control". It is obvious from these figures that the tracking control expressed by equation (4.146) consumes less control energy than the closed-loop control to provide the same controlled building response. In other words, the tracking control provides better controlled response than the closed-loop control for the same level of control energy used for the appendage movements. The figures show that the optimal design is the one which yields J_1 = 24.53, J_2 = 15.53, and J_3 = 8.25. The optimal designs are summarized in Table 4.8 where the values of J_1 and J_2 using equation (4.146) are 24.53 and 15.53 while the corresponding values are 29.20 and 17.80 when using equation (4.145).

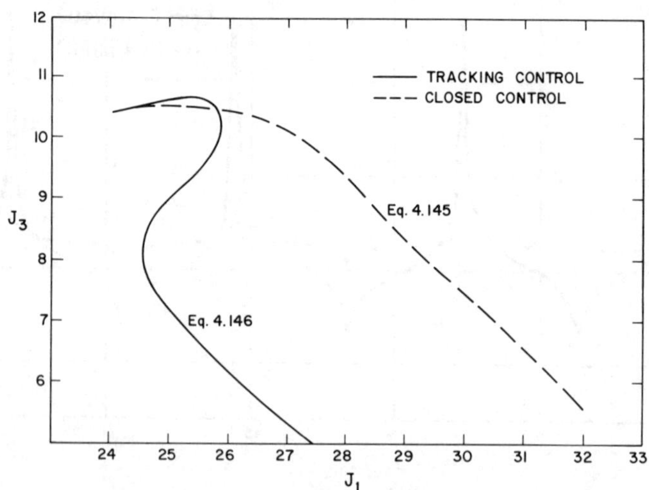

Figure 4.50 - Relationships of J_1-J_3 for Optimal Control Laws

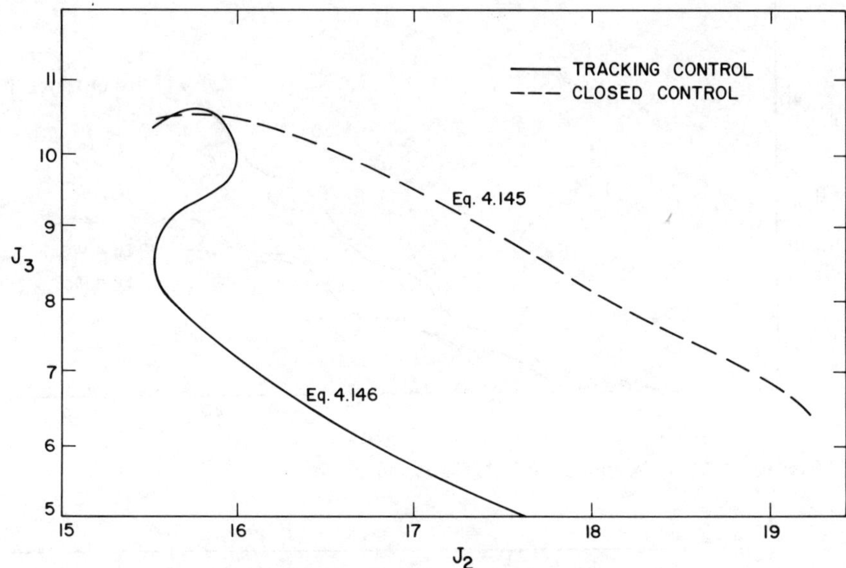

Figure 4.51 - Relationship of J_2-J_3 for Optimal Control Laws

Table 4.8 - Comparison of Cost Functions

Case	J_1	J_2	J_3
Uncontrolled	48.82	33.74	...
Equation (4.144)	32.25	19.15	12.01
Equation (4.145)	29.20	17.80	8.25
Equation (4.146)	24.53	15.53	8.25
Equation (4.145)	24.53	15.53	10.42
Equation (4.146)	24.53	15.53	8.25

Figure 4.52 shows the response of the deflection index J_1. The acceleration index J_1 is shown in Figure 4.53, and the control energy index J_3 is shown in Figure 4.54.

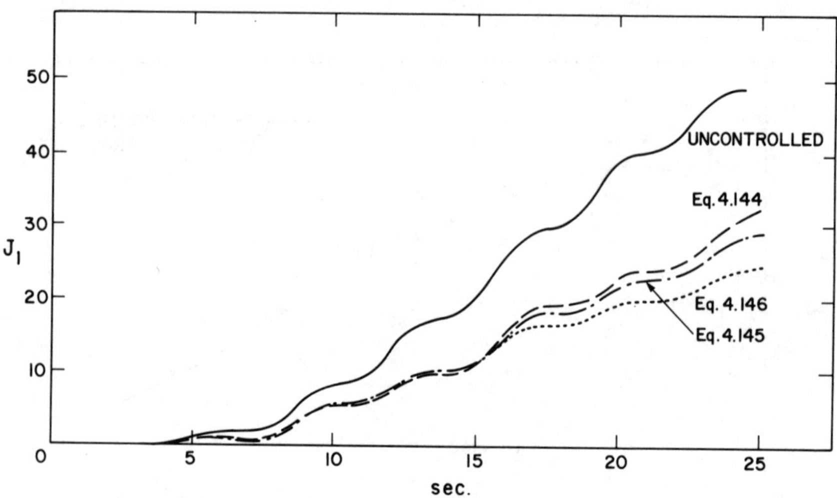

Figure 4.52 - Response of Deflection Index J_1

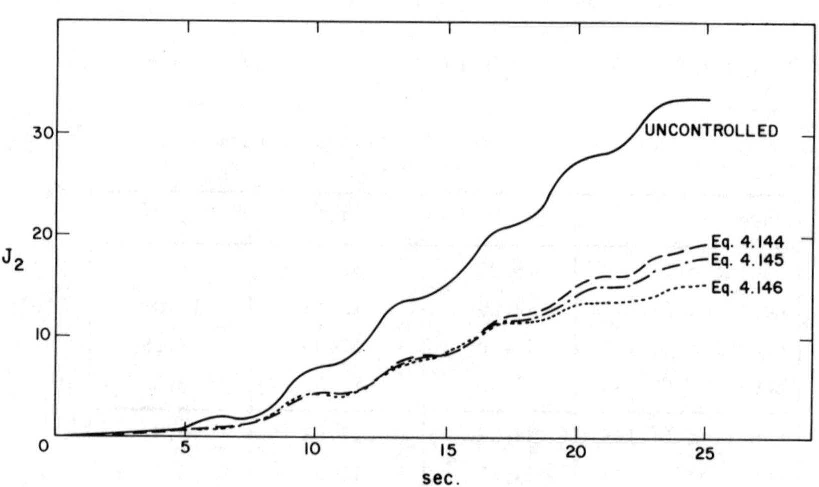

Figure 4.53 - Response of Acceleration Index J_2

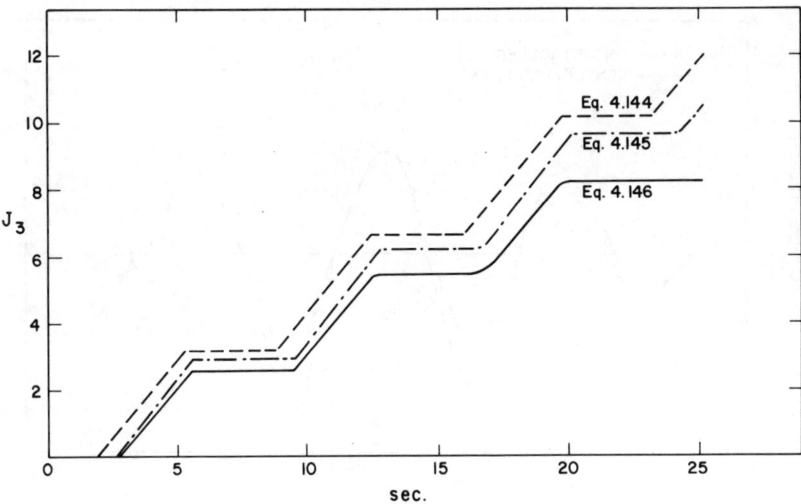

Figure 4.54 - Response of Control Energy Index J_3

The deflection and velocity responses of the uncontrolled and controlled building are shown in Figures 4.55 and 4.56. It is apparent that there is a significant reduction in the controlled response by the suggested method.

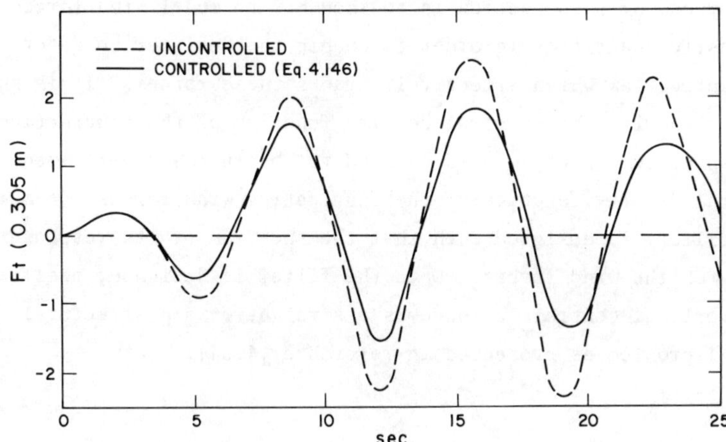

Figure 4.55 - Deflection Response

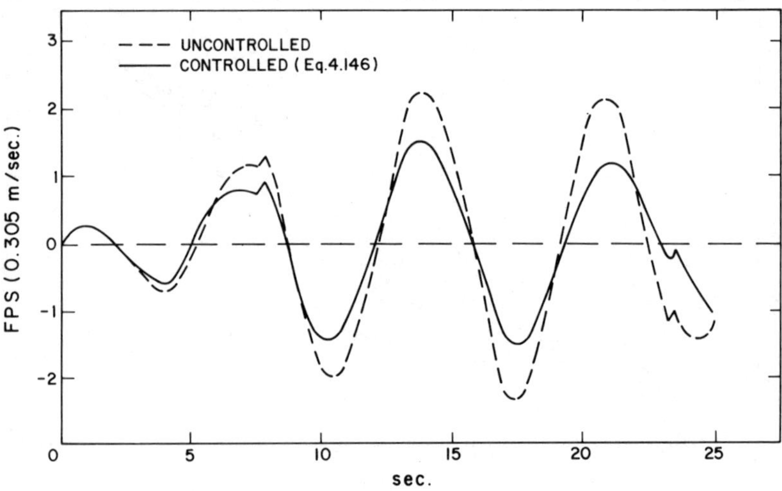

Figure 4.56 - Velocity Response

4.7 STOCHASTIC CONTROL OF TALL BUILDINGS

4.7.1 *Introduction*

The purpose of this section is to show how to model wind forces as
stochastic quantities in order to enable the designer to determine
the control law which effectively tracks these forces. It is shown
that by knowing the spectral density function of the fluctuating
wind speed component, one is enabled to obtain the generalized
spectral density function of the fluctuating wind forces. A model,
or a filter, is designed such that the spectrum of its response
simulates the wind forces. Once the filter is designed, the finding
of an optimal control law becomes a straightforward structural
control problem as presented in reference [4.34].

4.7.2 *Equations of Motion*

The tall building is subjected to drag wind forces and a concentrated control force at height x'. The equation of motion is again

$$EI \frac{\partial^4 y}{\partial x^4} + C \frac{\partial y}{\partial t} + m \frac{\partial^2 y}{\partial t^2} = D(x,t) + U(t)\delta(x-x') , \qquad (4.163)$$

where $D(x,t)$ are the drag wind forces, and $U(t)$ is the concentrated control force. The drag wind forces are expressed by [4.33],

$$D(x,t) = \frac{1}{2} \rho C_D B U^2 (x,t) , \qquad (4.164)$$

where ρ is the air density, C_D is the drag coefficient taken as a constant, B is the building width, and $U(x,t)$ is the wind speed at height x.

The wind speed can be expressed in terms of a mean component, $\overline{U}(x)$, and a fluctuating component, $u(x,t)$, as

$$U(x,t) = \overline{U}(x) + u(x,t) . \qquad (4.165)$$

Substituting equation (4.165) into equation (4.164) and neglecting the second order term, one obtains

$$D(x,t) = 0.5\rho C_D B\overline{U}^2 (x) + \rho C_D B\overline{U}(x)u(x,t) . \qquad (4.166)$$

In equation (4.166), the first term, which is the steady drag force, shall be denoted by $\overline{D}(x)$, and the second term, which is the random drag force, shall be denoted by $d(x,t)$.

The solution of equation (4.163) is assumed as

$$y(x,t) = \phi_1(x)b_1(t) \qquad (4.167)$$

in which $\phi_1(x)$ is the mode shape of the fundamental mode, and $b_1(t)$ is the generalized coordinate.

Substituting equation (4.167) into equation (4.163), and applying an integral transformation, one obtains the equation of motion of the first mode as

$$\ddot{b}_1 + 2\zeta_1\omega_1\dot{b}_1 + \omega_1^2 b_1 = \frac{1}{M_1} \int_0^L \phi_1(x)\overline{D}(x)\,dx + \frac{1}{M_1} \int_0^L \phi_1(x)d(x,t)\,dx$$

$$+ U(t)\phi_1(x') , \qquad\qquad (4.168)$$

in which M_1 is the first mode's generalized mass.

4.7.3 *Spectrum of Wind Forces*

The recommended wind speed spectrum for the fluctuating component
in the low frequency range at height x is given by

$$S_u(x,n) = \frac{U_*^2}{n} \frac{200f}{(1+50f)^{5/3}} , \qquad\qquad (4.169)$$

in which $S_u(x,n)$ is the spectral density function of the fluctuating
along wind speed, U_* is the shear velocity, $f = nx/\overline{U}(x)$, and n is
the frequency $= \omega/2\pi$.

The co-spectrum of the wind speed at points (x_1,z_1) and
(x_2,z_2) is given by

$$S_{u_1u_2}^c(x_1,x_2,n) = S_u^{1/2}(x_1,n)S_u^{1/2}(x_2,n)e^{-\hat{f}} , \qquad\qquad (4.170)$$

in which $\hat{f}$ is expressed as

$$\hat{f} = \frac{n[C_x^2(x_1-x_2)^2+C_z^2(z_1-z_2)^2]^{1/2}}{1/2[\overline{U}(x_1)+\overline{U}(x_2)]} . \qquad\qquad (4.171)$$

In equation (4.171), $C_x \simeq 10$, $C_z \simeq 16$ and z_1, z_2 are the horizontal
coordinates.

However, if the two points coincide, the co-spectrum of
equation (4.170) reduces to equation (4.169). The spectrum of the
generalized fluctuating wind forces (d_1) in the first mode is given by

$$E[d_1(t)d_1(t+\tau)] = \frac{1}{M^2}E\left[\int_0^L \int_0^L \phi_1(x_1)\phi_1(x_2)d(x_1,t)d(x_2,t+\tau)dx_1 dx_2\right]$$

$$= S^2\frac{C_D^2 B^2}{M_1^2} E \int_0^L \int_0^L \phi_1(x_1)\phi_1(x_2)\overline{U}(x_1)\overline{U}(x_2)S^c_{u_1 u_2}$$

$$(x_1,x_2,n)dx_1 dx_2] \; . \tag{4.172}$$

Thus one obtains

$$S_{d_1}(n) \simeq \frac{\rho^2 B^2 C_D^2}{M_1^2}\left[\int_0^L \phi_1(x_1)\overline{U}(x_1)S_u^{1/2}(x_1,n)dx_1\right]$$

$$\cdot \left[\int_0^L \phi_1(x_2)\overline{U}(x_2)S_u^{1/2}(x_2,n)dx_2\right] \; . \tag{4.173}$$

In order to make equation (4.173) workable, one has to determine the shear velocity U_* from equation (4.169). This quantity depends on the roughness of the terrain and the mean wind speed at the reference height 10 meters above ground. The mean wind speed $\overline{U}(x)$ is determined from

$$\overline{U}(x) = \overline{U}(33)\left(\frac{x}{33}\right)^\alpha , \tag{4.174}$$

in which α is the roughness coefficient. The shear velocity U_* is determined from

$$U_* = \frac{\overline{U}(x)}{2.5 \; \ln(x/x_0)} , \tag{4.175}$$

where x_0 depends on the roughness of the terrain.

The integrals of equation (4.173) can be determined numerically. The spectrum of the mean wind forces is constant and given by

$$\overline{S}_{D_1} = \frac{\rho^2 C_D^2 B^2}{4M_1^2}\left[\int_0^L \phi_1(x_1)\overline{U}^2(x_1)dx_1 \int_0^L \phi_1(x_2)\overline{U}^2(x_2)dx_2\right] . \tag{4.176}$$

4.7.4 *Design of Filters*

Various types of filters have been used in references [4.23,4.35],
in order to simulate random disturbances. The first-order and
second-order Markov processes are usually used for disturbances
having a variable spectrum with respect to frequencies as shown in
Figures 4.57 and 4.58. For those disturbances which have constant
power spectral functions, the white noise process is an appropriate
model. For uneven spectral density functions, as wind, one may
approximate the spectrum by an even function, or one may use the
trial approach to select the filter parameters which make its
spectrum fit the wind force spectrum.

The procedure is to plot the spectral density functions
of the generalized wind forces for every mode in the building.
The obtained diagram is compared with the diagram displaying the
spectral density functions of the standard filters [4.35]. The
filter with a spectrum closest to the actual spectrum is selected,
and the filter parameters are chosen such that its mean square
response is the same as the mean square response of the generalized
wind forces, i.e.,

$$\sigma_s^2 = \int_o^\infty S_d(\omega)\,d\omega \ .$$

(4.177)

If the filter equation has more than one parameter to be determined,
equation (4.177) will not be sufficient to determine the other
parameters. In this case, the designer must use some additional
significant points from the actual spectral density function to be
reflected in the filter spectral density function.

4.7.5 *Design of a Stochastic Control Law*

The filter equation of motion can be expressed in state form as

$$\dot{\underline{Z}} = \underline{A}_f \underline{Z} + \underline{W}_x \ , \qquad \underline{Z}_0 \ ,$$

(4.178)

in which $\underline{Z}$ is the filter state vector, $\underline{A}_f$ is the filter state matrix, and $\underline{W}_z$ is the white noise vector of intensity $\underline{V}_z$.

The relationship between the wind disturbance in equation (4.168) and the filter state matrix is

$$\underline{C}_f\underline{Z} = \frac{1}{M_1} \int_0^L \phi_1(x)d(x,t)dx \ , \qquad (4.179)$$

in which $\underline{C}_f$ is a constant matrix determined according to the dimension of the filter and the number of modes taken.

Writing equation (4.168) in state-space form as

$$\underline{\dot{X}} = \underline{A}\underline{X} + \underline{B}\underline{U} + \underline{d}_1 + \overline{D}_1 \ , \qquad (4.180)$$

where

$$\underline{A} = \begin{bmatrix} 0 & 1 \\ -\omega_1^2 & -2\zeta_1\omega_1 \end{bmatrix} \ , \qquad (4.181)$$

$$\underline{B} = [0 \quad \phi_1(L)]^T \ , \qquad (4.182)$$

$$d_1 = \frac{1}{M_1} \int_0^L \phi_1(x)d(x,t)dx \ , \qquad (4.183)$$

$$\overline{D}_1 = \frac{1}{M_1} \int_0^L \phi_1(x)\overline{D}(x)dx \ , \qquad (4.184)$$

equations (4.180) and (4.178) can be combined to read

$$\begin{bmatrix} \dot{X} \\ \dot{Z} \end{bmatrix} = \begin{bmatrix} \underline{A} & \underline{C}_f \\ 0 & \underline{A}_f \end{bmatrix} \begin{bmatrix} X \\ Z \end{bmatrix} + \begin{bmatrix} \underline{B} \\ 0 \end{bmatrix} U(t) + \begin{bmatrix} \underline{W}_x \\ \underline{W}_z \end{bmatrix} \ . \qquad (4.185)$$

In equation (4.185), $\underline{W}_x$ is the white noise vector to model $\overline{D}_1$.

Expressing equation (4.185) in a short state form, one has

$$\underline{\dot{X}}_c = \underline{A}_c\underline{X}_c + \underline{B}_c\underline{U} + \underline{W}_c \ , \qquad (4.186)$$

in which $\underline{X}_c = [\underline{X} \ \ \underline{Z}]^T$, $\underline{B}_c = [\underline{B} \ \ 0]^T$, $\underline{W}_c = [\underline{W}_x \ \ \underline{W}_z]^T$.

For a quadratic objective function of the form

$$J = \frac{1}{2} \int_o^{t_f} (\underline{X}_c^T \underline{Q} \underline{X}_c + \underline{U}^T \underline{R} \underline{U}) dt \, , \qquad (4.187)$$

in which t_f is the final control time, $\underline{Q}$ is a positive semi-definite weighing matrix, and $\underline{R}$ is again a positive definite weighing matrix, the optimal stochastic control law is given by

$$\underline{U} = -\underline{R}^{-1} \underline{B}_c^T \underline{P} \underline{X}_c \, , \qquad (4.188)$$

in which $\underline{P}$ is the Riccati matrix obtained from solving the following matrix differential equation:

$$-\underline{\dot{P}} = \underline{P} \underline{A}_c + \underline{A}_c^T \underline{P} + \underline{Q} - \underline{P} \underline{B}_c \underline{R}^{-1} \underline{B}_c^T \underline{P} \, , \qquad \underline{P}(t_f) = \underline{0} \, . \qquad (4.189)$$

The variance of the controlled building is obtained from solving the differential equation

$$\underline{\dot{K}} = [\underline{A}_c - \underline{B}_c \underline{R}^{-1} \underline{B}_c^T \underline{P}] \underline{K} + \underline{K} [\underline{A}_c - \underline{B}_c \underline{R}^{-1} \underline{B}_c^T \underline{P}]^T + \underline{V}_c \, , \qquad \underline{K}(t_0) \, , \qquad (4.190)$$

in which $\underline{K}$ is the covariance matrix, and $\underline{V}_c$ is the matrix which contains the intensities of the white noise vector $\underline{W}_c$.

The variance of the control force U is obtained from [4.35]

$$\underline{K}_{u,u} = \underline{R}^{-1} \underline{B}_c^T \underline{P} \underline{K} \underline{P} \underline{B}_c \underline{R}^{-1} \, . \qquad (4.191)$$

4.7.6 *Numerical Investigations for Tendon Control*

The building used in Section 4.3 shall be investigated here as an example. The equation of motion of the first mode is

$$\begin{bmatrix} \dot{b}_1 \\ \ddot{b}_1 \end{bmatrix} = \begin{bmatrix} 0 & 1 \\ -4.971 & -0.04 \end{bmatrix} \begin{bmatrix} b_1 \\ \dot{b}_1 \end{bmatrix} + \begin{bmatrix} 0 \\ \dfrac{-1.32}{mL} \end{bmatrix} U + \begin{bmatrix} 0 \\ d_1 \end{bmatrix} + \begin{bmatrix} 0 \\ \overline{D}_1 \end{bmatrix} \, . \qquad (4.192)$$

It is assumed that the wind tunnel tests have yielded
the roughness variables of the terrain as x_0 = 1.98 ft, and α = 0.4.
These values indicate that the building is built in a large city.
Assuming the mean wind speed to be $\overline{U}(33)$ = 66 fps, then, U_* = 9.39
fps is obtained by using equation (4.175). Substituting into
equation (4.173), the spectral density function of the generalized
fluctuating wind forces is given by

$$S_d(n) = \frac{\rho^2 C_C^2 B^2}{(mL)^2}\left[\int_0^L \phi_1(x)\overline{U}(x)\frac{U_*\sqrt{200f}}{n(1+50f)^{5/6}}\,dx\right]^2, \qquad (4.193)$$

in which $f = nx/\overline{U}(x)$, $\phi_1(x) = \cosh\frac{\lambda x}{L} - \cos\frac{\lambda x}{L} - \sigma\left(\sinh\frac{\lambda x}{L} - \sin\frac{\lambda x}{L}\right)$,
σ = 0.73409, λ = 1.875 and $n = \omega/2\pi$.

Using the following data: drag coefficient C_D = 1.3
and air density ρ = 0.0763 pcf, the numerical integration of equation (4.193) has yielded the curve plotted in Figure 4.59. Comparing this diagram with Figures 4.57 and 4.58, one concludes that
the first order Markov filter is the best model to present fluctuating wind forces.

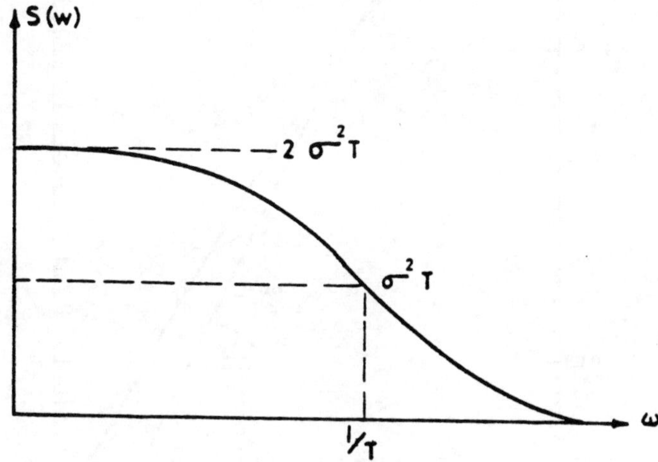

Figure 4.57 - First-Order Markov Filter

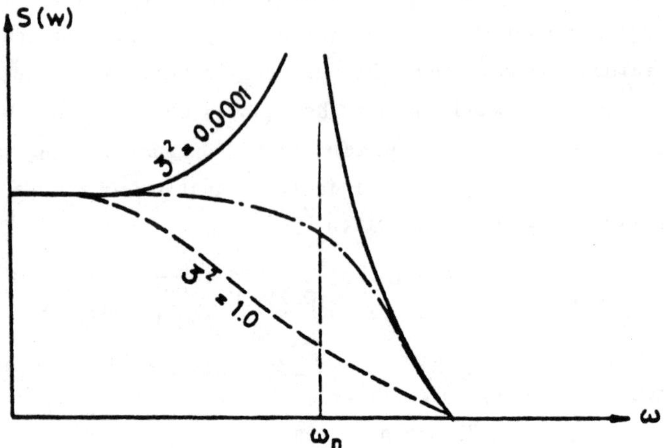

Figure 4.58 - Second-Order Markov Filter

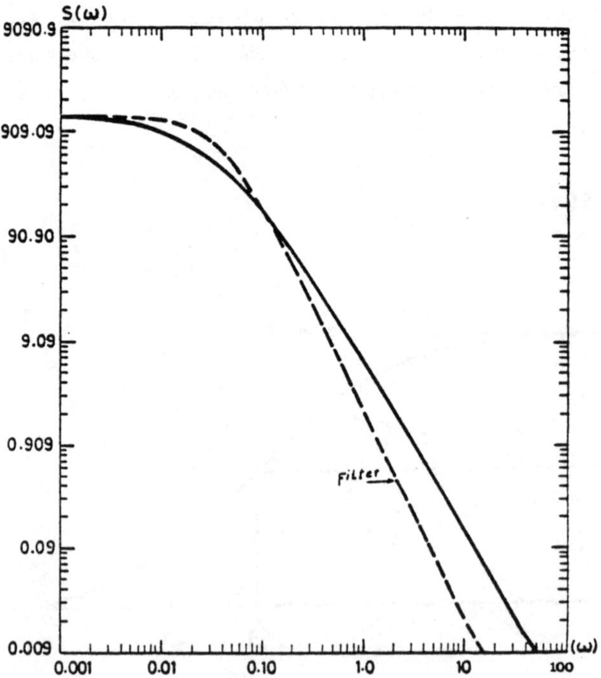

Figure 4.59 - Design of a Filter

The first order Markov filter spectral density function is given [4.35] as

$$S(\omega) = \frac{2\sigma^2/T}{\omega^2 + (1/T)^2} \; ,$$

(4.194)

in which the parameters σ^2 and T need to be determined. Selecting two points from the graph of Figure 4.59, say at $\omega = 0.001$ and $\omega^2 = 1.0$, one can determine the parameters σ^2 and T from equation (4.194). These parameters are then defined in order to let the areas under both curves be equal. In this example, the mean square response of the spectrum, equation (4.177), was 82.50, and the parameters $T = 25.6$ and $\sigma^2 = 26.637$ have been determined. With these parameters, the filter spectrum is also plotted in Figure 4.59.

In order to study the controlled response of the building, the filter equation in time domain has been determined as

$$\dot{\underline{Z}} = \frac{-1}{T} \; Z + W_z(t) \; ,$$

(4.195)

in which $W_z(t)$ is white noise of intensity $(2\sigma^2/T)$ and $Z(t)$ is equal to $d_1(t)$.

According to equation (4.185) and (4.186), the variables are

$$\underline{X}_c^T = [b_1 \quad \dot{b}_1 \quad Z] \; ,$$

(4.196)

$$\underline{A}_c = \begin{bmatrix} 0 & 1 & 0 \\ -4.971 & -0.04 & 1 \\ 0 & 0 & -0.039 \end{bmatrix} \; ,$$

(4.197)

$$\underline{B}_c^T = \begin{bmatrix} 0 & \dfrac{-1.32}{mL} & 0 \end{bmatrix} \; .$$

(4.198)

The optimal stochastic control law was obtained for the weighing factors

$$Q = \begin{bmatrix} 25 & 25 & 0 \\ 25 & 2 & 0 \\ 0 & 0 & 0 \end{bmatrix} , \qquad (4.199)$$

$$R = 0.6 \times 10^{-14} . \qquad (4.200)$$

Assuming $t_f = 30$ seconds, the Riccati matrix $\underline{P}$ was obtained as

$$\underline{P} = \begin{bmatrix} -11.11 & 1.764 & -1.768 \\ 1.764 & 1.5 & 0.199 \\ -1.768 & 0.199 & P_6(t) \end{bmatrix} , \qquad (4.201)$$

in which $P_6(t)$ is a time varying element.

From equations (4.198) and (4.201), the gain matrix is obtained as a constant, and is given by

$$\underline{R}^{-1}\underline{B}_c^T\underline{P} = 10^6 [35.27 \quad 30.02 \quad 3.98] . \qquad (4.202)$$

Substituting equation (4.202) into equation (4.190), the closed-loop augmented state matrix is

$$\underline{A}_c - \underline{B}_c \underline{R}^{-1}\underline{B}_c^T\underline{P} = \begin{bmatrix} 0 & 1 & 0 \\ -9.203 & -3.642 & 0.52 \\ 0 & 0 & -0.039 \end{bmatrix} . \qquad (4.203)$$

One important observation to be made here is that the eigenvalues of the passive controlled system are $(-0.02 \pm J2.22, -0.039)$ whereas the active controlled system has the eigenvalues $(-1.82 \pm J2.426, -0.039)$. This indicates the possibility of determining the closed-loop gain elements in equation (4.202) by the pole assignment method but the open-loop element (last element in equation (4.202)) cannot be determined.

This is so because the augmented system is not completely controllable due to the arrangement of $\underline{B}_c$ in equation (4.198).

However, one may find the open-loop elements which improve the
building response by trial and error.

Considering the turbulent wind component only, the matrix
$\underline{V}_c$ is

$$\underline{V}_c = E[\underline{W}_c^T \, \underline{W}_c] = \begin{bmatrix} 0 & 0 & 0 \\ 0 & 0 & 0 \\ 0 & 0 & 2.081 \end{bmatrix}, \qquad (4.204)$$

in which $E[\underline{W}_z^T \, \underline{W}_z] = 2\sigma^2/T = 2.081$.

The variance of the controlled and uncontrolled responses
are determined from the solution of equation (4.190). The variance
of the deflection at the top of the building, is shown in Figure
4.60. The variance of the velocity is shown in Figure 4.61. The
variance of the total control force is given in Figure 4.62, and the
open-loop control force is also given in Figure 4.62.

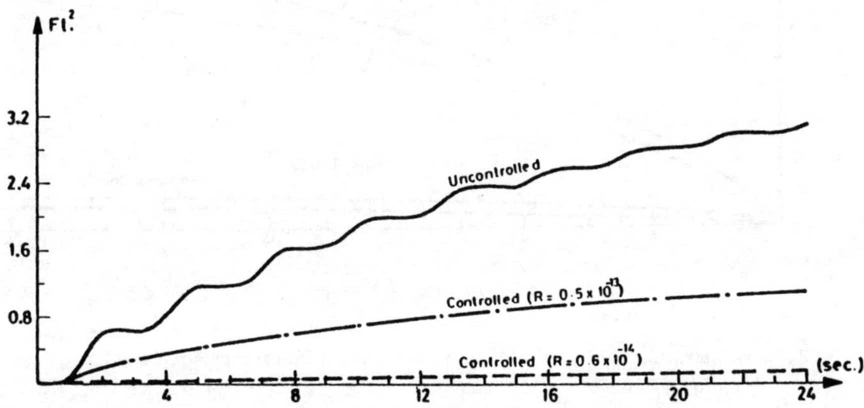

Figure 4.60 - Variance of Deflection Response

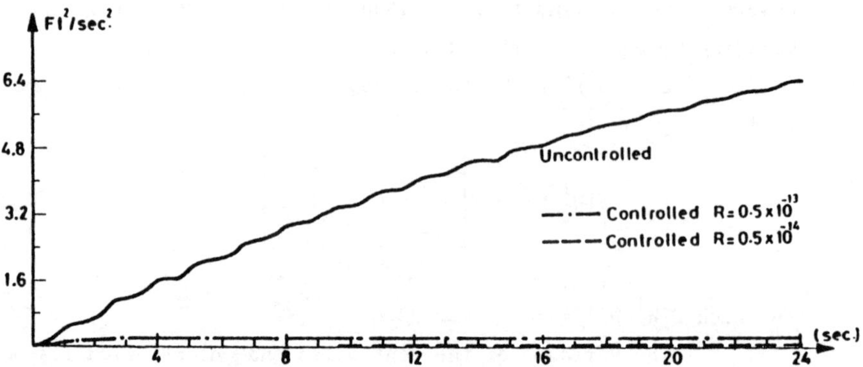

Figure 4.61 - Variance of Velocity Response

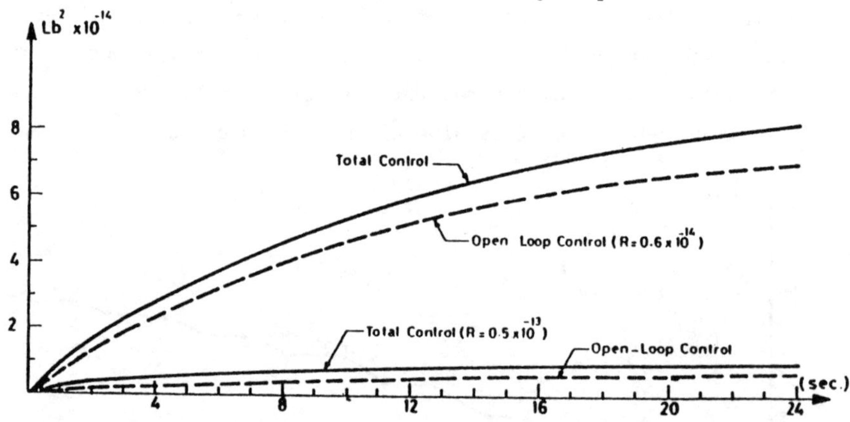

Figure 4.62 - Variance of Control for a Response

4.7.7 *Numerical Investigation of Tuned Mass Damper Control Application*

A tall building which is controlled by an optimal active tuned
mass damper is considered in this section. The building's first
mode natural frequency is assumed to be $\omega_1 = 1.228$ rps, the
damping ratio $\zeta_1 = 0.01$, and the generalized mass $M_1 = 13.42 \times 10^6$
lb. The optimal passive tuned mass damper parameters are $M_2 = 0.02M_1$,
$\omega_2 = 1.204$ rps, and $\zeta_2 = 0.07$.

Considering the first mode and TMD, the equations of motion can be expressed in a matrix state form as

$$\dot{\underline{X}} = \underline{A}\underline{X} + \underline{B}\underline{U} + \overline{\underline{D}}_1 + \underline{d}_1, \tag{4.205}$$

where

$$\underline{A} = \begin{bmatrix} 0 & 1 & 0 & 0 \\ -\omega_1^2 & -2\zeta_1\omega_1 & 0.04\omega_2^2 & 0.08\zeta_2\omega_2 \\ 0 & 0 & 0 & 1 \\ 2\omega_1^2 & 4\zeta_1\omega_1 & -1.08\omega_2^2 & -2.16\zeta_2\omega_2 \end{bmatrix}, \tag{4.206}$$

$$\underline{B}^T = \begin{bmatrix} 0 & \dfrac{-2}{M_1} & 0 & \left(\dfrac{4}{M_1} + \dfrac{1}{M_2} \right) \end{bmatrix}, \tag{4.207}$$

$$\overline{\underline{D}}_1^T = \begin{bmatrix} 0 & \dfrac{1}{M_1} \int\limits_0^L \phi_1(x)\overline{D}(x)\,dx & 0 & \dfrac{-2}{M_1} \int\limits_0^L \phi_1(x)\overline{D}(x)\,dx \end{bmatrix}, \tag{4.208}$$

$$\underline{d}_1^T = \begin{bmatrix} 0 & \dfrac{1}{M_1} \int\limits_0^L \phi_1(x)d(x,t)\,dx & 0 & \dfrac{-2}{M_1} \int\limits_0^L \phi_1(x)d(x,t)\,dx \end{bmatrix}. \tag{4.209}$$

By adding the filter equation (4.195) to equation (4.205), one obtains equation (4.186), in which $\underline{A}_c$, $\underline{B}_c$, $\underline{W}_c$ are given by

$$\underline{A}_c = \begin{bmatrix} 0 & 1 & 0 & 0 & 0 \\ -\omega_1^2 & -2\zeta_1\omega_1 & 0.04\omega_2^2 & 0.08\zeta_2\omega_2 & 1 \\ 0 & 0 & 0 & 1 & 0 \\ 2\omega_1^2 & 4\zeta_1\omega_1 & -1.08\omega_2^2 & -2.16\zeta_2\omega_2 & -2 \\ 0 & 0 & 0 & 0 & -\dfrac{1}{T} \end{bmatrix}, \tag{4.210}$$

$$\underline{B}_c^T = \begin{bmatrix} 0 & \dfrac{-2}{M_2} & 0 & \left(\dfrac{4}{M_1} + \dfrac{1}{M_2} \right) & 0 \end{bmatrix}, \tag{4.211}$$

$$\underline{W}_c^T = \begin{bmatrix} 0 & 0 & 0 & 0 & w_z \end{bmatrix}. \tag{4.212}$$

In equation (4.212), w_z is the input white noise to the filter, its intensity is $2\sigma^2/T$, and the steady state wind force has been disregarded.

Using the same filter parameter of the previous section, one finds that the eigenvalues of the passively controlled system

are $(-0.35 \pm J1.392, -0.35 \pm J1.06, -0.039)$, where the last eigenvalue remains as it is because of the uncontrollability of the
filter, one has to use the following gain matrix

$$\underline{G} = 10^6 [2.769 \quad -1.69 \quad 0.05 \quad -0.352 \quad g] \,, \tag{4.213}$$

where $\underline{G}$ has been determined for $\omega_2 = 1.375$ rps [4.29] using the
pole assignment method, and g is the corresponding open-loop element which has a degree of arbitrariness in its determination as
has been concluded in the previous section. In this equation, $\underline{G}$
replaces the term $(\underline{R}^{-1}\underline{B}_c^T\underline{P})$ in equation (4.188).

The value of g shall be determined by trial and error.
One way to guess the range of this value is to let the elements
of $\underline{C}_f$ in the closed loop matrix, equation (4.203), be zero. By doing
so, we attempt to eliminate completely the effect of the wind forces.
In this example, to let these elements be zero, one has to assume

$$\frac{-2}{M} g + 1 = 0 \,, \tag{4.214}$$

$$\left(\frac{4}{M_1} + \frac{1}{M_2}\right) g - 2 = 0 \,. \tag{4.215}$$

From equation (4.214) the value of g follows as
$g = 0.49 \times 10^6$. From equation (4.215) one obtains $g = 6.7 \times 10^6$.
The first value eliminates the generalized wind force on the building but it has an effect on the tuned mass damper. The second value
eliminates the generalized wind force on the tuned mass damper but
provides forces on the building. In fact, the detrimental coupling
between the building and the tuned mass damper causes difficulties
in getting a better control than the closed-loop control.

Figure 4.63 shows the building controlled by active TMD,
Figure 4.64 gives the variance of the deflection, and Figure 4.65
gives the variance of the velocity at the top of the building.
Figure 4.66 shows the variance of the control force for the assumed
values of g. It is obvious that ATMD is not as effective as
active tendon because of the loss of stiffness.

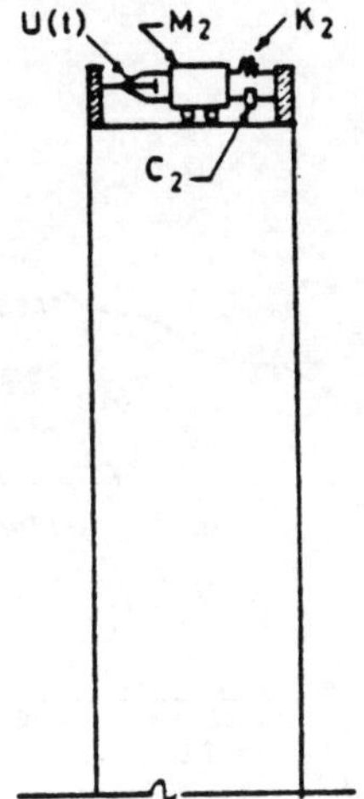

Figure 4.63 - ATMD Mechanism

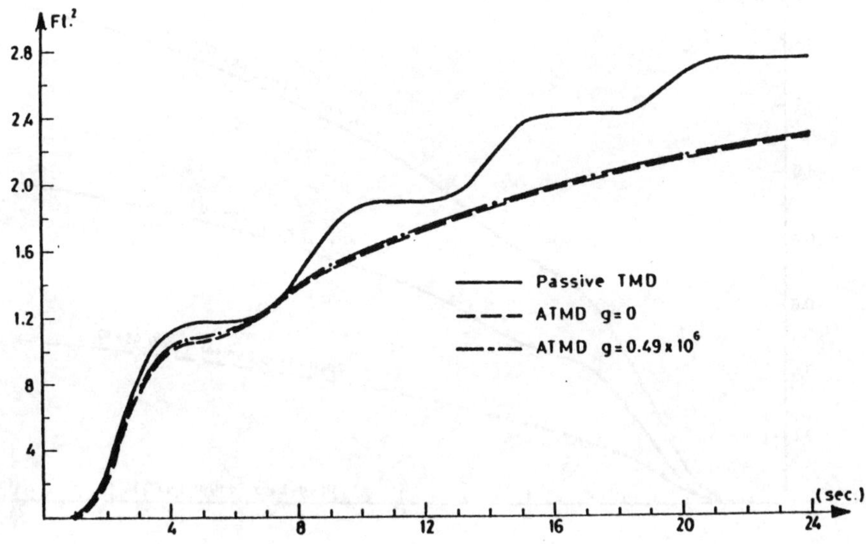

Figure 4.64 - Variance of Deflection Response

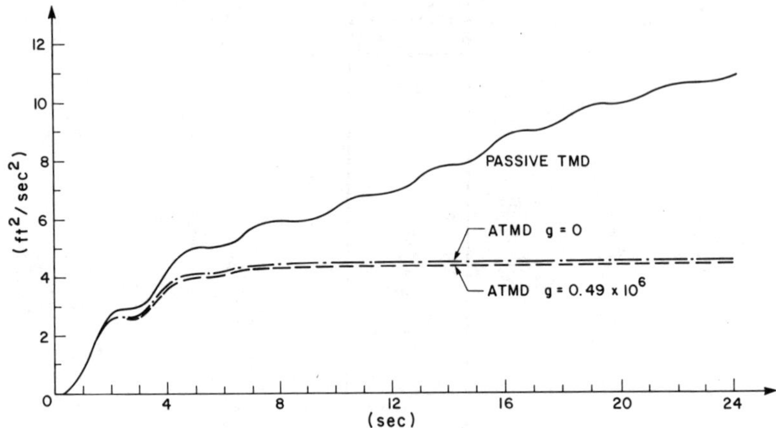

Figure 4.65 - *Variance of Velocity Response*

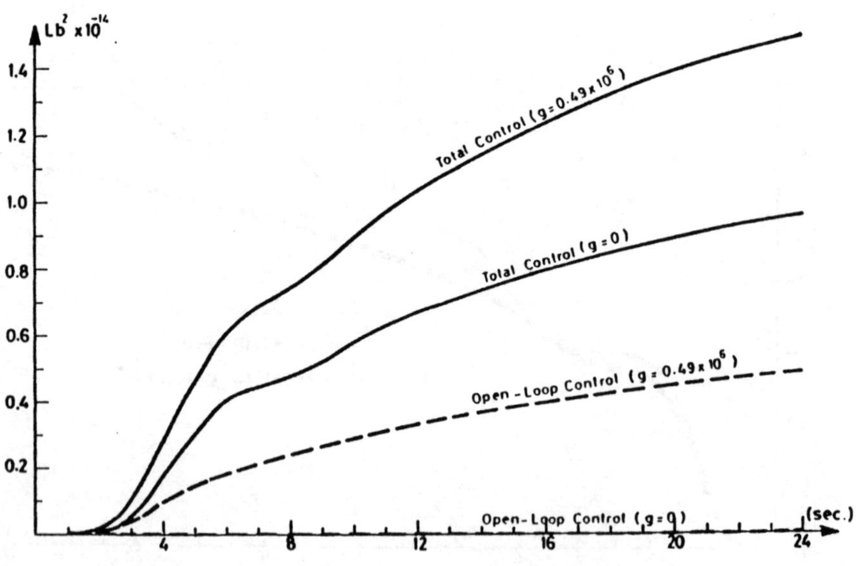

Figure 4.66 - *Variance of Control Force Response*

4.7.8 *Control by Appendages*

The technique of control by appendages, in which one transforms
the stochastic wind into deterministic wind using Fourier series,
has been shown earlier. The procedure of modelling the wind by
a filter is the same. The only change occurs when determining
the Riccati matrix for the augmented system. However, this part
will not be shown, because in principle it is equal to what has
been said in the previous two sections.

4.7.9 *Conclusions*

It has been shown that by modelling the stochastic wind forces by
a proper filter driven by white noise makes the design of the
stochastic optimal control law a straightforward control problem.
The control law can be found by using optimal control theory.
However, one may also use the pole assignment method, but then
the determination of the open-loop gain elements must be by a trial
and error process.

4.8 CONTROL OF TALL BUILDINGS TAKING TIME DELAY INTO ACCOUNT

It has been shown in the previous sections that the control of a
building starts with measuring the structure's response, then estimating the structure's state variables from the available measurements, and it ends by applying the design active control forces
which are functions of the estimated state variables. This process
involves a time lag between the instant of measuring the structural
response and that of applying the control forces which are designed
to control the measured response. The time lag may affect the
stability of the structure, or increase the controlled structural
response as compared to the one which is expected according to the
original design. This lag effect is present in every control process and the designer should actually design the control forces
taking the delay into account.

In the previous sections, the time delay effect has been disregarded. This may be justified as follows: a flexible tall building usually has a large natural period as compared with the time delay and thus its effect is small. One can also claim that the previous study was only intended to indicate the possibility of using active control forces for the suppressing of the vibrations in tall buildings. In this section, the time delay effect on the controlled building response will be studied. It is shown how to compensate for this effect, and how much this compensation costs. It will also be shown that the time delay effect could be used as a criterion for selecting that control mechanism which is less than all the others affected by the time delay.

4.8.1 *Equations of Motion*

In previous sections, the equations of motion of the building have been expressed in state-space form by

$$\dot{\underline{X}} = \underline{A}\underline{X} + \underline{B}\underline{U} + \underline{d} \; ; \quad \underline{X}(t_0) = \underline{\alpha} \; , \tag{4.216}$$

in which $\underline{X}$ is the n-state vector, $\underline{U}$ the r-control vector, $\underline{d}$ the n-dimensional disturbance vector, $\underline{X}(t_0)$ the n-dimensional initial conditions for $\underline{X}$, $\underline{A}$ the state matrix containing the building characteristics, $\underline{B}$ the control matrix indicating the locations and types of the control forces $\underline{U}$, and $\underline{d}$ the generalized wind forces on the building.

In most cases, the state variables $\underline{X}$ are estimated from the available measurements using an observer (filter). The settling times of the devices used in measuring the building response, estimating state variables, and operating the control mechanism cause delay in the control and state variables. Thus, equation (4.216) should be expressed, in reality, as follows:

$$\dot{\underline{X}} = \sum_{i=o}^{n} \underline{A}_i \underline{X}(t-\tau_{xi}) + \sum_{j=o}^{r} \underline{B}_j \underline{U}(t-\tau_{uj}) + \underline{d} \; , \tag{4.217}$$

in which τ_{xi} is the delay in ith state variable; and τ_{uj} is the delay in jth control variable.

The initial conditions become

$$\underline{X}(t_0) = \underline{X}_0 , \qquad -\underline{\alpha}_x \le t_0 \le \underline{0} , \tag{4.218}$$

$$\underline{U}(t_0) = \underline{U}_0 , \qquad -\underline{\alpha}_u \le t_0 \le \underline{0} , \tag{4.219}$$

in which the n-vector $\underline{\alpha}_x$ contains all α_{xi}, and the r-vector $\underline{\alpha}_u$ contains all α_{uj}.

4.8.2 *Design of Control Laws*

Various methods have been proposed in the literature, e.g., [4.36-4.40], for finding the proper control law for time delayed systems. The computations of an optimal control law for time delay systems are very complicated and time-consuming. Even the implementation of the control laws become very difficult because they require the feedback of the current states, the delayed states, and the control variables. For these reasons, and in order to remain practical in the implementation of the control law, designers have paid their attention to finding suboptimal control laws able to overcome the time-delay effect.

One of the simple methods frequently used in the literature [4.36,4.37,4.40] is the Taylor-Series expansion method. The accuracy of this method depends on the time delay being small as compared with the natural period of the structure. This method shall be used in this section because of its simplicity and the ease in the implementation of the control law.

For the delays τ_{xi} in the state variables, and τ_{uj} in the control variables of equation (4.217), the Taylor-Series expansion of $\underline{X}(t-\tau_{xi})$ and $\underline{U}(t-\tau_{uj})$ are, respectively,

$$\underline{X}(t-\tau_{xi}) = \underline{\dot{X}}(t) - \tau_{xi}\underline{\dot{X}}(t) + \frac{\tau_{xi}^2}{2!}\underline{\ddot{X}}(t) - \frac{\tau_{xi}^3}{3!}\underline{\dddot{X}}(t) + \cdots , \tag{4.220}$$

$$\underline{U}(t-\tau_{uj}) = \underline{U}(t) - \tau_{uj}\underline{\dot{U}}(t) + \frac{\tau_{uj}^2}{2!}\underline{\ddot{U}}(t) - \frac{\tau_{uj}^3}{3!}\underline{\dddot{U}}(t) + \cdots . \tag{4.221}$$

The optimal control for equation (4.217) can be obtained by taking the full series in equations (4.220) and (4.221). For practical purposes, these series are truncated and terms only up to a certain order are retained.

Another way of expressing the time delay is as follows:

$$\underline{X}(t) = \underline{X}(t-\tau_{xi}+\tau_{xi}) = \underline{X}(t-\tau_{xi}) + \tau_{xi}\underline{\dot{X}}(t-\tau_{xi}) + \frac{\tau_{xi}^2}{2!}\underline{\ddot{X}}(t-\tau_{xi}) + \cdots .$$
$$\tag{4.222}$$

Whatever the way of expressing the delay, the corresponding series are substituted into equation (4.217). One can then form an augmented state-space system containing the variables $\underline{X}(t)$, $\underline{X}(t-\tau_{xi})$, $\underline{U}(t)$, $\underline{U}(t-\tau_{uj})$, etc. The augmented system can be expressed as

$$\underline{\dot{X}}_c = \underline{A}_c\underline{X}_c + \underline{B}_c\underline{U}_c + \underline{d}_c , \qquad \underline{X}_0(t_0) , \tag{4.223}$$

which is in a similar form as equation (4.216) but the dimensions are different. Therefore, one can design $\underline{U}_c$ by any suitable method in order to control the structure and the time delay effect.

4.8.3 *Numerical Applications to Tendon Control*

The building used in Section 4.3 shall be investigated and designed taking the time delay effect into account. Considering the first mode and that the delay is in the control force, the equation of motion can be expressed as

$$\ddot{b}_1 + 2\zeta_1\omega_1\dot{b}_1 + \omega_1^2 b_1 = F_1(t) - \frac{1.32}{mL} U(t-\tau) , \tag{4.224}$$

in which $b_1(t)$ is the generalized coordinate of the first mode, ζ_1 is the damping ratio in the first mode, ω_1 is the natural frequency of the first mode; $U(t-\tau)$ is the actual control force on

the building with a time delay τ, m is the mass of the building
per unit height, L is the height of the building, and $F_1(t)$ is the
generalized wind force.

The control $U(t-\tau)$ can be replaced using the Taylor-
Series expansion of equation (4.222) for $U(t)$ as follows:

$$U(t) = U(t-\tau) + \tau\dot{U}(t-\tau) + \frac{\tau^2}{2}\ddot{U}(t-\tau) , \qquad (4.225)$$

i.e.,

$$\ddot{U}(t-\tau) = \frac{2}{\tau^2} U(t) - \frac{2}{\tau^2} U(t-\tau) - \frac{2}{\tau}\dot{U}(t-\tau) . \qquad (4.226)$$

For simplicity, the series is truncated to the second
order. Thus, one has another second order differential equation
in terms of $U(t-\tau)$. Denoting the state variables $X_1 = b_1$, $X_2 = \dot{b}_1$,
$X_3 = U(t-\tau)$, and $X_4 = \dot{U}(t-\tau)$, equation (4.224) and (4.225) can be
expressed in the state form of equation (4.223), where,

$$\underline{A}_c = \begin{bmatrix} 0 & 1 & 0 & 0 \\ -\omega_1 & -2\zeta_1\omega_1 & \frac{-1.32}{mL} & 0 \\ 0 & 0 & 0 & 1 \\ 0 & 0 & \frac{-2}{\tau^2} & \frac{-2}{\tau} \end{bmatrix} , \qquad (4.227)$$

$$\underline{B}_c^T = \begin{bmatrix} 0 & 0 & 0 & \frac{2}{\tau^2} \end{bmatrix} , \qquad (4.228)$$

$$\underline{d}_c^T = \begin{bmatrix} 0 & F_1(t) & 0 & 0 \end{bmatrix} . \qquad (4.229)$$

The initial conditions for X_3 and X_4 are assumed to be zero for
$t \leq \tau$. The design of $U(t)$ can now be done by any suitable method.
For simplicity, the pole assignment method shall be used [4.9,4.24].
The design objective is to determine the control law which takes
the time delay into account and the design which gives acceptable
results.

Considering the numerical data L = 1000 ft, m = 11000 lb/ft,
ω_1 = 2 rps, ζ_1 = 0.01, F_1 = 20 sin 2.2t, and τ = 0.2 sec, which is
small as compared to the first mode natural period (3.14 sec).

The eigenvalues of the matrix $\underline{A}_c$ of equation (4.227) are $(-0.02 \pm J2.0)$ and $(-5 \pm J5)$. The first two eigenvalues represent the building characteristics, and the other two concern the control with time delay. The controlled building characteristics are specified to be $(-0.5 \pm J2.0)$ in order to introduce active damping to the building. The other two eigenvalues are assumed to be $(\alpha \pm J5)$, where α is found by trial and error and provides the desirable controlled response. In order to ease the process of design, the following indices are used:

$$J_D = \int_o^T X_1^2 dx ~, \tag{4.230}$$

$$J_V = \int_o^T X_2^2 dx ~, \tag{4.231}$$

$$J_u = \int_o^T U^2 dx ~, \tag{4.232}$$

$$J_T = \int_o^T U^2(t-\tau) dx ~, \tag{4.233}$$

in which J_D involves the deflection response, J_V involves the velocity response, J_u involves the applied control response, and J_T involves the final control with time delay.

For various values of α, the relationships between J_u-J_D and J_u-J_V are shown, respectively, in Figures 4.67 and 4.68. This relationship is also shown for $\tau = 0.0$. It is obvious from these figures that in order to obtain the same response obtained from the design corresponding to $\tau = 0.0$, one has to apply a larger control force which may affect the stability of the system. However, using $\alpha = -18$ provides, approximately, the same control force but only little increase in the building response which occurs for $\tau = 0.0$. In this case, the applied control force is given by

$$U(t) = [-8.63 \times 10^6 \quad 56.58 \times 10^6 \quad -6.671 \quad -0.53] \underline{X}_c ~.$$

$$\tag{4.234}$$

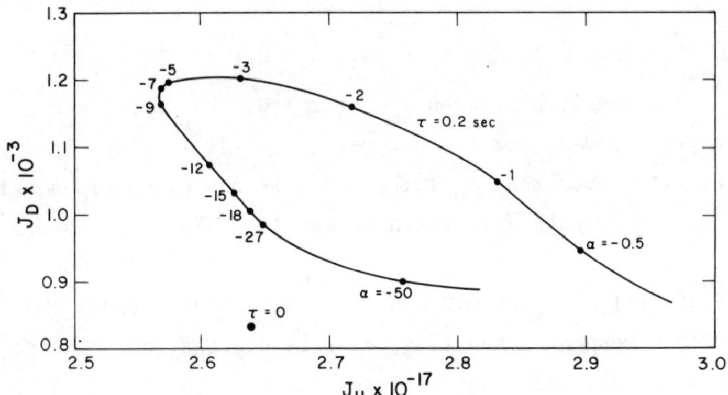

Figure 4.67 - Relationship Between Deflection and Control Indices

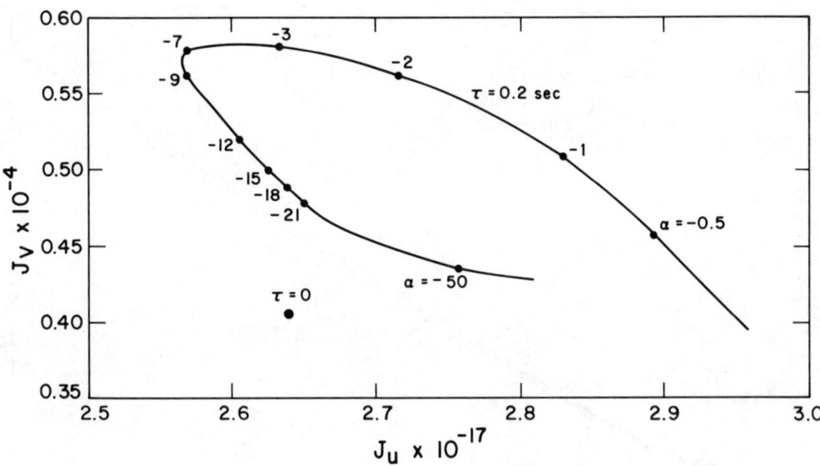

Figure 4.68 - Relationship Between Velocity and Control Indices

This is different from the design for $\tau = 0.0$ which is

$$U(t) = [2.08\times10^6 \quad 8.02\times10^6]\underline{X} \; , \tag{4.235}$$

in which $\underline{X}^T = [b_1 \quad \dot{b}_1]$, and $\underline{X}_c^T = [b_1 \quad \dot{b}_1 \quad U(t-\tau) \quad \dot{U}(t-\tau)]$.

A comparison between J_u for $\tau = 0.0$, J_u for $\tau = 0.2$ using $\alpha = -18$, and J_T for $\tau = 0.2$ using $\alpha = -18$ is given in Figure 4.69. It is obvious that J_u for $\tau = 0.0$ is approximately equal to that for $\tau = 0.2$, when α is taken to be -18. Also, the values of J_T are little less than the values of J_u. Figure 4.70 shows the response of $U(t)$ when $\tau = 0.0$ and $\tau = 0.2$. It also shows the control force response when time delay is disregarded. This case represents the designed control force for $\tau = 0.0$ when actually a time lag of 0.2 seconds exists in the feedback process. Figures 4.71 and 4.72 show, respectively, the deflection and velocity responses for the uncontrolled building, the controlled building for $\tau = 0.0$ using current feedback, the controlled building for $\tau = 0.0$ using lag feedback of 0.2 seconds, and the controlled building for $\tau = 0.2$ seconds and $\alpha = -18$.

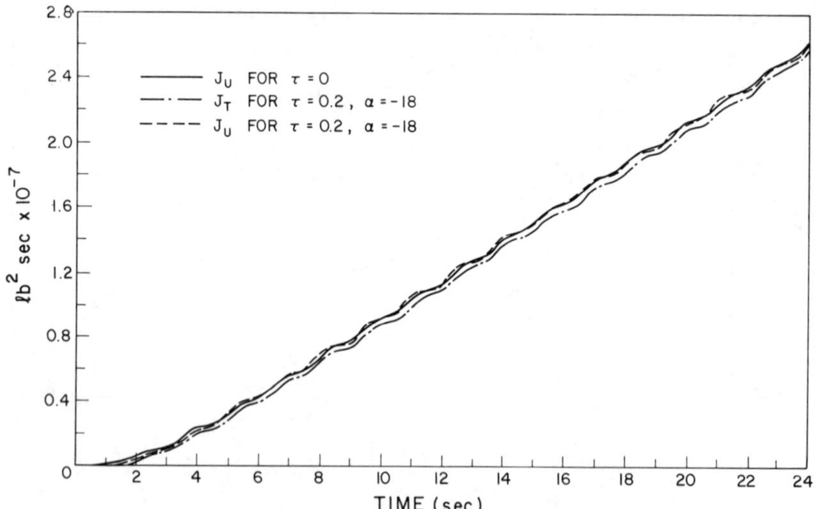

Figure 4.69 - Response of Control Indices

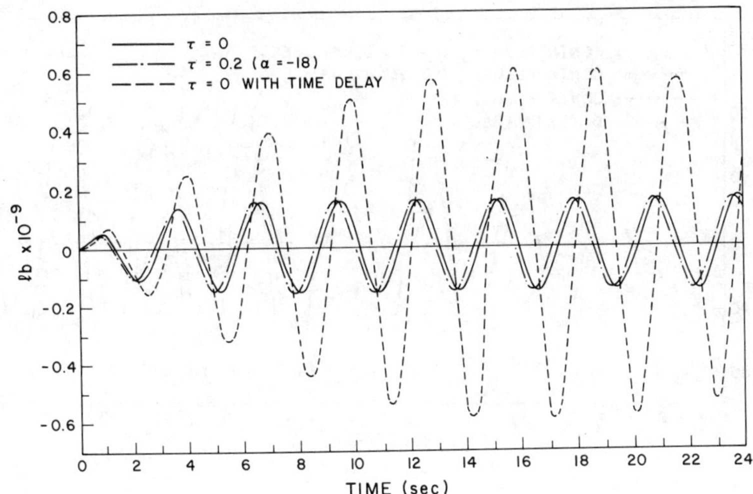

Figure 4.70 - *Control Force Response*

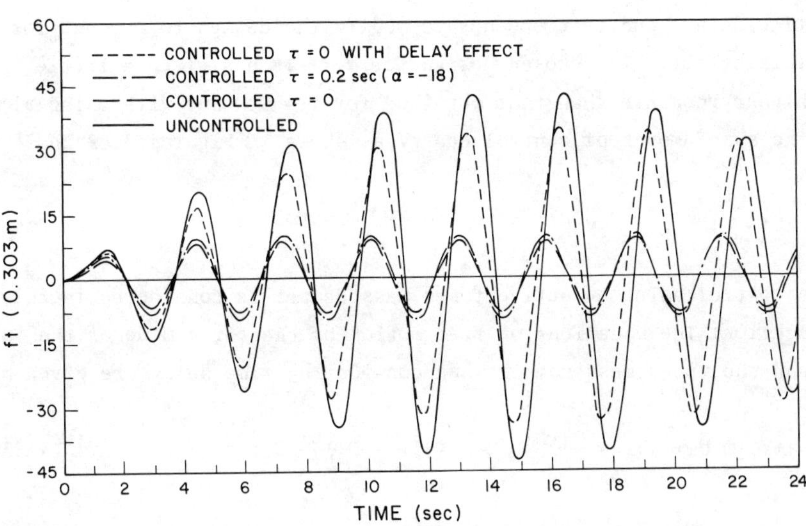

Figure 4.71 - *Deflection Response*

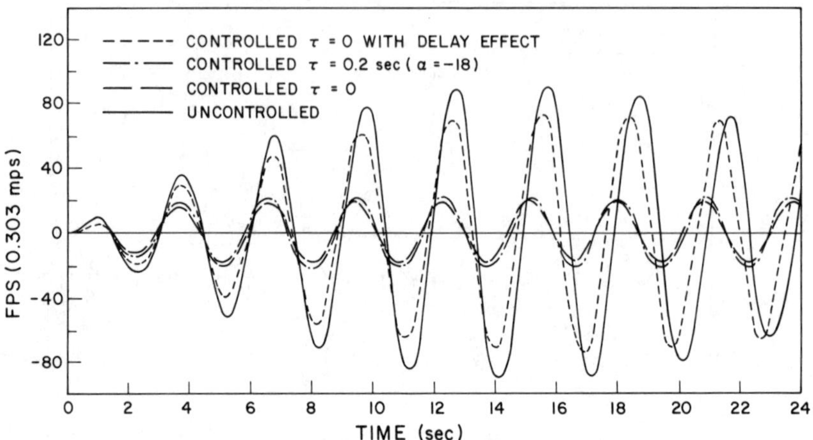

Figure 4.72 - Velocity Response

It is obvious that the last case is a little higher than
the ideal case for $\tau = 0.0$ using current feedback. One also observes
the increase in the controlled response if the lag feedback is considered. This indicates the sensitivity of the tendon mechanism to
time delay, and that one has to modify the design to account for
this effect. The chosen design for $\alpha = -18$ has given a little
higher response than that obtained for $\tau = 0.0$ but with using almost
the same amount of control energy as shown in Figures 4.69 and 4.70.

4.8.4 *Numerical Application to TMD Control*

A control using an active tuned mass damper is considered in this
section. The equations of free motion of the first mode of the building
and the tuned mass damper when considering time delay are given by

$$\ddot{b}_1+2\zeta_1\omega_1\dot{b}_1+\omega_1^2 b_1 = \frac{\phi_1(L)}{mL} (C_2\dot{Z}+K_2Z) - \frac{\phi_1(L)}{mL} U(t-\tau) , \qquad (4.236)$$

$$\ddot{z}+2\zeta_2\omega_2\dot{z}+\omega_2^2 z = -\ddot{b}_1\phi_1(L) + \frac{1}{M_2} U(t-\tau) , \qquad (4.237)$$

in which z is the relative displacement between the damper and the
building, ζ_2 is the damping ratio in the damper, ω_2 is the natural
frequency of the damper, M_2 is the mass of the damper, $\phi_1(L)$ is the
first mode shape at top of the building; and $U(t-\tau)$ is the actual
control force between the building and the tuned mass damper.

Using equation (4.226) for the delayed control force,
one can write equations (4.236) and (4.237) in state space form as
equation (4.223) where $\underline{A}_c$ and $\underline{B}_c$ are respectively given as

$$\underline{A}_c = \begin{bmatrix} 0 & 1 & 0 & 0 & 0 & 0 \\ -\omega_1^2 & -2\zeta_1\omega_1 & \dfrac{2K_2}{mL} & \dfrac{2C_2}{mL} & \dfrac{-2}{mL} & 0 \\ 0 & 0 & 0 & 1 & 0 & 0 \\ 2\omega_1^2 & 4\zeta_1\omega_1 & -a & -b & C & 0 \\ 0 & 0 & 0 & 0 & 0 & 1 \\ 0 & 0 & 0 & 0 & \dfrac{-2}{\tau^2} & \dfrac{-2}{\tau} \end{bmatrix} , \qquad (4.238)$$

$$\underline{B}_c^T = \begin{vmatrix} 0 & 0 & 0 & 0 & 0 & \dfrac{2}{\tau^2} \end{vmatrix} , \qquad (4.239)$$

in which a = $(2K_2/mL+K_2/M_2)$, b = $(4C_2/mL+C_2/M_2)$, C = $(4/mL+1/M_2)$,
and the initial conditions are chosen to be X_1 = 2.5, X_2 = -2.5,
X_3 = 10, X_4 = -4 and X_5 = X_6 = 0 for t $\leq$ τ. The initial conditions
are specified to be non-zero because it has been pointed out that
active tuned mass dampers are only effective for free and steady
state vibrations.

In this section, the optimal TMD parameters which have
been determined in Section 4.5.5.3 are used. Data considered are
L = 1000 ft, m = 13420 lb/ft, ω_1 = 1.228 rps; ζ_1 = 0.01, τ = 0.2
seconds, M_2 = 0.02 mL, ω_2 = 1.375 rps, and ζ_2 = 0.07. The eigenvalues of $\underline{A}_c$ using these numerical values are (-0.03 $\pm$ J1.518),
(-0.05 $\pm$ J1.109), and (-5 $\pm$ J5). The actively controlled building
is assumed to have eigenvalues of (-0.35 $\pm$ J1.518), (-0.35 $\pm$ J1.109)
and (α $\pm$ J5), where α is determined by trial and error as in the
previous section.

Using the same indices of equations (4.230) to (4.233), the relationships between J_D-J_u, and between J_v-J_u for various values of α are given in Figures 4.73 and 4.74. In these figures also the points are shown which represent the design for $\tau = 0.0$. One notices that when $\alpha \geq -29$, the controlled response is almost unchanged when increasing the control energy. For the sake of comparison, $\alpha = -35$ consumes, approximately, the same energy as in the case where $\tau = 0.0$. In this case, the gain matrix is obtained as

$$\underline{K} = [0.67 \times 10^8 \quad -0.78 \times 10^8 \quad -0.108 \times 10^7 \quad -0.103 \times 10^8 \quad -25.64 \quad -1.22] \ , \tag{4.240}$$

whereas the gain matrix for $\tau = 0.0$ is given by

$$\underline{K} = [0.246 \times 10^7 \quad -0.323 \times 10^7 \quad -0.77 \times 10^5 \quad -0.409 \times 10^6] \ . \tag{4.241}$$

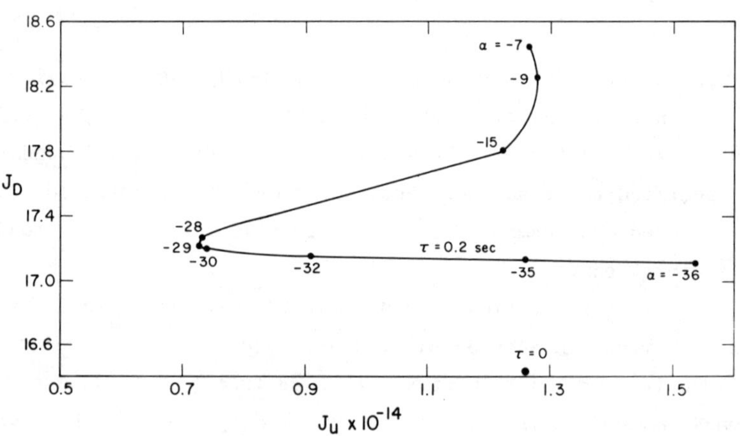

Figure 4.73 - *Relationship Between Deflection and Control Indices*

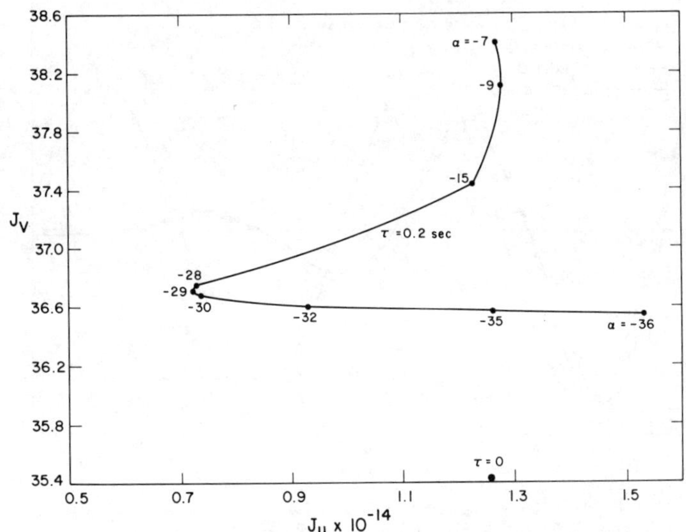

Figure 4.74 - Relationship Between Velocity and Control Indices

Comparison between the uncontrolled response, the controlled response with neglected time delay, the controlled response using equation (4.241) considering a time lag of 0.2 seconds in the feedback, and the controlled response considering a time delay using equation (4.240) and using $\alpha = -35$ are shown in Figure 4.75 for deflection response, and in Figure 4.76 for velocity response. It is obvious from these figures that the time lag has no significant effect on the free vibration response of the building in contrast to the previous example. Figure 4.76 gives a comparison between the response of the control force for three cases, that is $\tau = 0.0$, $\tau = 0.0$ with a lag of 0.2 seconds, and $\tau = 0.2$ seconds using $\alpha = -35$. One observes a difference in the magnitudes during the first two seconds after which the three responses are almost equal. These results indicate the efficiency of an active tuned mass damper against time delay effect.

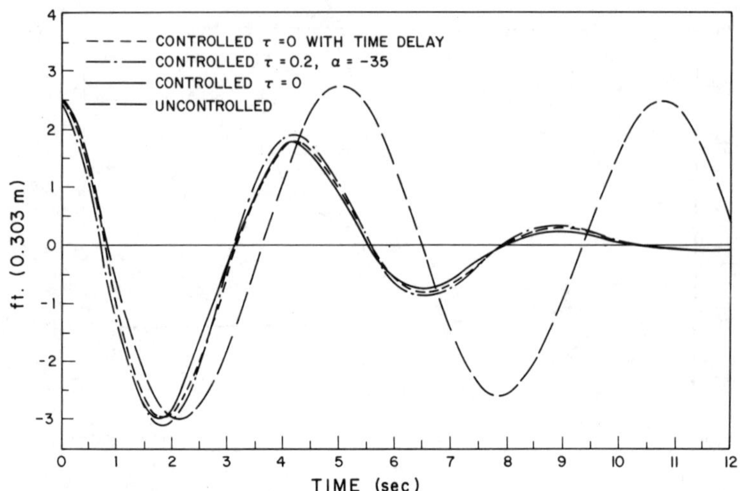

Figure 4.75 - Deflection Response

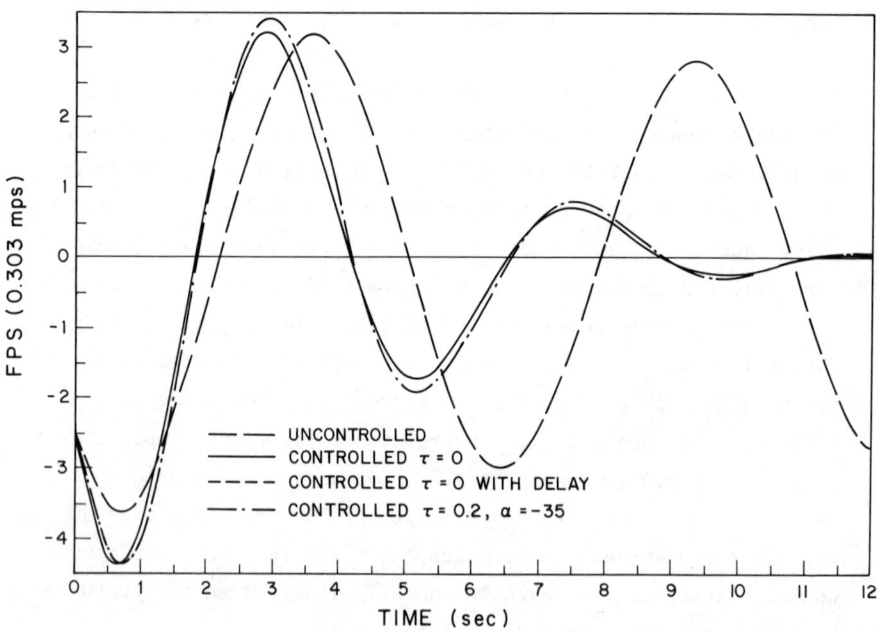

Figure 4.76 - Velocity Response

4.9   FEASIBILITY OF ACTIVE CONTROL OF TALL BUILDINGS

4.9.1 *Introduction*

In this section, the feasibility of active control of tall buildings
by the previous solutions shall be studied. The feasibility shall
be measured in terms of the magnitude of the control forces, and
the rate of flow of fluids in the actuators which generates the
control forces. This shall assist in pointing out the advantages
and disadvantages of each control mechanism.

4.9.2 *Feasibility of Active Tendon Control*

The control force generated by the tendon is proportional to the
magnitude of the ram displacement of the actuator. If the ram
displacement if $y(t)$, then the control force $U(t)$ is given by

$$U(t) = \frac{y(t)}{L} EA \tag{4.242}$$

in which L = tendon lengths, E = tendon modulus of elasticity,
and A = tendon cross section areas.

In order to determine the rate of flow in the actuator,
it is assumed that the cross section area of the piston is A_p.
The rate of flow of the fluid in and out of the actuator is obtained
from

$$Q(t) = \dot{y}(t)A_p \ . \tag{4.243}$$

The feasibility of using tendon control is measured in
terms of the rate of flow of the fluid. This reflects the volumes
of the fluid used to generate the control forces.

In order to determine $Q(t)$, one has

$$Q(t) = A_p\dot{y}(t) = A_p \frac{\dot{U}(t)L}{EA} \ , \tag{4.244}$$

in which $\dot{U}(t)$ is the rate of change of the control force $U(t)$ with respect to time.

As an index for volume changes, the following performance criteria are chosen

$$J = \int_o^T \left[\frac{Q(t)}{A_p}\right]^2 dt \ ,$$

$$J_1 = \int_o^T \dot{U}^2 \left[\frac{L}{EA}\right]^2 dt \ . \tag{4.245}$$

Equation (4.245) represents the volume of the fluid per unit cross section area of the pistons. It indicates the total volume changes in time T.

The tall building considered in the previous sections shall now be used as an example. The equation of motion of the first mode with the control force due to active tendon is given by

$$\ddot{A}(t)+2\zeta_1\omega_1\dot{A}(t)+\omega_1^2 A(t) = \frac{F(t)}{mL} - \frac{1.32}{mL} U(t)-\omega_{1p}^2 A(t) \ , \tag{4.246}$$

in which $F(t)$ is the wind force in the first mode, 1.32 is the constant due to tendon length and modulus of elasticity, ω_{1p}^2 is the stiffness due to passive control.

The building parameters are again $\omega_1 = 1.228$ rps, $m = 13.42 \times 10^3$ lb.sec^2/ft^2, $L = 1000$ ft. The pretension in the passive tendon introduces more stiffness to the building by:

$$\omega_{1p}^2 = \frac{0.35}{mL} \times \frac{2EA}{D} \cos^2\theta = 0.7933 \text{ rps}, \tag{4.247}$$

in which D is the width of the building, and θ is the angle of slope of the tendon with respect to the vertical as presented in Section 4.4

In order to let the building have eigenvalues of (-0.38 ± 1.5175), the following gain matrix must be used:

$$\underline{K} = 10^6 [1.466 \quad 7.477] \ . \tag{4.248}$$

The wind forces used in Sections 4.6.3 shall be used here too. The parameters used in calculating wind forces are C_d = 1.3, B = D = 100 ft, ρ = 0.0763, $\Delta\omega$ = 40/25, $\overline{U}$ = 66 fps and K = 0.04. The total generalized wind forces and the generalized turbulent wind forces are, respectively, plotted in Figures 4.77 and 4.78.

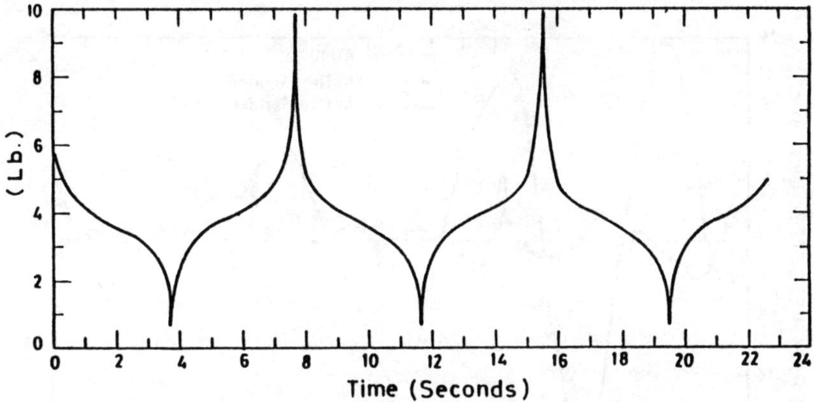

Figure 4.77 - Total Generalized Wind Forces

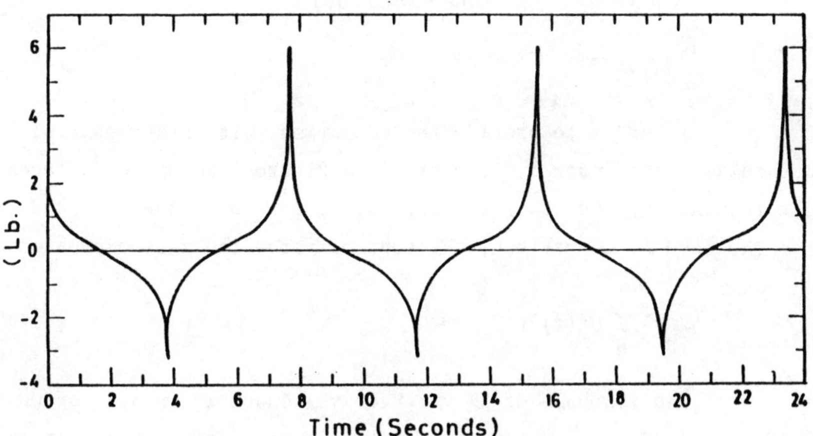

Figure 4.78 - Generalized Turbulent Wind Forces

The rate of flow per unit piston area which represents
the ram velocity, $\dot{y}(t)$, during the control process is given in
Figure 4.79 and is denoted by "active tendon 1". This has been
calculated by considering the total tendon length as 1100 ft,
$E = 30,000$ ksi, and cross-section areas of tendons $= 1.0$ ft^2.

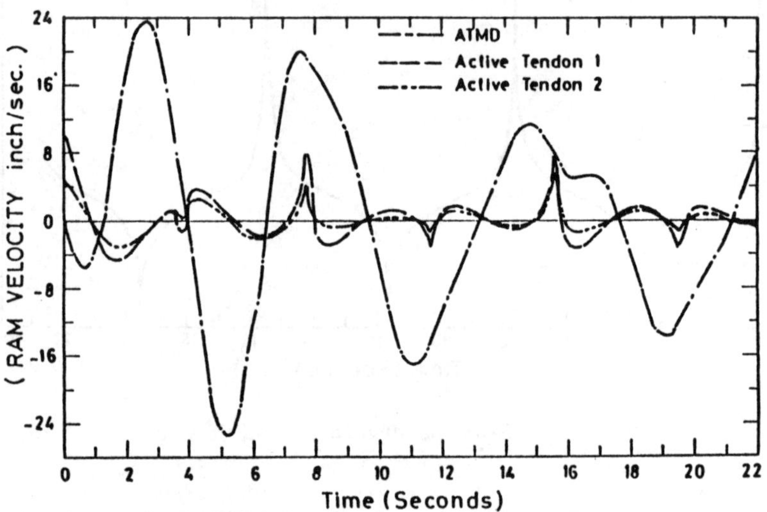

Figure 4.79 - RAM Velocity

In order to provide the comparison with other control
mechanisms, the index J_1 is plotted in Figure 4.80 up to 20 seconds.
The response of the control force, $U(t)$, is shown in Figure 4.81.
For the sake of comparison, the control force index is expressed by

$$J_u = \int_o^T U^2(t)dt \ . \tag{4.249}$$

The response of J_u for "active tendon 1" is also shown
in Figure 4.82. The deflection response, velocity response at the
top of the building are shown, respectively, in Figures 4.83 and
4.84. For the sake of comparison, the deflection and velocity are
measured in terms of

$$J_D = \int_o^T A^2(t)dt \; , \qquad (4.250)$$

$$J_V = \int_o^T \dot{A}^2(t)dt \; . \qquad (4.251)$$

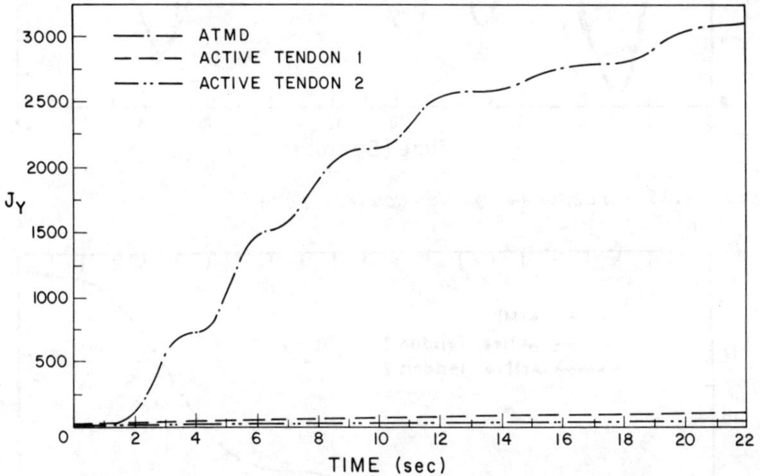

Figure 4.80 - Response of the Volume of Fluid in the Actuator per unit Piston Area

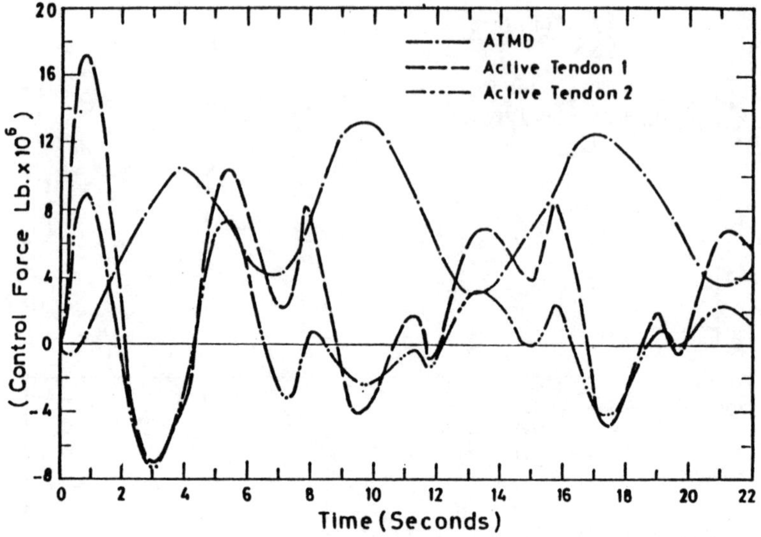

Figure 4.81 - Response of the Control Force

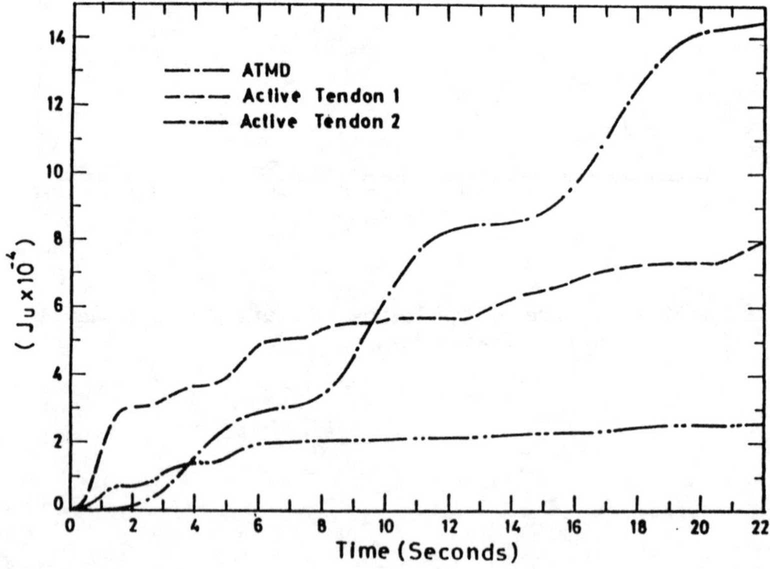

Figure 4.82 - Response of Control Index

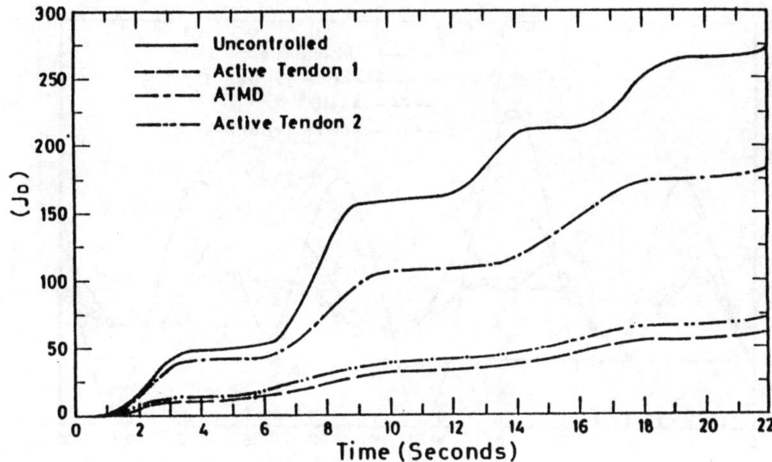

Figure 4.83 - Deflection Response, J_D

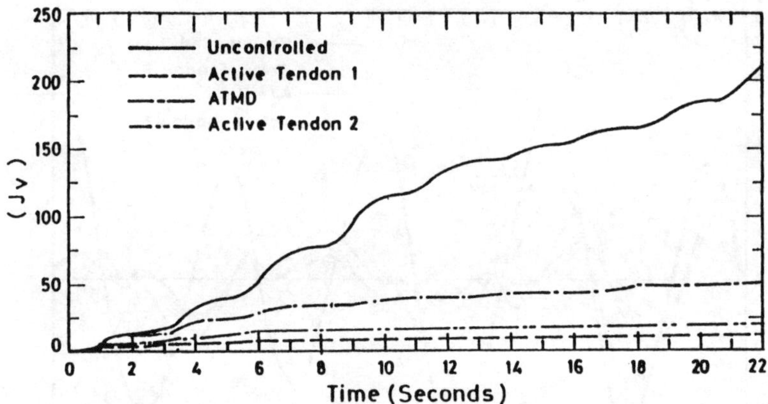

Figure 4.84 - Velocity Response, J_v

The responses of J_D and J_V for "active tendon 1" are shown, respectively, in Figures 4.85 and 4.86.

 H.H.E. *Leipholz and M. Abdel-Rohman*

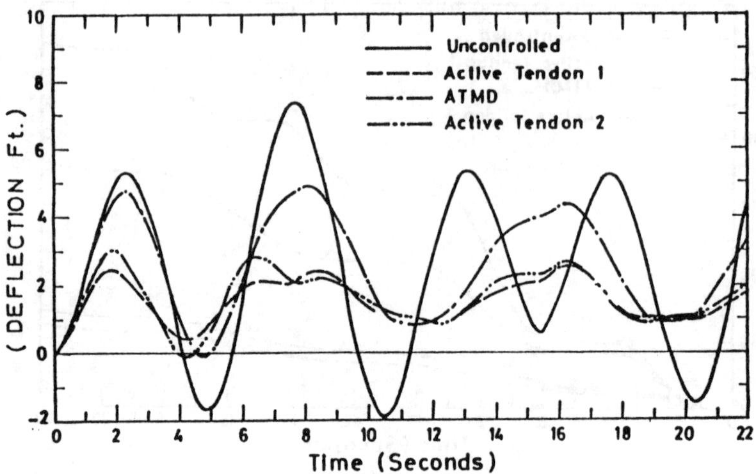

Figure 4.85 - Deflection Response

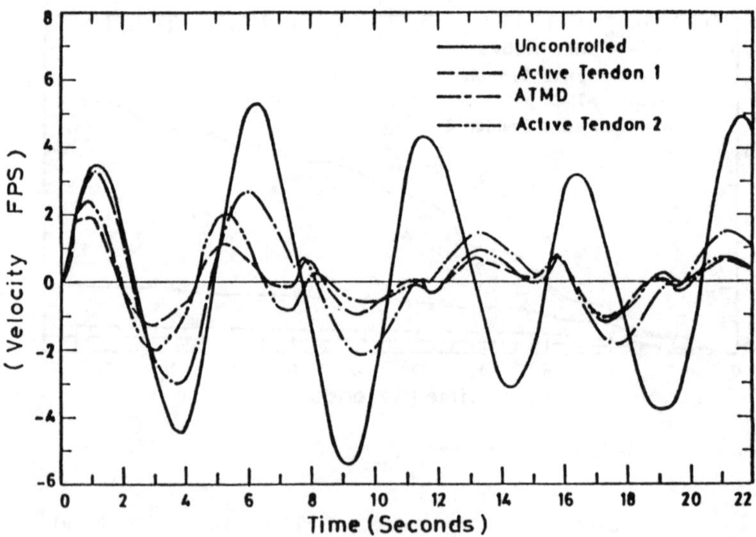

Figure 4.86 - Deflection Response at the Top

Another design denoted by "active tendon 2" has also been studied. This design makes the building eigenvalues to be $(-0.19 \pm J1.5056)$. The gain matrix in this case is

$$\underline{K} = [-6.979 \times 10^2 \quad 3.614 \times 10^6] \ . \tag{4.252}$$

Comparison between "active tendon 1" and "active tendon 2" are given in Figures 4.79 to 4.86. It is obvious that the control "active tendon 1" is better than "active tendon 2", but on account of the control energy and the actuator's motion. However, the feasibility of using both systems is obvious because the response of the ram is practical and does not exceed 10 in/sec. Also, the magnitudes of the control forces are within practical limits. For "active tendon 1" the maximum value is about 16,000 kips, and for "active tendon 2" it is about 8,000 kips.

4.9.3 *Feasibility of Active Tuned Mass Damper Control*

In this case, the actuator movement is represented by the relative sway between the damper and the top of the building. One, thus, writes

$$y(t) = W_T(t) - \phi(L)A(t) \ , \tag{4.253}$$

in which $y(t)$ is the ram displacement, $\phi(L)$ is the mode shape at the top of the building, $A(t)$ is the generalized coordinate of the first mode, and $W_T(t)$ is the damper displacement. The equations of motion of the building and TMD are, again,

$$\ddot{A} + 2\zeta_1\omega_1\dot{A} + \omega_1^2 A = \frac{\phi_1(L)}{mL}(C_2\dot{Z} + K_2 Z) - \frac{\phi_1(L)}{mL}U(t) + \frac{F_1}{mL} \ , \tag{4.254}$$

$$\ddot{Z} + 2\zeta_2\omega_2\dot{Z} + \omega_2^2 Z = -\ddot{A}\phi_1(L) + \frac{U(t)}{M_2} \ , \tag{4.255}$$

in which F_1 is the wind force on the building, $Z(t)$ is the relative displacement between TDM and building.

Therefore, the ram displacement in equation (4.253) can be determined as shown in Figure 4.87 from

$$y(t) = z(t) \ .$$
(4.256)

The rate of flow per unit piston area is thus

$$\frac{Q(t)}{A_p} = \dot{y}(t) = \dot{z}(t) \ .$$
(4.257)

As an index for the amount of fluid consumption for the actuator moment, equation (4.245) becomes

$$J_1 = \int_o^T \left(\frac{Q(t)}{A_p}\right)^2 dt = \int_o^T \dot{y}^2 dt \ .$$
(4.258)

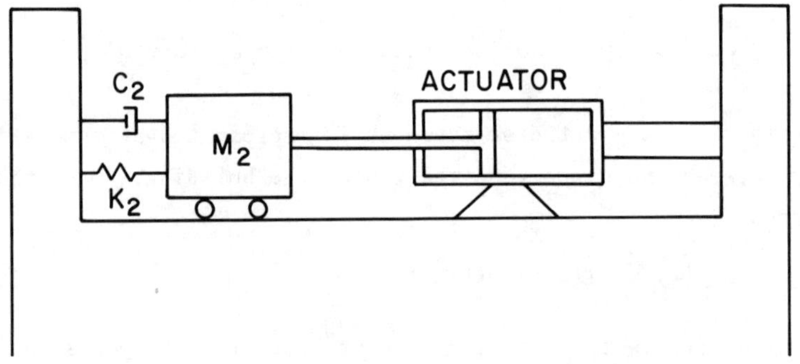

Figure 4.87 - Active Tuned Mass Damper Mechanism

The same building parameters of Section 4.9.2 were taken here. The tuned mass damper parameters are those which provide optimal design, i.e., $M_2 = 0.02$ mL, $\omega_2 = 1.375$ rps, $\zeta_2 = 0.07$. The eigenvalues of the passive controlled building are $(-0.068 \pm J1.392)$ and $(-0.035 \pm J1.06)$. In order to let the controlled building have the eigenvalues $(-0.35 \pm J1.392)$ and $(-0.35 \pm J1.106)$, the following gain matrix is used:

$$\underline{K} = 10^6 [2.769 \quad -1.69 \quad 0.0505 \quad -0.353] \ .$$
(4.259)

For the same wind forces of Figures 4.77 and 4.78, the
rate of flow of fluid in the actuator is given in Figure 4.79 as compared with active tendon control. In Figure 4.80, the flow index
of equation (4.258) is also plotted as compared with active tendon
control. Figures 4.81 and 4.82 show the control force response
and the energy index as compared with active tendon control.
Figures 4.81 to 4.86 give a comparison of the building response
using active TMD with that using active tendon control.

It is obvious, as concluded previously, that active TMD
is not feasible and not effective for forced vibration control.
It consumes more energy and power but does not provide acceptable
control. This is attributed to the lack of stiffness. Perhaps, when
an ATMD were coupled with passive tendon, it may provide acceptable
results.

The maximum ram velocity, as in Figure 4.79, is 24 in/sec,
the maximum control force is 14,000 kips as shown in Figure 4.81.
These results prove again the uneffectiveness of ATMD for forced
vibration control [4.30].

4.9.4 *Feasibility of Active Control by Appendages*

One main disadvantage in using appendages for tall buildings control is that it depends on the mean wind direction. It is not
that easy to make the appendage direction always perpendicular to
the mean wind direction. In this section, the feasibility of
using appendage for forced vibration control is studied, using
the same wind model considered in Section 4.9.2.

The appendage is a steel plate of dimension 100×40×(2/12)ft.
It is hinged along the building width (100 ft) and can be raised
up or down using compression members and hydraulic actuators. The
process of operation of this mechanism is shown in Figure 4.88. In
this process, the appendage may be fully folded, fully deployed, or in
an in between state. The purpose of this study is to investigate the
response of the actuator force which causes the appendage movement.

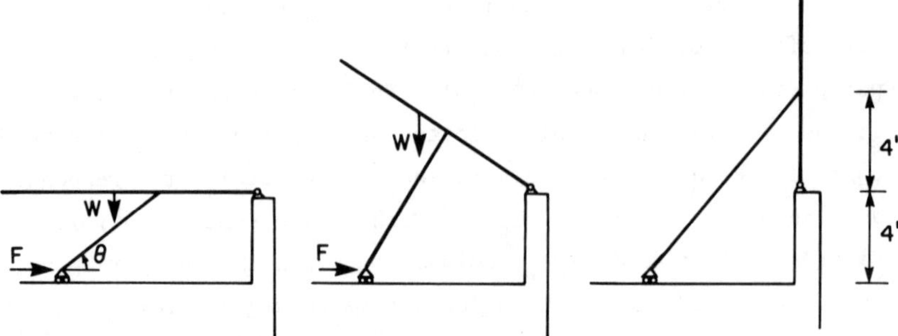

Figure 4.88 - Appendage Operating Limits

The same building parameters used in the previous section are used here where ω_1 = 1.228 rps, m = 13.42 × 10^3 lb. mass/ft, L = 1000 ft, B = 100 ft. The appendage parameters are A_p = 0.04A_s, C_D = 1.3, Q_{11} = Q_{22} = 0.2D5, and η = 0.05D-4.

Using these parameters, the Riccati matrix for t_f = 25 seconds is

$$\underline{P} = \begin{bmatrix} 89339.63 & 9761.22 \\ 9761.22 & 84995.84 \end{bmatrix} . \qquad (4.260)$$

In order to determine the force which operates the actuator, consider the general appendage position of Figure 4.89.

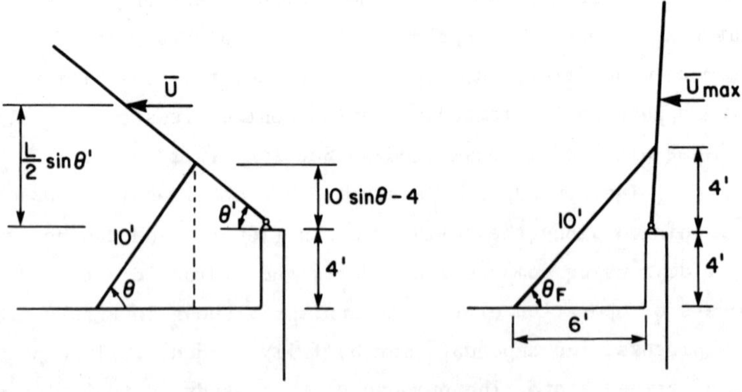

Figure 4.89 - Appendage Mechanism Dimensions

From Figure 4.89 we have the following relationships:

$$\sin\theta' = \frac{10\sin\theta-4}{4} = 2.5\sin\theta-1 \ , \tag{4.261}$$

$$\overline{U}(t) = \sin\theta'\overline{U}_{max} \ . \tag{4.262}$$

From Figure 4.90, one may obtain the following relation: taking moment about A, gives

$$F'\cos\theta(10\cos\theta-4)+F'\sin\theta(4\cos\theta')-w\ \frac{L}{2}\ \cos\theta'-\overline{U}\ \frac{L}{2}\ \sin\theta' = J_A\alpha \ , \tag{4.263}$$

where α is the angular acceleration, J_A is the mass moment of inertia about A, and $\overline{U}(t)$ is the wind force on the appendage.

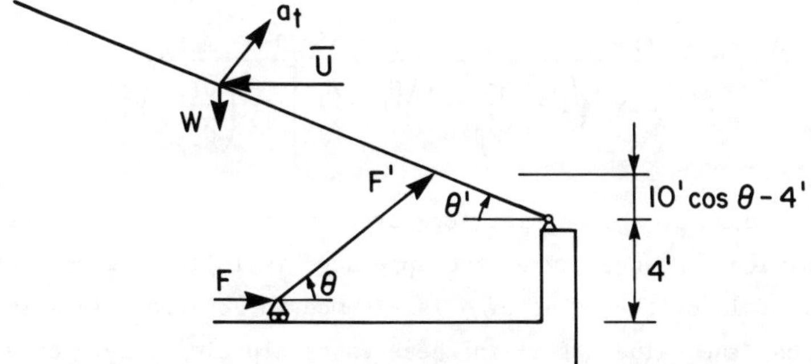

Figure 4.90 - Applied Forces on Appendage Mechanism

The applied force F' can now be determined as

$$F'(10\cos^2\theta-4\cos\theta+4\sin\theta\cos\theta') = J_A\alpha + \frac{wL}{2}\ \cos\theta' + \frac{\overline{U}L}{2}\ \sin\theta' \ .$$

Since

$$F'(\cos\theta[4\sin\theta']+4\sin\theta\cos\theta') = 4F'\sin(\theta+\theta') \ ,$$

one has

$$F' = \frac{J_A\alpha+w\ \frac{L}{2}\ \cos\theta'+\overline{U}\ \frac{L}{2}\ \sin\theta'}{4\sin(\theta+\theta')} \ , \tag{4.264}$$

in which w is the appendage weight, $J_A = mL^2/3$, m is the appendage mass, $\overline{U}$ is the control force generated by the appendage.

In order to determine the angular acceleration, one follows the following approach:

$$\alpha = \ddot{\theta} \, , \qquad (4.265)$$

$$\theta' = \sin^{-1} \frac{U(t)}{U_{max}} \, , \qquad (4.266)$$

$$\dot{\theta}' = \frac{d}{dt} \sin^{-1}\left(\frac{\overline{U}(t)}{\overline{U}_{max}}\right) = \frac{1}{\sqrt{1 - \left(\frac{\overline{U}(t)}{\overline{U}_{max}}\right)^2}} \frac{d\overline{U}(t)}{\overline{U}_{max}\,dt} \, , \qquad (4.267)$$

$$\ddot{\theta}' = \frac{d}{dt} \dot{\theta}' = \frac{\ddot{\overline{U}}(t)}{\overline{U}_{max}} \frac{1}{\sqrt{1 - \frac{\overline{U}^2}{\overline{U}^2_{max}}}} + \left(\frac{\dot{\overline{U}}(t)}{\overline{U}_{max}}\right)^2 \frac{\frac{\overline{U}(t)}{\overline{U}_{max}}}{\left[1 - \left(\frac{\overline{U}(t)}{\overline{U}_{max}}\right)^2\right]^{3/2}} \, . \qquad (4.268)$$

This should be the angular acceleration in the general operating condition. However, when the appendage is fully developed or fully folded, the value of α is assumed to be zero. This is so because the values of θ' in these cases are either zero or $90°$.

The actuator force F can be determined from the equilibrium in the horizontal direction, i.e.,

$$F = F'\cos\theta \, . \qquad (4.269)$$

This value of F' represents the total force including the static force which keeps the appendage in the horizontal position. This static force is evaluated as

$$F'_s = \frac{WL}{8\sin\theta_0} = \frac{16.33D4\times2\times40}{8\times0.4} = 408.325 \times 10^4 \text{ lb} \, . \qquad (4.270)$$

The actuator static force is thus,

$$F_{sh} = F'_s \cos\theta_s = F'_s \times 0.9165 = 374.209 \times 10^4 \ \text{lb} \ . \tag{4.271}$$

The force applied by the actuator during the appendage motion, which shall be denoted by F_2, is

$$F_2 = F' \cos\theta - 374.208 \times 10^4 \ . \tag{4.272}$$

For the same wind forces considered in Section 4.6 and shown again in Figure 4.91, the response of the actuator force F_2 is given in Figure 4.92. The maximum value of this force is about 2.8×10^5 kips, which is much higher than active tendon and active TMD control. The comparison between using appendage and not using it for deflection control is given in Figure 4.93 and for velocity in Figure 4.94. The energy response is indicated by the deployment factor $U(t)$ in Figure 4.95. From the deflection response and velocity response of Figures 4.96 and 4.97, one may conclude that active appendages are not very useful for tall buildings' control. The magnitudes of the actuator force was higher than expected. However, these control forces are applied for a very small period. This indicates that the consumed energy is much less than for the other control mechanisms, but the feasibility is still questionable as the reduction in the response depends on the flexibility of the building.

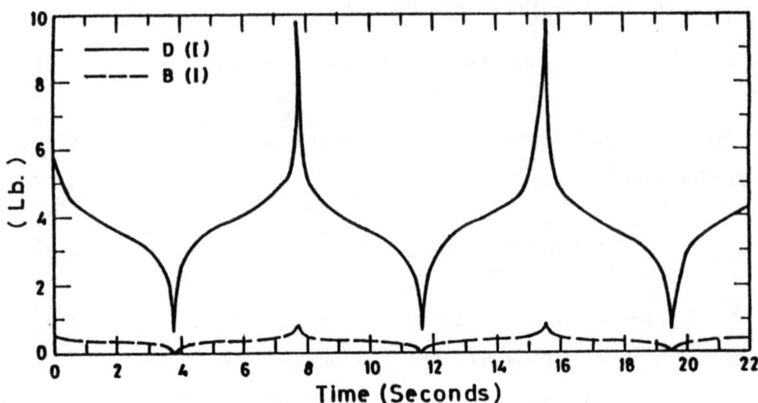

Figure 4.91 - Generalized Wind Forces

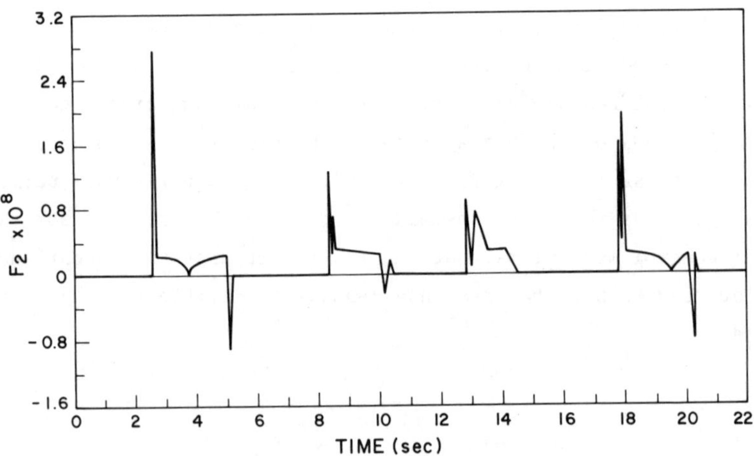

Figure 4.92 - Actuator Force Response

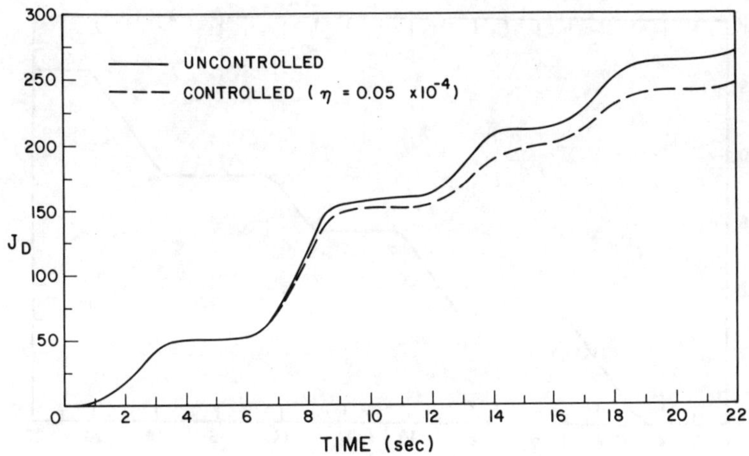

Figure 4.93 - Deflection Response, J_D

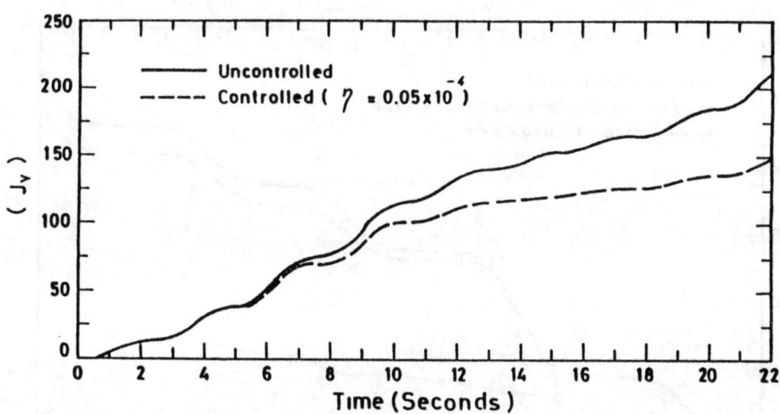

Figure 4.94 - Velocity Response, J_v

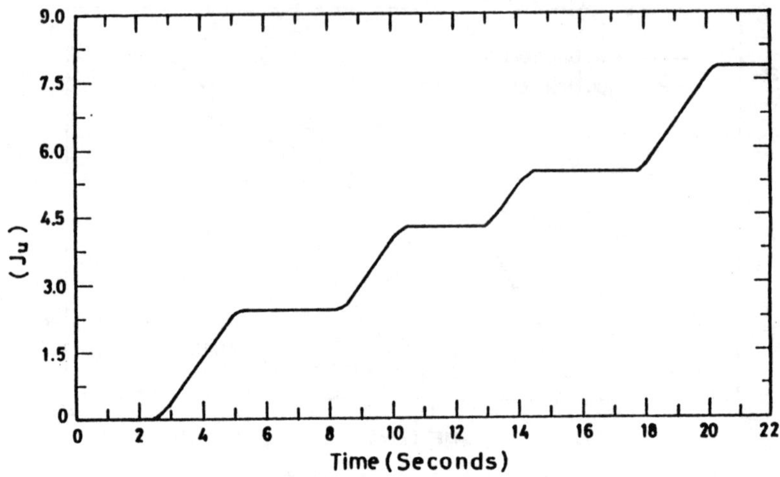

Figure 4.95 - Control Response, J_u

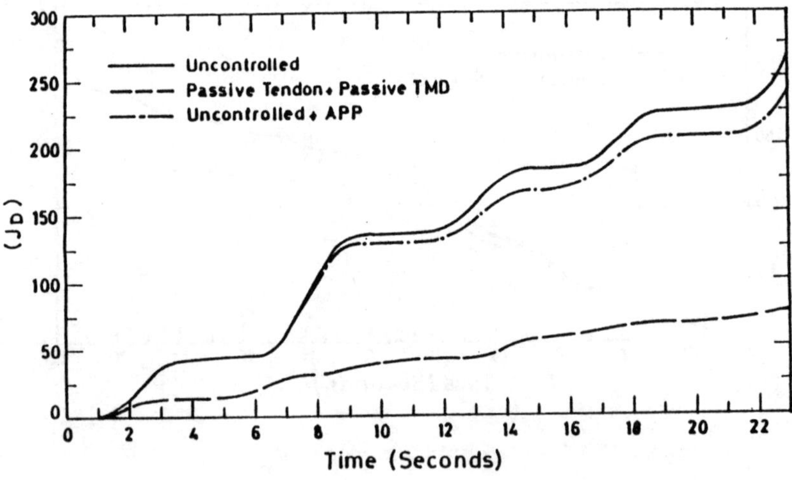

Figure 4.96 - Building Deflection Response, J_D

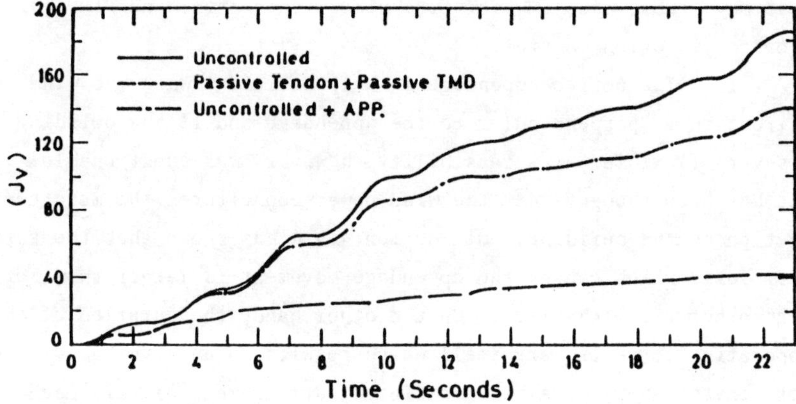

Figure 4.97 - Building Velocity Response, J_v

4.10   FINDING A PRACTICAL CONTROL MECHANISM FOR TALL BUILDINGS

4.10.1   *Introduction*

The work presented in the previous sections has shown the advantages and disadvantages of the control mechanisms which can be
used for tall buildings' control. The active tendon control was
very efficient in providing active damping and active stiffness
to the building. It has shown the possibility for a practical application. The active tendon control, however, was sensitive to time
delay effect. If the time delay is considered in the design, then,
the consumed control energy in this case can almost be the same as if
time delay is neglected.

The active tuned mass damper (ATMD) was only effective
for free vibration control because it introduces active damping
to the building. However, ATMD was not effective for forced vibration control because of the lack in active stiffness. The ATMD
has shown nonsensitivity for time delay effect, but it also requires
huge amounts of fluids which are needed for the actuator's motion.

One may, thus, use ATMD for free vibration control and PTMD for forced vibration control.

The active appendage is only efficient when the wind direction is perpendicular to the appendage and if the building is very flexible. Its feasibility, however, was questionable. It has been thought that the wind power constitutes the main control action on the building, but Section 4.9.4 has shown that the actuator force which causes the appendage movement is larger than for the other two mechanisms. On the other hand, the duration of the operating force is very small which results in an acceptable amount of control energy. Maybe, a different arrangement of this mechanism would make it more feasible.

Using these results, we shall investigate the possibility of combining any of the previous control mechanisms in order to obtain an efficient and practical control mechanism. It is hoped that the disadvantages of some control mechanisms shall be compensated by the advantages of the others. The combinations considered in this study are as follows:

    (1) Combined passive tendon and passive TMD

    (2) Combined active tendon and passive TMD

    (3) Combined passive tendon and active TMD

    (4) Combined active tendon and active TMD

    (5) Combined appendage with (1)

    (6) Combined appendage with (2)

    (7) Combined appendage with (3)

    (8) Combined appendage with (4)

4.10.2 *The Uncontrolled Building Response*

The tall building considered in the investigations is the same building treated in Section 4.9. The building parameters are $m = 13.42 \times 10^3$ lb.s^2/ft^2, $D = 100$ ft, $L = 1000$ ft, $\omega_1 = 1.228$ rps, and $\zeta_1 = 0.01$. The building is subjected to wind forces as in Section 4.9. The wind data are again, $C_D = 1.3$, $\rho = 0.0763$ pcf,

$\Delta\omega$ = 40/25, $\overline{U}$ = 66 fps, K = 0.04 and Φ = 4000 ft. In order to
judge the building response, the following mean square quantities
are used, respectively, for deflection response and velocity
response

$$J_D = \int_o^T A^2(t)dt \; ,$$
(4.273)

$$J_V = \int_o^T \dot{A}^2(t)dt \; .$$
(4.274)

The uncontrolled building responses are given in Figures 4.96 and
4.97. These figures are used as ceilings when looking for an
effective control mechanism.

4.10.3 *Combined Passive Tendon and PTMD*

Passive tendon mechanism provides passive stiffness to the building.
Passive TMD introduces passive damping and negative stiffness to
the building. By combining the two mechanisms together, one is
able to obtain passive stiffness and passive damping. It has been
pointed out in the previous sections that the passive stiffness
introduced by the tendons is 0.7933 $(rps)^2$, and the damping ratio
becomes 10 percent by virtue of PTMD. Figures 4.96 and 4.97 show
the building response when using this mechanism. It is obvious
that this mechanism does not require any control energy and it
provides a reduction in the deflection of 70 percent, and in the
velocity of 75 percent.

The TMD displacement and velocity responses are measured,
respectively, by

$$J_{TMDF} = \int_o^T WT^2(t)dt \; ,$$
(4.275)

$$J_{TMDV} = \int_o^T \dot{W}T^2(t)dt \; .$$
(4.276)

Figures 4.100 and 4.101 show PTMD responses using equations (4.275) and (4.276) for T = 23 seconds. The PTMD response indices were 1034 and 546.4, respectively, for deflection and velocity as shown in Table 4.9. The eigenvalues of this case are (-0.0417 ± J1.5985) and (-0.065 ± J1.1412).

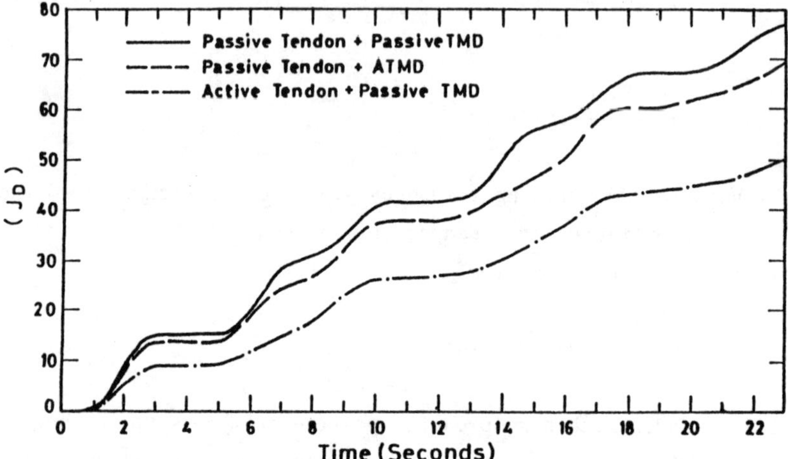

Figure 4.98 - Building Deflection Response, J_D

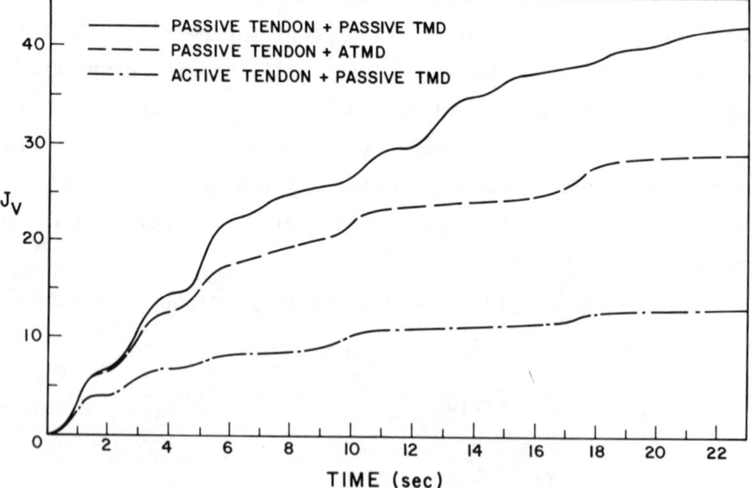

Figure 4.99 - Building Velocity Response, J_v

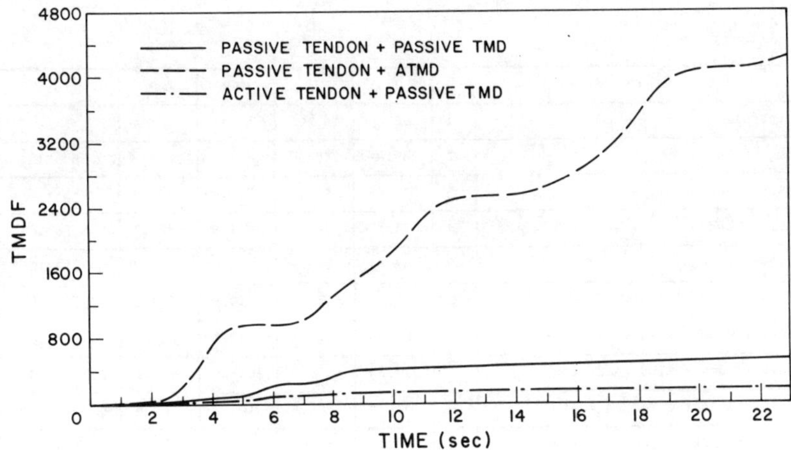

Figure 4.100 - TMD Deflection Response, J_{TMDF}

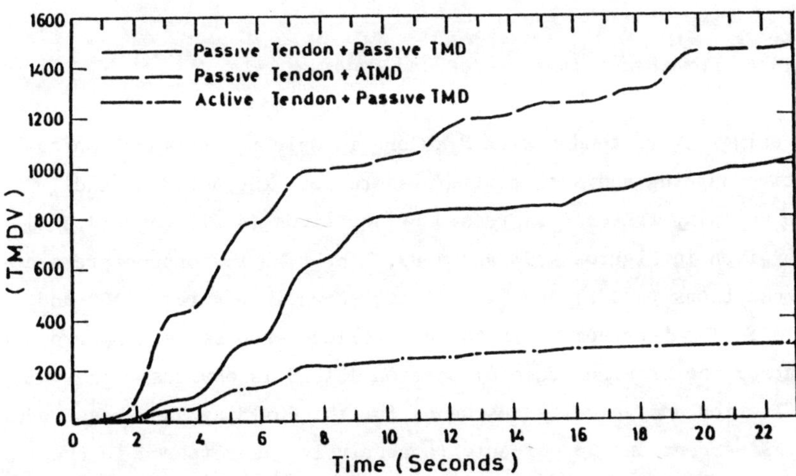

Figure 4.101 - TMD Velocity Response, J_{TMDV}

Table 4.9 - Performance of Practical Control Mechanisms

Case		J_D	J_V	TMDF	TMD	TU
Uncontrolled	1	365.0	185.6	....	....	....
Uncontrolled + App.	2	239.99	140.52	....	....	0.56×10^{17}
Passive Tendon + Passive TMD	3	77.45	41.81	546.4	1034.0	....
Passive Tendon + Passive TMD + App.	4	74.87	34.06	451.9	840.7	0.8398×10^{14}
Passive Tendon + ATMD	5	69.81	28.79	4258.0	1477.0	0.5287×10^{15}
Active Tendon + Passive TMD	6	50.7	13.22	179.6	295.5	0.6932×10^{5}
Active Tendon + Passive TMD + App.	7	51.01	12.11	158.3	255.3	0.7337×10^{15}
Active Tendon + ATMD (Addition)	8	49.34	11.58	3039.0	609.3	0.1064×10^{16}
Similar to Active Tendon + ATMD	9	61.44	16.28	105.2	169.6	0.2118×10^{15}
Similar to Active Tendon + ATMD + App.	10	61.47	15.01	99.1	155.1	0.2379×10^{15}
Similar to Passive Tendon + ATMD	11	71.23	23.57	600.2	678.1	0.324×10^{14}
Similar to Passive Tendon + ATMD'	12	72.42	27.12	299.2	431.9	0.1653×10^{14}

4.10.4   *Combined Active Tendon and Passive TMD*

By using active tendon with PTMD one is able to introduce more
active damping and active stiffness to both the building and TMD.
The building response expressed by equations (4.273) and (4.274)
are given in Figures 4.98 and 4.99. The PTMD response expressed
by equations (4.275) and (4.276) are shown in Figures 4.100 and
4.101. The improvement in the controlled response as compared to
that of the previous case of Section 4.10.3 is obvious. The
deflection and velocity responses for the building have been reduced
by 35 percent and 68 percent, respectively, as compared to those
in the case of the mechanism of Section 4.10.3. Also, the deflection and velocity responses of TMD have been reduced by 72 and
67 percent, respectively, of those in case of passive tendon and
PTMD. The energy required to gain those responses can be represented by

$$J_u = \int_o^T U^2(t)dt \; , \tag{4.277}$$

in which U(t) is the control force used for active tendon.

The response of the energy consumption is shown in Figure 4.102. This response enables the building and TMD to have the eigenvalues (-0.2099 ± J1.6709) and (-0.0811 ± J1.1501).

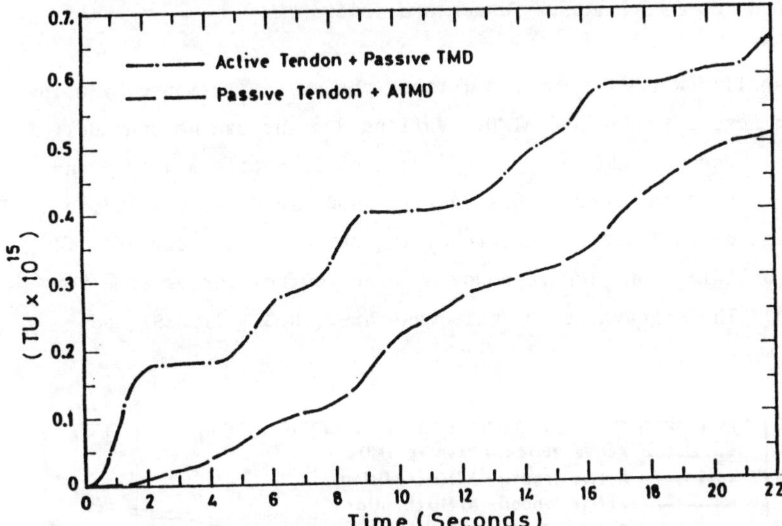

Figure 4.102 - Energy Response, J_u

4.10.5 Combined Passive Tendon and Active TMD

The third mechanism is to use passive tendon with ATMD. This may reduce the energy consumed in the previous mechanism. The ATMD was chosen to be the one of Section 4.5 with gain $\underline{K} = 10^6 [2.769 \; -1.69 \; 0.050 \; -0.352]$. However, in the presence of passive tendon, the eigenvalues become (0.118 ± J1.674) and (-0.5677 ± J0.843). At T = 23 seconds, the indices for the deflection and velocity response for the building were, respectively, 69.5 and 28.79. The reduction in the responses as compared to those in the case of using

passive tendon and PTMD are, respectively 10 and 30 percent. The
TMD response has, however, increased to be 4258 for the deflection
index and 1477 for the velocity index. The reduction in energy
was only 23 percent. It is obvious that this mechanism does not
improve much the previous response of case 4.10.4. The response
indices are given in Table 4.9 and Figures 4.98 and 4.102.

4.10.6 *Combined Active Tendon and Active TMD*

We shall now try to improve further the system response by using
both active tendon and ATMD. Various designs can be considered
here: One may add the previous designs in Sections 4.10.4 and
4.10.5 to each other. This case is shown in Figures 4.103 to 4.107
and given in Table 9. The resulting response was very good for
the building, but TMD response and the control forces were very
high. The eigenvalues of this case are $(-0.3 \pm J1.785)$ and
$(-0.57 \pm J0.821)$.

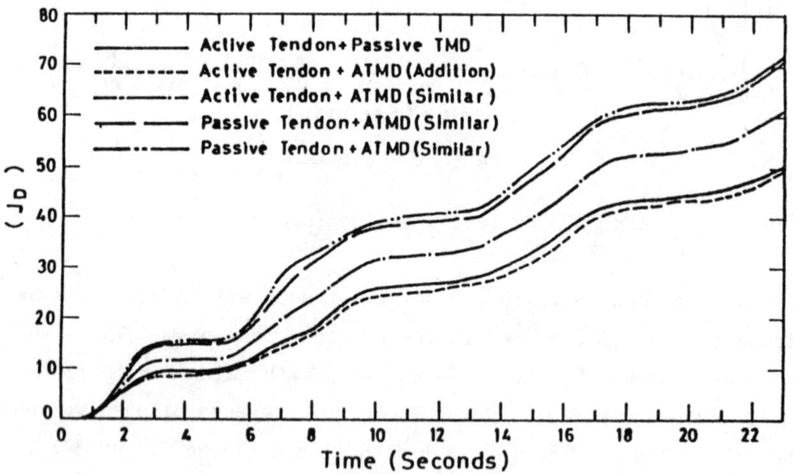

Figure 4.103 - Building Deflection Response, J_D

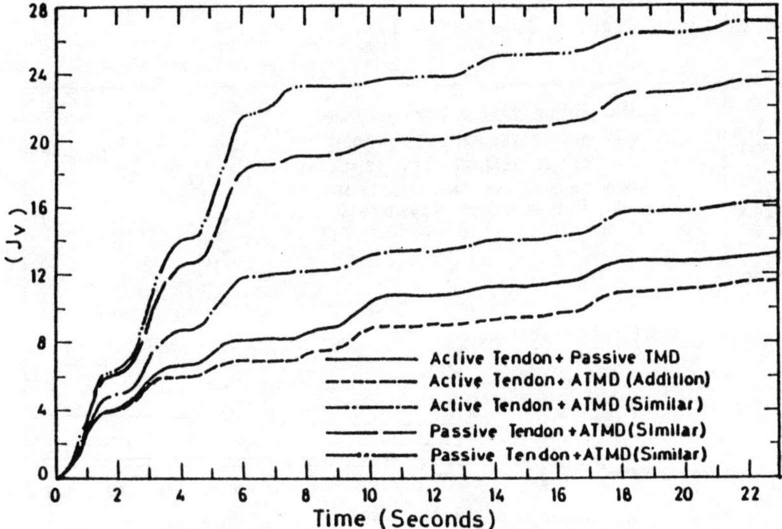

Figure 4.104 - Building Velocity Response, J_v

Other designs have been made considering a multi-control
problem. Cases 9 and 11 in Table 4.9 show the obtained responses
due to these designs. Case 9 provides a reduction in the values
of equations (4.273) to (4.276) as compared to those for passive
tendons and PTMD by 20, 61, 80 and 82 percent, respectively. The
consumed energy was only 30 percent of case 6. Case 11 consumes
less energy but the controlled response is higher than in case 9.
It is thus concluded that a proper design for the control forces
in this case may provide favourable results. Cases 9 and 11 were
designed to provide eigenvalues of $(-0.2 \pm J1.6)$ and $(-0.2 \pm J1.2)$,
but they have different gain matrices because the problem here is
of the multicontrol type, which does not provide a unique control gain.

Another design has been made by considering a semi-active
TMD. The building response was almost the same as in case 11 but
the TMD responses have decreased significantly. Also, the control
energy was about 50 percent of that in case 11. This case is

number 12 in Table 4.9 and denoted by [Passive Tendon + ATMD (similar)] in Figures 4.103 to 4.107.

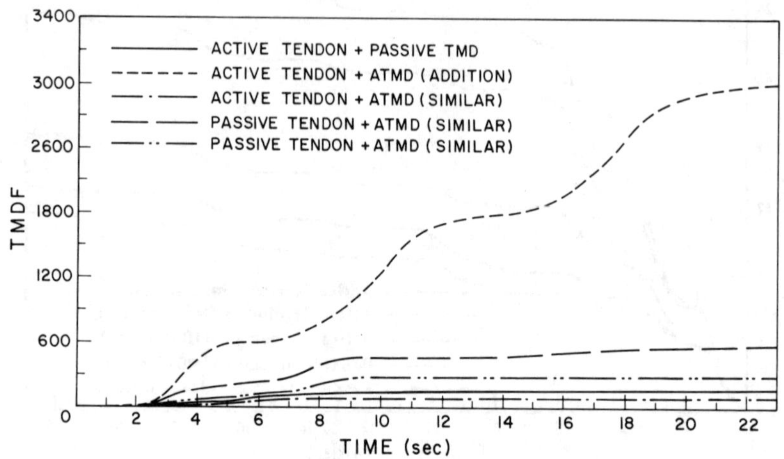

Figure 4.105 - TMD Deflection Response, J_{TMDF}

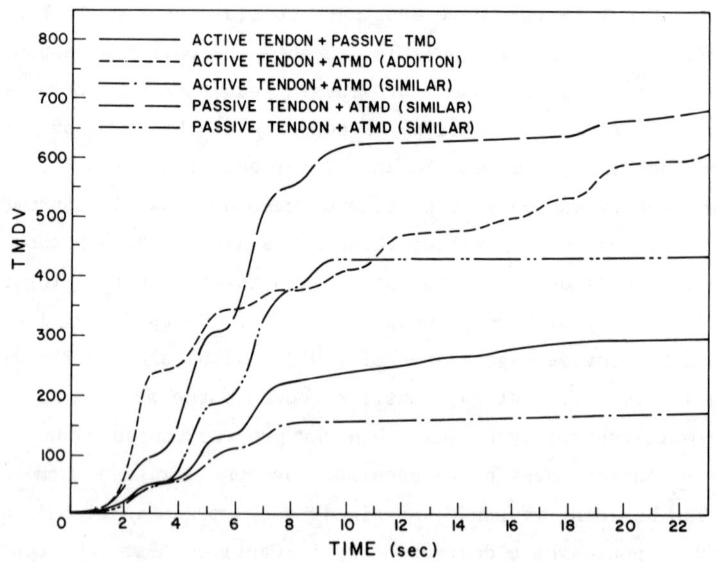

Figure 4.106 - TMD Velocity Response, J_{TMDV}

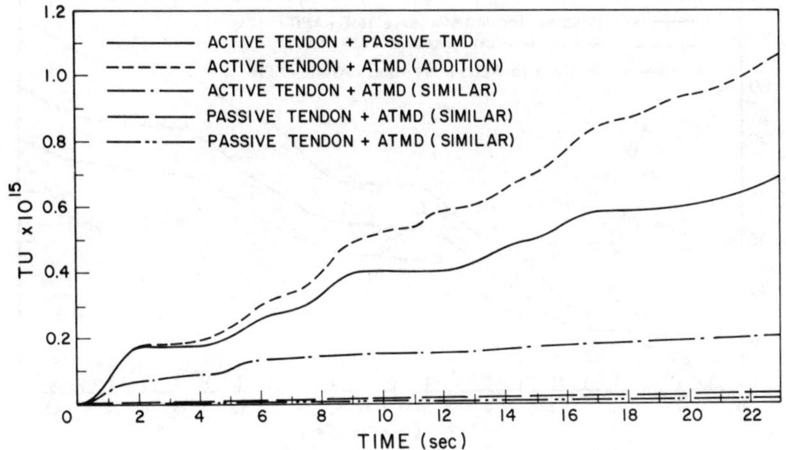

Figure 4.107 - Control Force Response, J_u

4.10.7 *Combined Appendages, Passive Tendon and PTMD*

By adding an appendage to case 3 of Table 4.9, one obtains additional small reductions in the building and TMD responses. The reductions obtained in case 4 as compared to case 3 according to equations (4.273) to (4.276) are, respectively 3, 18, 17 and 18 percent. However, the energy required is 2.6 times the energy used in case 11. Thus, the mechanism is not very beneficial and can be neglected. The responses following from using this mechanism are given in Figures 4.108 to 4.112.

4.10.8 *Combined Appendage, Active Tendon and PTMD*

This mechanism represents case 7 in Table 4.9. It should provide additional improvement as compared to case 6. However, it only assists in decreasing the TMD response by 11 percent for displacement and 13 percent for velocity on the account of increasing the energy by 5 percent as compared to case 6. This case can also be rejected.

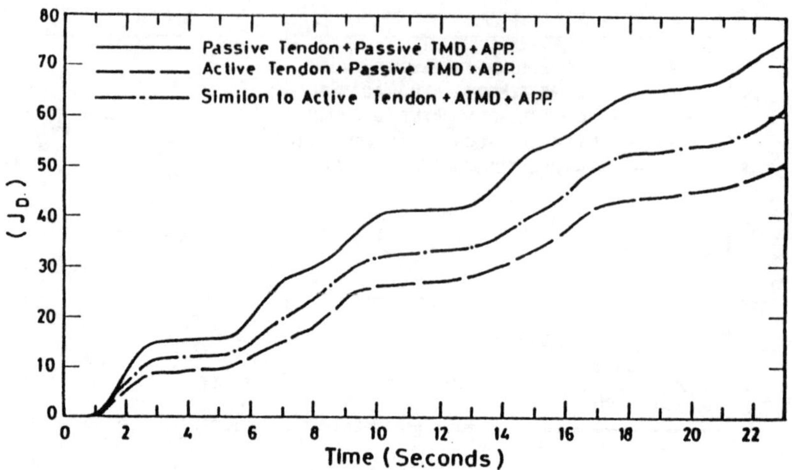

Figure 4.108 - Building Deflection Response, J_D

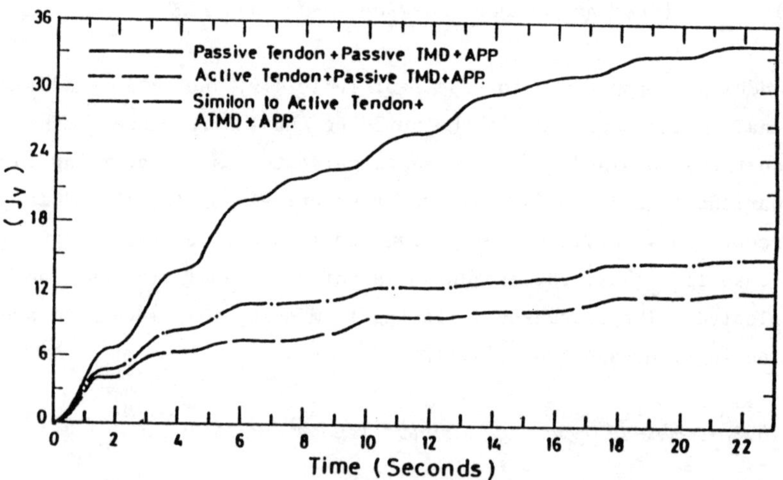

Figure 4.109 - Building Velocity Response, J_v

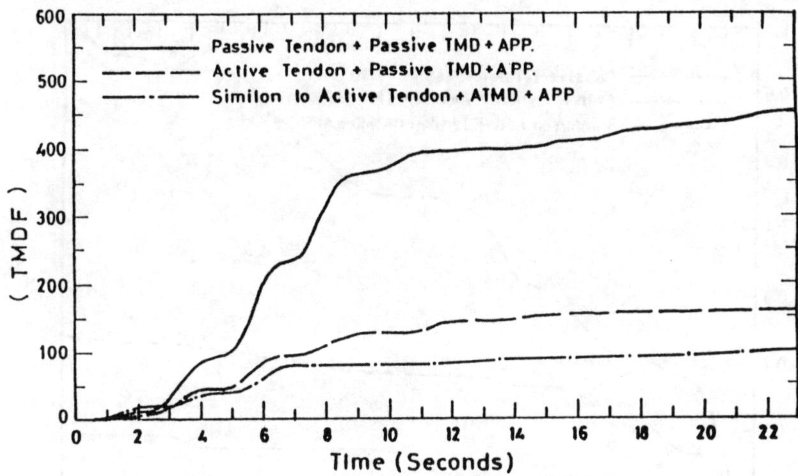

Figure 4.110 - TMD Deflection Response, J_{TMDF}

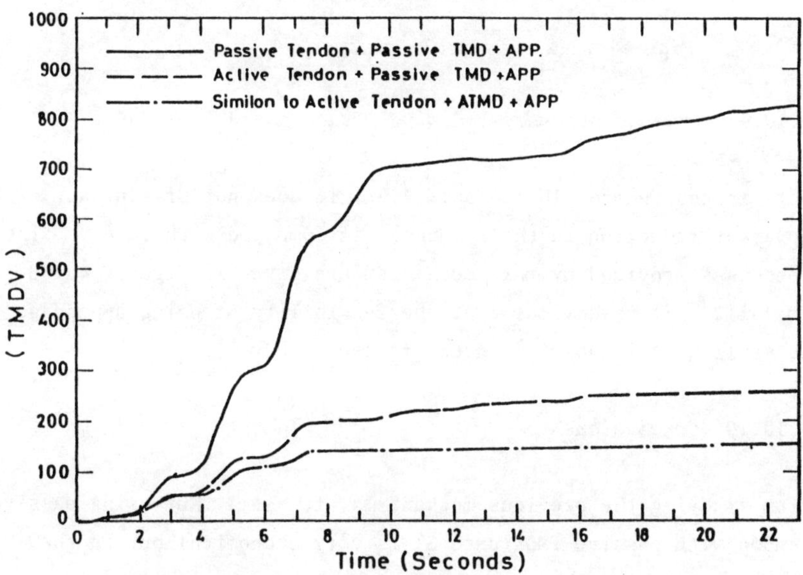

Figure 4.111 - TMD Velocity Response, J_{TMDv}

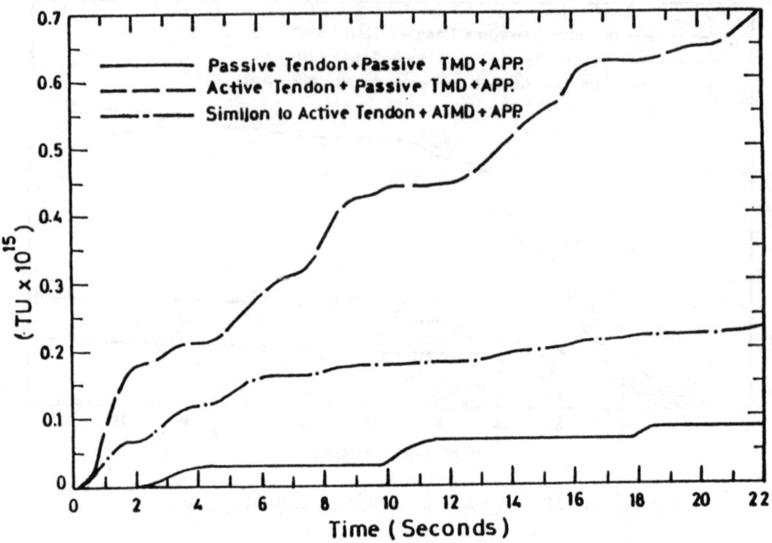

Figure 4.112 - Control Response, J_u

4.10.9 *Combined Appendage, Active Tendon and ATMD*

This is case number 10 in Table 4.9. It does not present any
apparent reduction in the responses if compared with case 9. The
responses provided by this mechanism are given in Figures 4.108
to 4.112. It is obvious that the feasibility of using appendage
is still questionable and needs further study.

4.10.10 *Conclusions*

From studying the previous mechanisms, it seems that using passive
tendon with passive TMD (case 3) is very economical but on the
account of violent TMD response. A better mechanism is using
active tendon and active TMD. If this mechanism is designed properly, it could provide significant reduction in the building and
TMD response but consumes a moderate control energy only. Case 9 of

Table 4.9 supports this conclusion. Another mechanism that can
be used, is case 12, which provides moderate controlled response
but consumes very little energy. As a general conclusion, it can
be stated that by combining a tendon control mechanism with a TMD
control mechanism, one is able to obtain a very acceptable controlled response using minimum control energy.

4.11   REFERENCES

[4.1]   ZUK, W., "Kinetic Structures", *ASCE, Civil Engineering*,
        Vol. 39, No. 12, December, 1968, pp. 62-64.

[4.2]   YAO, J.T.P., "Concept of Structural Control", *ASCE, Journal
        of the Structural Division*, Vol. 98, No. ST7, Proc. Paper
        9048, July, 1972, pp. 1567-1574.

[4.3]   ZUK, W. and CLARK, R.H., *Kinetic Architecture*, Van Nostrand
        Reinhold Co., New York, New York, 1970.

[4.4]   ROORDA, J., "Tendon Control in Tall Structures", *ASCE, Journal
        of the Structural Division*, Vol. 101, No. ST3, Proc. Paper
        1168, March, 1975, pp. 505-521.

[4.5]   YANG, J.N. and GIANNOPOULAS, F., "Active Control of Two-
        Cable Stayed Bridge", *ASCE, Journal of the Engineering
        Mechanics Division*, Vol. 105, No. EM5, October, 1979,
        pp. 795-810.

[4.6]   YANG, J.N. and GIANNOPOULAS, F., "Active Tendon Control of
        Structures", *ASCE, Journal of the Engineering Mechanics
        Division*, Vol. 104, No. EM3, June, 1978, pp. 551-568.

[4.7]   ABDEL-ROHMAN, M. and LEIPHOLZ, H.H.E., "Automatic Active
        Control of Structures", *ASCE, Journal of the Structural
        Division*, Vol. 106, No. ST3, March, 1980, pp. 663-677.

[4.8]   ABDEL-ROHMAN, M. and LEIPHOLZ, H.H.E., "General Approach to
        Active Structural Control", *ASCE, Journal of the Engineering
        Mechanics Division*, Vol. 105, No. EM6, December, 1979,
        pp. 1007-1023.

[4.9]   ABDEL-ROHMAN, M. and LEIPHOLZ, H.H.E., "Structural Control
        by Pole Assignment Method", *ASCE, Journal of the Engineering
        Mechanics Division*, Vol. 104, No. EM5, October, 1978,
        pp. 1159-1175.

[4.10] KLEIN, R.E., CUSANO, C. and SLUKEL, J.V., "Investigation of
        Method to Stabilize Wind Induced Oscillations in Large
        Structures", Presented at the *1972 ASME Winter Annual Meeting*,
        Paper No. 72-WT/AUT-11, New York, November, 1972.

[4.11] CHANG, J.C. and SOONG, T.T., "The Use of Aerodynamic Appendages for Tall Building Control", *Proceedings of the IUTAM Symposium on Structural Control,* held at the University of Waterloo, 1979, Editor: H. Leipholz, North Holland Publishing Co., 1980, pp. 199-210.

[4.12] SOONG, T.T. and SKINNER, G.T., "Experimental Study of Active Structural Control", *ASCE, J. of Engineering Mechanics Division,* Vol. 107, No. EM6, December, 1981, pp. 1057-1068.

[4.13] LUND, R.A., "Active Damping of Large Structures in Winds", *Proceedings of the IUTAM Symposium on Structural Control,* held at the University of Waterloo, 1979, Editor: H. Leipholz, North Holland Publishing Co., 1980, pp. 459-471.

[4.14] PETERSON, N.R., "Design of Large Scale Tuned Mass Dampers", presented at the *1979 ASCE Convention and Exposition,* held in Boston, Massachusetts, April, 1979, reprint 3578.

[4.15] CHANG, J.C. and SOONG, T.T., "Structural Control Using Active Tuned Mass Dampers", *ASCE, Journal of the Engineering Mechanics Division,* Vol. 106, No. EM6, December, 1980, pp. 1091-1098.

[4.16] ABDEL-ROHMAN, M., QUINTANA, V.H. and LEIPHOLZ, H.H.E., "Optimal Control of Civil Engineering Structures", *ASCE, Journal of the Engineering Mechanics Division,* Vol. 106, No. EM1, February, 1980, pp. 57-73.

[4.17] ABDEL-ROHMAN, M. and LEIPHOLZ, H.H.E., "Active Control of Flexible Structures", *ASCE, Journal of the Structural Division,* Vol. 104, No. ST8, August, 1978, pp. 1251-1266.

[4.18] YANG, J.N. and GIANNOPOULAS, F., "Active Control and Stability of Cable-Stayed Bridge", *ASCE, Journal of the Engineering Mechanics Division,* Vol. 105, No. EM4, August, 1979, pp. 677-694.

[4.19] SAE-UNG, S.V. and YAO, J.T.P., "Active Control of Building Structures", *ASCE, Journal of the Engineering Mechanics Division,* Vol. 104, No. EM2, April, 1978, pp. 335-350.

[4.20] ABDEL-ROHMAN, M. and LEIPHOLZ, H.H.E., "Active Control of Tall Buildings", *ASCE, Journal of the Structural Division,* Vol. 109, No. 3, March, 1983, pp. 628-645.

[4.21] LIN, Y.K., "Structural Response under Turbulent Flow Excitation", offprint from "Random Excitation of Structures by Earthquakes and Atmospheric Turbulence", *CISM Courses and Lectures No. 225,* International Center for Mechanical Sciences, Springer-Verlag, Wien, New York, 1977, pp. 239-307.

[4.22] VALCAITIS, R., SHINOZUKA, M. and TAKENO, N., "Response Analysis of Tall Buildings to Wind Loading", *ASCE, J. of the Struc. Division*, Vol. 101, No. ST3, March, 1975, pp. 585-600.

[4.23] KWAKERNAAK, H. and SIVAN, R., *Linear Optimal Control Systems*, Wiley-Interscience, New York, New York, 1972.

[4.24] ABDEL-ROHMAN, M., "Active Control of Large Structures", *ASCE, J. of the Eng. Mech. Division*, Vol. 108, No. EM5, October, 1982, pp. 719-720.

[4.25] BRYSON, A.E. and HO, Y.C., *Applied Optimal Control*, Hemisphere Publishing Co., Washington, D.C., 1975.

[4.26] ABDEL-ROHMAN, M., "Design of Optimal Observers for Structural Control", *IEEE, Division D, J. of Control Theory and Applications*, Vol. 131, No. 4, 1984, pp. 158-163.

[4.27] McNAMARA, R.J., "Tuned Mass Dampers for Buildings", *ASCE, J. of the Struc. Div.*, Vol. 103, No. ST9, September, 1977, pp. 1785-1798.

[4.28] MAHMOODI, P., "Structural Dampers", *ASCE, J. of the Struc. Div.*, Vol. 95, No. ST8, August, 1969, pp. 1661-1672.

[4.29] ABDEL-ROHMAN, M., "Design of Active TMD for Tall Building Control", *Int. J. of Buildings and Environments*, Vol. 19, No. 3, 1984, pp. 191-195.

[4.30] ABDEL-ROHMAN, M., "Effectiveness of Active TMD in Tall Buildings", *Trans. of the Canadian Society of Mechanical Engineers*, Vol. 8, No. 4, 1985.

[4.31] CRANDALL, S.H. and MARK, W.D., *Random Vibration in Mechanical Systems*, Academic Press, New York, 1973.

[4.32] DAVENPORT, A.G., "The Application of Statistical Concepts to Wind Loading of Structures", *Proc. Inst. of Civil Engineers*, Vol. 19, 1961, pp. 449-472.

[4.33] SIMIU, E. and SCANLAN, R.H., *Wind Effects on Structures - An Introduction to Wind Engineering*, John Wiley and Sons Inc., New York, New York, 1978.

[4.34] ABDEL-ROHMAN, M. and LEIPHOLZ, H.H.E., "Stochastic Control of Structures", *ASCE, Journal of the Structural Division*, Vol. 107, No. ST 7, 1981, pp. 1313-1325.

[4.35] MAYBECK, P.S., *Stochastic Models, Estimation and Control*, Vol. 1, Academic Press, New York, 1979.

[4.36] HAMMARSTROM, I.G. and GROS, K.S., "Adaptation of Optimal Control Theory to Systems with Time Delays", *International Journal of Control*, Vol. 32, No. 2, 1980, pp. 320-357.

[4.37] MUTHARASAN, R. and LUUS, R., "Analysis of Time-Delay Systems by Series Approximations", *American Institute of Chemical Engineers Journal*, Vol. 21, 1975, pp. 567-572.

[4.38]   OGUNNAIKE, B.A., "A New Approach to Observer Design for
         Time-Delay Systems", *International Journal of Control*,
         Vol. 33, No. 3, 1981, pp. 519-542.

[4.39]   OH, S.E. and LUUS, R., "Optimal Feedback Control of Time-
         Delay Systems", *American Institute of Chemical Engineers
         Journal*, Vol. 22, No. 1, January, 1976, pp. 140-147.

[4.40]   SOLIMAN, M.A. and RAY, W.H., "Optimal Feedback Control for
         Linear-Quadratic Systems having Time Delays", *International
         Journal of Control*, Vol. 15, 1972, pp. 609-615.

[4.41]   ABDEL-ROHMAN, M., "Optimal Control of Tall Buildings by
         Appendages", *ASCE, Journal of the Structural Division*,
         Vol. 110, No. ST5, May, 1984, pp. 937-947.

[4.42]   ABDEL-ROHMAN, M., "Stochastic Control of Tall Buildings
         Against Wind", *Journal of Wind Engineering and Industrial
         Aerodynamics*, Vol. 17, 1984, pp. 251-264.

Chapter V

A Review of Methods in the Control of Continuous Systems

5.1 INTRODUCTION

As pointed out in Chapter 2, control of continuous systems represents a specific situation which requires a specific approach.
Mainly, if the differential operator of the control problem has
become nonselfadjoint, one is faced with the following difficulty:
the possible occurrence of spillover. Spillover is the incidence
of instability by virtue of higher, unstable modes, which have not
been considered in the design of the control.

In view of the danger of spillover, a modal approach is
not advisable, and new methods may have to be developed. How that
can be done, will be shown to a certain extent in this chapter.
Essentially, three possibilities will be discussed: In the first
case, the operators characterizing the problem will be considered
directly, and one will try to come to conclusions on the behaviour
of the control system, taking the properties of the operators into
account. In the second case, a closed form solution to the problem
will be sought, thus enabling one to avoid the modal approach. In
the third case, a Liapunov like approach will be presented for continuous systems with nonselfadjoint operators.

5.2   CONSIDERATIONS OF EXISTENCE AND PERFORMANCE OF A CONTROL

Consider the control of a plate subjected to loading transversal to
the plate's surface. Let first the motion of the uncontrolled plate
be investigated. Assuming that the plate is a thin one, the equation of motion for this plate reads

$$\rho h \ddot{w} + N \nabla^4 w = q(x,y,t) \; .$$
(5.1)

In (5.1), ρ is the density per unit volume, h the plate's constant
thickness, N the flexural rigidity, $w(x,y,t)$ the transversal deflection of the plate's middle surface, q the distributed load, x and
y the spatial coordinates, and t the time. Moreover,

$$\ddot{w} = \frac{\partial^2 w}{\partial t^2} \; , \qquad \nabla^4 w = \frac{\partial^4 w}{\partial x^4} + 2 \frac{\partial^4 w}{\partial x^2 \partial y^2} + \frac{\partial^4 w}{\partial y^4} \; .$$
(5.2)

In addition to (5.1), boundary and initial conditions must be
specified as to define the problem of the plate's motion definitely.
In order to simplify the situation, let the initial conditions be
homogeneous ones. The boundary conditions may be indicated by the
expression

$$U[w]_B = 0 \; ,$$
(5.3)

where U is a "vector" of operators on w with respect to x and y
(not t), and where the resulting expression $U[w]$ is to be taken at
the boundary B (contour) of the plate.

It is known that the operator in (5.1), i.e.,

$$D = \rho h \frac{\partial^2}{\partial t^2} + N \nabla^4 \; ,$$
(5.4)

has an inverse

$$D^{-1} = \frac{1}{\rho h} \int_o^t \int_S \int \left[\sum_i \sum_j \phi_{ij}(x,y) \phi_{ij}(\xi,\eta) \sin \frac{\omega_{ij}(t-\tau)}{\omega_{ij}} \right] \ldots d\rho d\eta d\tau \; ,$$
(5.5)

in which the ϕ_{ij} are the eigenfunctions and the ω_{ij} the eigenvalues
of the auxiliary problem

$$N\nabla^4\phi = \rho h\omega_{ij}^2\phi \ , \quad U[\phi]_B = 0 \ . \tag{5.6}$$

A detailed derivation of (5.5) will be presented later.

By means of (5.3), (5.4) and (5.5), the plate's motion
can be described concisely in the form

$$Dw = q \ , \quad U[w]_B = 0 \ , \ \text{plus homogeneous initial} \tag{5.7}$$
$$\text{conditions}$$

$$w = D^{-1}q \ , \tag{5.8}$$

respectively. The advantage of (5.8) is that the operator D^{-1}
implies already fulfillment of boundary and initial conditions,
as will be seen later.

Let now the idea of a control be brought into play.
Then, (5.1) changes into

$$\rho h\ddot{w} + N\nabla^4 w + u = q \ , \tag{5.9}$$

where

$$u = Aw \tag{5.10}$$

is the control, A being a yet to be specified operator.

Hence, (5.9) is identical to

$$Dw + Aw = q \ . \tag{5.11}$$

By virtue of the existence of D^{-1}, (5.11) can be transformed into

$$w + D^{-1}Aw = D^{-1}q \ . \tag{5.12}$$

With

$$D^{-1}q = f \ , \tag{5.13}$$

which is a known quantity, one has the "integral equation"

$$w = -D^{-1}Aw + f \; ,$$

(5.14)

as an expression of the control problem.

Depending on the nature of the operator A, (5.14) might be a complicated integro-differential equation requiring basically further development of integral equation theory in the context of continuous systems' control theory. In order to keep things simple here, set

$$A \equiv cE \; ,$$

where c is a positive constant and E the unit operator. Then,

$$w(x,y,t) = -c \int_o^t \int \int_S G_0(x,y,\xi,\eta,t,\tau)w(\xi,\eta,\tau)d\xi d\eta d\tau + f(x,y,t) \; ,$$

(5.15)

where

$$G_0(x,y,\xi,\eta,t,\tau) = \sum_i \sum_j \left(\phi_{ij}(x,y)\phi_{ij}(\xi,\eta)\sin \frac{\omega_{ij}(t-\tau)}{\omega_{ij}} \right) ,$$

(5.16)

and $\int \int_S \ldots d\rho d\eta$ is the double integral over the plate's surface.

Let for (5.15) the "abbreviated" form

$$w = -c \int_\Omega G_0 w d\Omega + \int_\Omega G_0 q d\Omega$$

(5.17)

be used in the following, where (5.13) has also been applied. Moreover, let it be observed that (5.8) and (5.13) together imply that

$$\int_\Omega G_0 q d\Omega = f \equiv w_0 \; ,$$

(5.18)

if w_0 has been chosen as the notation for the *uncontrolled* response of the plate following from (5.1), (5.3). Hence, also

$$w = c \int_\Omega G_0 w d\Omega + w_0 \; .$$

(5.19)

The first problem arising is the *existence problem:* does (5.19) have a solution? This question can be answered in the positive, using a theorem from integral equation theory: if c in (5.19) is not an eigenvalue λ of the homogeneous problem

$$w = -\lambda \int_\Omega G_0 w d\Omega, \quad U[w]_B = 0, \tag{5.20}$$

then there exists a solution. In other words, the problem

$$\left.\begin{array}{l} w + c \int_\Omega G_0 w d\Omega = Tw = w_0 \ , \\[2em] T = E + c \int_\Omega G_0 \cdots d\Omega \ , \end{array}\right\} \tag{5.21}$$

has a "resolvent" T^{-1}, so that $w = T^{-1}w_0$.

The next problem to be tackled is the *performance problem:* is it possible to choose a value for c so that the response w of the controlled system (5.19) is a "better" one than that of the uncontrolled system (5.8)?

Let it be assumed that "better" means

$$w = \varepsilon w_0, \quad 0 < \varepsilon < 1 \ . \tag{5.22}$$

Then, one has to require in (5.21) that

$$Tw \equiv T\varepsilon w_0 = w_0 \tag{5.23}$$

holds true. Hence,

$$||w_0|| = ||T\varepsilon w_0||$$

shall hold.

That means,

$$||w_0|| \leq ||T|| \ ||\varepsilon|| \ ||w_0|| \ ,$$

$$1 \leq ||T|| \ ||\varepsilon|| \ ,$$

$$||T|| \geq \frac{1}{||\varepsilon||} > 1 \tag{5.24}$$

is required to hold. Yet,

$$||T|| = \sup\left(1 + c \max_{x,y,t} \int_\Omega |G_0| d\Omega\right) ,$$

as known in the theory of integral equations and functional analysis [5.1]. Therefore, the performance requirement is satisfied for any $0 < c$, since then

$$||T|| = \sup\left(1 + c \max_{x,y,t} \int_\Omega |G_0| d\Omega\right) > 1 .$$

No question that the larger c has been chosen the better the performance of the controlled plate will be. This will be shown by means of a numerical example, which will be presented later.

5.3   A CLOSED FORM SOLUTION TO THE CONTROL PROBLEM

Let, as the second approach, a closed form solution of the same problem be sought. Then, the boundary-initial value problem for the controlled plate reads

$$\rho h \ddot{w} + N\nabla^4 w + cw = q \ , \quad U[w]_B = 0 \ , \quad w(0,x,y) = 0 \ , \quad \dot{w}(0,x,y) = 0 \ .$$
$$(5.25)$$

Applying Laplace transformation, i.e., setting

$$\bar{w}(x,y,p) = \int_o^\infty e^{-pt} w(x,y,t)dt \ , \tag{5.26}$$

$$\bar{q}(x,y,p) = \int_o^\infty e^{-pt} q(x,y,t)dt \ , \tag{5.27}$$

the differential equation in (5.25) changes into

$$\rho h p^2 \bar{w} + N\nabla^4 \bar{w} + c\bar{w} = \bar{q} \ . \tag{5.28}$$

Let the auxiliary problem

$$N\nabla^4 \phi_{ij}(x,y) = \rho h \omega_{ij}^2 \phi_{ij}(x,y) \ , \quad U[\phi_{ij}]_B = 0 \tag{5.29}$$

be solved. It yields the orthonormal eigenfunctions ϕ_{ij} of the plate's free vibration, which satisfy the condition

$$\int_S \int \phi_{ij}\phi_{k\ell}\,dS = \delta_{ik}\delta_{j\ell} \ , \tag{5.30}$$

where the integration is over the plate's surface S.

Use the expansions

$$\overline{w} = \sum_{m,n} A_{mn}\phi_{mn} \ , \qquad \overline{q} = \sum_{m,n} B_{mn}\phi_{mn} \ . \tag{5.31}$$

By virtue of the orthonormality of the ϕ_{ij}, one has

$$B_{mn} = \int_S \int \overline{q}\phi_{mn}\,dxdy \ . \tag{5.32}$$

Using (5.31) in (5.28) yields

$$\sum_{m,n} (\rho h p^2 + c)A_{mn}\phi_{mn} + \sum_{m,n} A_{mn} N\nabla^4\phi_{mn} = \sum_{m,n} B_{mn}\phi_{mn} \ . \tag{5.33}$$

By means of (5.29), equation (5.33) can be rewritten as

$$\sum_{m,n} A_{mn}(\rho h p^2 + [c + \rho h \omega_{mn}^2])\phi_{mn} = \sum_{m,n} B_{mn}\phi_{mn} \ . \tag{5.34}$$

As a consequence of (5.34), one has

$$A_{mn}\rho h\left(p^2 + \left[\frac{c}{\rho h} + \omega_{mn}^2\right]\right) = B_{mn} \ ,$$

and with

$$c/\rho h = c^*$$

as well as (5.32),

$$A_{mn} = \frac{\dfrac{1}{\rho h}\displaystyle\int_S \int \overline{q}\phi_{mn}(\xi,\eta)\,d\xi d\eta}{p^2 + [\omega_{mn}^2 + c^*]} \ . \tag{5.35}$$

Using (5.35) in the first equation in (5.31) yields

$$\bar{w}(x,y) = \sum_{m,n} \frac{1}{p^2+[\omega_{mn}^2+c^*]} \phi_{mn}(x,y) \int\int_S \frac{\bar{q}}{\rho h} \phi_{mn}(\xi,\eta)d\xi d\eta \ . \qquad (5.36)$$

Inverting the Laplace transforms in (5.36) results in

$$w(x,y,t) = \sum_{m,n} \int_o^t \int\int_S \frac{\phi_{mn}(\xi,\eta)\phi_{mn}(x,y)}{\rho h[\omega_{mn}^2+c^*]^{1/2}} \sin[\omega_{mn}^2+c^*]^{1/2}$$

$$\cdot \ (t-\tau)q(\xi,\eta,\tau)d\xi d\eta d\tau \ . \qquad (5.37)$$

Equation (5.37) can be changed into

$$w(x,y,t) = \sum_{m,n} \left\{ \int\int_S \left[\int_o^t \frac{\sin[\omega_{mn}^2+c^*]^{1/2}(t-\tau)}{[\omega_{mn}^2+c^*]^{1/2}} \ q(\xi,\eta,\tau)dt \right] \right.$$

$$\left. \cdot \ \frac{\phi_{mn}(\xi,\eta)\phi_{mn}(x,y)}{\rho h} \ d\xi d\eta \right\} \ .$$

Integration by parts with respect to τ yields

$$w(x,y,t) = \sum_{m,n} \left\{ \int\int_S \frac{\phi_{mn}(\xi,\eta)\phi_{mn}(x,y)}{\rho h[\omega_{mn}^2+c^*]} \ q(\xi,\eta,t)d\xi d\eta \right.$$

$$- \int\int_S \frac{\phi_{mn}(\xi,\eta)\phi_{mn}(x,y)}{\rho h[\omega_{mn}^2+c^*]} \cdot \cos[\omega_{mn}^2+c^*]^{1/2}tq(\xi,\eta,0)d\xi d\eta$$

$$\cdot \int\int_o^t\int_S \frac{\phi_{mn}(\xi,\eta)\phi_{mn}(x,y)}{\rho h[\omega_{mn}^2+c^*]} \ \cos[\omega_{mn}^2+c^*]^{1/2}(t-\tau) \cdot \frac{\partial}{\partial\tau} q(\xi,\eta,\tau)d\xi d\eta d\tau \left. \right\} \ .$$

$$(5.38)$$

Let the following assumptions be made:

(i) $\qquad |\phi_{mn}| < M \ ,$

(ii) $\qquad |q| < K_1 \ , \qquad \left| \int_o^t \frac{\partial q}{\partial\tau} \ d\tau \right| < K_2 \ .$

$$(5.39)$$

Then,

$$
|w(x,y,t)| \leq \sum_{m,n} \left\{ \frac{|q(\xi,\eta,t)|_{max}}{[\omega_{mn}^2+c^*]} \frac{M^2 S}{\rho h} \right\} \frac{|q(\xi,\eta,0)|_{max}}{[\omega_{mn}^2+c^*]} \frac{M^2 s}{\rho h}
$$

$$
+ \frac{\left|\int_o^t \frac{\partial q}{\partial \tau} d\tau\right|}{[\omega_{mn}^2+c^*]} \frac{M^2 S}{\rho h} \leq \frac{M^2 S}{\rho h} (2K_1+K_2) \sum_{m,n} \frac{1}{\omega_{mn}^2} \leq \frac{M^2 S}{\rho h} (2K_1+K_2) \sum_{m,n} \frac{1}{\omega_{mn}^2}
$$

$$
\leq \frac{M^2 S}{\rho h} (2K_1+K_2)K_3 < \infty \; ,
$$

if one sets

$$
\sum_{m,n} \frac{1}{\omega_{mn}^2} = K_3 \; , \tag{5.40}
$$

which is obviously a convergent series.

But then, it has been shown that

$$
w(x,y,t) = \sum_{m,n} \int\!\!\int_S \int_o^t \frac{\phi_{mn}(\xi,\eta)\phi_{mn}(x,y)}{\rho h[\omega_{mn}^2+c^*]^{12}} \sin[\omega_{mn}^2+c^*]^{1/2}(t-\tau)
$$

$$
\cdot \; q(\xi,\eta,\tau)d\xi d\eta d\tau \; ,
$$

is an absolutely and uniformly convergent series. Therefore,

$$
w(x,y,t) = \sum_{m,n}^{k,\ell} \int\!\!\int_S \int_o^t \frac{\phi_{mn}(\xi,\eta)\phi_{mn}(x,y)}{\rho h[\omega_{mn}^2+c^*]^{1/2}} \sin[\omega_{mn}^2+c^*]^{1/2}(t-\tau)
$$

$$
\cdot \; q(\xi,\eta,\tau)d\xi d\eta d\tau + r \; ,
$$

where r is a remainder for which the condition

$$
r \to 0 \qquad \text{if} \qquad k,\ell \to \infty
$$

holds. Hence,

$$\left| w(x,y,t) - \sum_{m,n}^{k,\ell} \int\limits_{o}^{t} \iint\limits_{S} \int \cdots \right| = |r| < \varepsilon , \tag{5.41}$$

for sufficiently large k and ℓ. The series on the left side of (5.41) is a finite one. Therefore, integration and summation can be interchanged. Consequently,

$$\left| w(x,y,t) - \int\limits_{o}^{t} \iint\limits_{S} \int \sum_{m,n}^{k,\ell} \frac{\phi_{mn}(\xi,\eta)\phi_{mn}(x,y)}{\rho h [\omega_{mn}^2 + c^*]^{1/2}} \sin[\omega_{mn}^2 + c^*]^{1/2}(t-\tau) \right.$$

$$\left. \cdot q(\xi,\eta,\tau)d\xi d\eta d\tau \right| < \varepsilon$$

and therefore

$$w(x,y,t) = \int\limits_{o}^{t} \iint\limits_{S} \int G(x,y,t,\xi,\eta,\tau)q(\xi,\eta,\tau)d\xi d\eta d\tau \tag{5.42}$$

holds, where the notation

$$\sum_{m,n}^{\infty} \frac{\phi_{mn}(\xi,\eta)\phi_{mn}(x,y)}{\rho h [\omega_{mn}^2 + c^*]^{1/2}} \sin[\omega_{mn}^2 + c^*]^{1/2}(t-\tau) = G(x,y,t,\xi,\eta,\tau) \tag{5.43}$$

has been used.

Moreover, it can be claimed that there exists the equivalent solution

$$w(x,y,t) = - \int\limits_{o}^{t} \iint\limits_{S} \int \tilde{G}(x,y,t,\xi,\eta,\tau) \frac{\partial}{\partial \tau} q(\xi,\eta,\tau)d\xi d\eta d\tau$$

$$- \iint\limits_{S} \int \tilde{G}(x,y,t,\xi,\eta,0)q(\xi,\eta,0)d\xi d\eta + \iint\limits_{S} \int \tilde{G}(x,y,t,\xi,\eta,t)q(\xi,\eta,t)d\xi d\eta , \tag{5.44}$$

with

$$\tilde{G}(x,y,t,\xi,\eta,\tau) = \sum_{m,n} \frac{\phi_{mn}(\xi,\eta)\phi_{mn}(x,y)}{\rho h [\omega_{mn}^2 + c^*]} \cos[\omega_{mn}^2 + c^*]^{1/2}(t-\tau) . \tag{5.45}$$

This is true owing to the fact that in (5.38) integration and summation can be interchanged by virtue of the earlier established absolute and uniform convergence of the series involved. This form of the solution will be used later.

There is still a need to show that assumptions (5.39,i) and (5.40) can indeed be justified from case to case. Take, for example, a freely supported, rectangular plate with the sidelengths a and b. For such a plate,

$$\phi_{mn} = \sin \frac{m\pi x}{a} \sin \frac{n\pi y}{b} \quad \text{with} \quad |\phi_{mn}| \leq M \equiv 1 \ ,$$

confirming (5.39,i) and

$$\omega_{mn}^2 = \left[\left(\frac{m}{a}\right)^2 + \left(\frac{n}{b}\right)^2\right]^2 \pi^2 \frac{N}{\rho h} \ .$$

Therefore, the series

$$\sum_{m,n} \frac{1}{\omega_{mn}^2} = \frac{\rho h}{N\pi^2} \sum_{m,n} \left[\left(\frac{m}{a}\right)^2 + \left(\frac{n}{b}\right)^2\right]^{-2}$$

is indeed convergent, confirming (5.40).

For an assessment of the performance requirement (5.22), one may realize that (5.28) can be written as

$$\rho h \ddot{w} + N\nabla^4 w = q - cw = f \ . \tag{5.46}$$

Then, proceeding as before, one has by Laplace transformation

$$\rho h p^2 \bar{w} + N\nabla^4 \bar{w} = \bar{f} \ . \tag{5.47}$$

Using the expansions

$$\bar{w} = \sum_{m,n} A_{mn} \phi_{mn} \ , \qquad \bar{f} = \sum_{m,n} C_{mn} \phi_{mn} \ , \tag{5.48}$$

one arrives by substituting (5.48) in (5.47) at

$$\sum_{m,n} \rho h p^2 A_{mn} \phi_{mn} + \sum_{m,n} A_{mn} N \nabla^4 \phi_{mn} = \sum_{m,n} C_{mn} \phi_{mn} \; . \qquad (5.49)$$

By means of (5.29), equation (5.49) can be rewritten as

$$\sum_{m,n} A_{mn} (\rho h p^2 + \rho h \omega_{mn}^2) \phi_{mn} = \sum_{m,n} C_{mn} \phi_{mn} \qquad (5.50)$$

yielding

$$A_{mn} = \frac{C_{mn}}{\rho h (p^2 + \omega_{mn}^2)} \; . \qquad (5.51)$$

Yet,

$$C_{mn} = \int \int_S \overline{f} \phi_{mn} \, dx dy \; . \qquad (5.52)$$

Therefore, also

$$A_{mn} = \frac{\dfrac{1}{\rho h} \displaystyle\int \int_S \overline{f} \phi_{mn} \, d\xi d\eta}{p^2 + \omega_{mn}^2} \; . \qquad (5.53)$$

Using (5.53) for the first expansion in (5.48) yields

$$\overline{w}(x,y) = \sum_{m,n} \frac{1}{p^2 + \omega_{mn}^2} \phi_{mn}(x,y) \int \int_S \frac{\overline{f}}{\rho h} \phi_{mn}(\xi,\eta) d\xi d\eta \; . \qquad (5.54)$$

Inverting the Laplace transforms in (5.54) results in

$$w(x,y,t) = \sum_{m,n} \int_o^t \int \int_S \frac{\phi_{mn}(x,y) \phi_{mn}(\xi,\eta)}{\rho h \omega_{mn}} \sin\omega_{mn}(t-\tau)$$

$$\cdot [-cw(\xi,\eta,\tau) + q(\xi,\eta,\tau)] d\xi d\eta d\tau \; . \qquad (5.55)$$

Rearranging and making use of the established fact that integration and summation can be interchanged leads to

$$w(x,y,t) + c \int_o^t \iint_S G_0(x,y,t,\xi,\eta,\tau)w(\xi,\eta,\tau)d\xi d\eta d\tau$$

$$= \int_o^t \iint_S G_0(x,y,t,\xi,\eta,\tau)q(\xi,\eta,\tau)d\xi d\eta d\tau \ , \qquad (5.56)$$

where

$$G_0 = \sum_{m,n} \frac{\phi_{mn}(x,y)\phi_{mn}(\xi,\eta)}{\rho h \omega_{mn}} \sin\omega_{mn}(t-\tau) \ . \qquad (5.57)$$

It is to be noted that solutions (5.56) and (5.42) are equivalent.

Obviously, the response w_0 of the uncontrolled plate follows from (5.56) for $c \equiv 0$ as

$$w_0(x,y,t) = \int_o^t \iint_S G_0(x,y,t,\xi,\eta,\tau)q(\xi,\eta,\tau)d\xi d\eta d\tau \ . \qquad (5.58)$$

Hence, (5.56) can be rewritten as

$$w(x,y,t) + c \int_o^t \iint_S G_0(x,y,t,\xi,\eta,\tau)w(\xi,\eta,\tau)d\xi d\eta d\tau = w_0(x,y,t) \ . \qquad (5.59)$$

Setting

$$w(x,y,t) + c \int_o^t \iint_S G_0(x,y,t,\xi,\eta,\tau)w(\xi,\eta,\tau)d\xi d\eta d\tau = T_0 w \ , \qquad (5.60)$$

one has from (5.59)

$$T_0 w = w_0 \ .$$

As shown earlier, if $||T_0|| > 1$, then $w = \varepsilon w_0$, $0 < \varepsilon < 1$, i.e., the performance requirement is met.

Yet,

$$||T_0|| = \sup_{s,t} \left(1 + c \max \int_o^t \iint_S |G_0| d\xi d\eta d\tau \right) > 1 \ .$$

That means, the performance requirement is indeed satisfied.

Although further considerations may be left to the reader, it goes already now without saying that this second approach, leading to solution (5.42), is much more powerful than the first one, as it provides more detailed information on the behaviour of the controlled system, especially on its stability, without resort to the questionable modal approach.

Based on formula (5.42), a numerical calculation has been carried out assuming

$$q(x,y,t) = 40 \sin 533.5t \text{ psi} ,$$

and for the plate the following data have been chosen:

$$L \equiv a = b = 50 \text{ inches}, \quad h = 1 \text{ inch},$$

$$\rho = 0.00073 \text{ lb.sec}^2/\text{in}^3,$$

$$\nu = 0.25, \quad E = 30 \times 10^6 \text{ psi}.$$

It may be pointed out that the load was chosen as to be in resonance with the first natural frequency of the plate. This has been done in order to show that even in an extremely disadvantageous case the control works very efficiently.

In order to assess the efficiency of the control, not deflections and corresponding control forces themselves have been determined, but their norms

$$J_D = \int_o^t w^2\left(\frac{L}{2} , \frac{L}{2} , t\right) dt \tag{5.61}$$

and

$$J_u = \int_o^t u^2\left(\frac{L}{2} , \frac{L}{2} , t\right) dt \tag{5.62}$$

for $0 \leq t \leq T = 0.5$ seconds.

The results are depicted in Figures 5.1 and 5.2. It becomes quite obvious from these figures that the larger the value of c, the better the effect of the control is, as predicted.

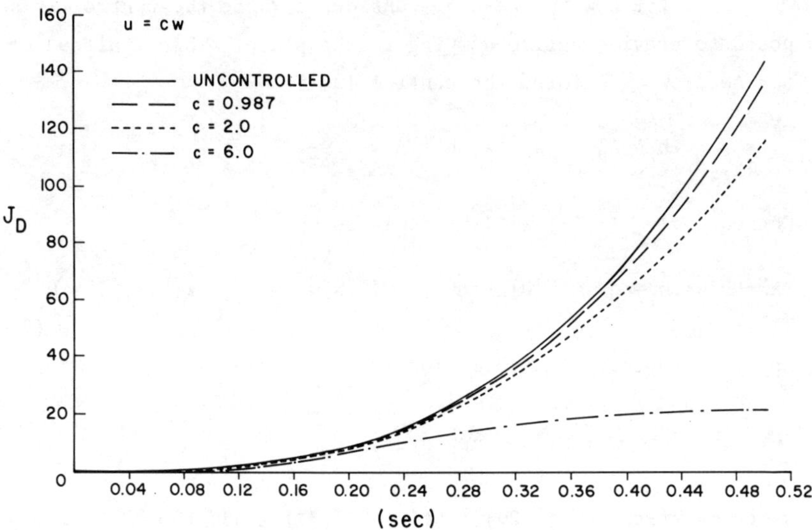

Figure 5.1 - Deflection Mean Square Response at Plate Centre

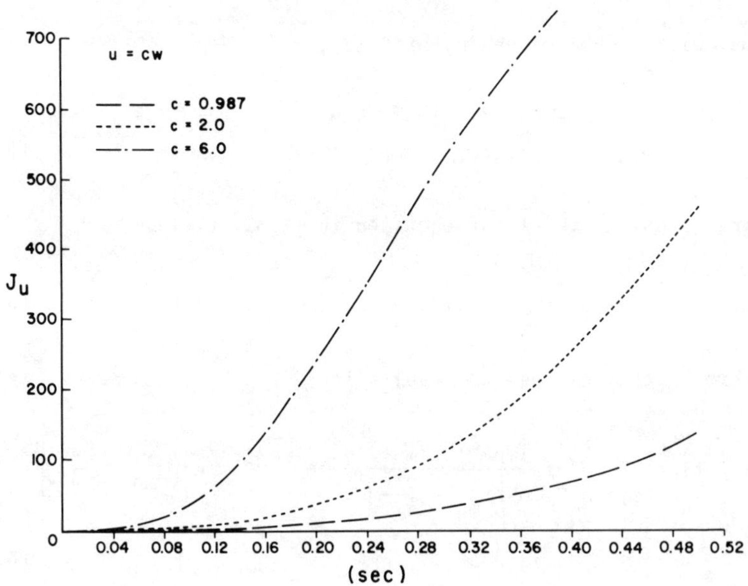

Figure 5.2 - Control Force Mean Square Response at Plate Centre

Let now the case be considered where the control is supposed to provide active damping to the plate. Then, instead of $u = Aw \equiv cw$ as before, the control law is

$$u = A\dot{W} \equiv c\dot{w} \quad \text{with} \quad A = c\,\frac{\partial}{\partial t} \,.$$

Consequently, (5.25) will be replaced by

$$\rho h\ddot{w} + N\nabla^4 w + c\dot{w} = q \,, \quad U[w]_B = 0 \,, \quad w(0,x,y) = 0 \,, \quad \dot{w}(0,x,y) = 0 \,.$$
$$(5.63)$$

Applying Laplace transformation to (5.63) yields

$$\rho h p^2 \overline{w} + N\nabla^4 \overline{w} + cp\overline{w} = \overline{q} \,, \tag{5.64}$$

which by virtue of (5.29), (5.30), (5.31) leads to

$$\sum_{m,n} (\rho h p^2 + cp + \rho h\omega_{mn}^2) A_{mn}\phi_{mn} = \sum_{m,n} B_{mn}\phi_{mn} \,. \tag{5.65}$$

Thus, with $c/\rho h = c^*$ and with (5.32),

$$A_{mn} = \frac{\dfrac{1}{\rho h} \int\int_S \overline{q}\phi_{mn}(\xi,\eta)\,d\xi d\eta}{p^2 + c^* p + \omega_{mn}^2} \,. \tag{5.66}$$

Using (5.65) in the first equation in (5.31) yields

$$\overline{w}(x,y) = \sum_{m,n} \frac{1}{p^2 + c^* p + \omega_{mn}^2} \,\phi_{mn}(x,y) \int\int_S \frac{\overline{q}}{\rho h}\,\phi_{mn}(\xi,\eta)\,d\xi d\eta \,. \tag{5.67}$$

Inverting this expression results in

$$w(x,y,t) = \sum_{m,n} \int\int_S \int_o^t \frac{\phi_{mn}(\xi,\eta)\phi_{mn}(x,y)}{\rho h\left[\omega_{mn}^2 - \dfrac{c^{*2}}{4}\right]^{1/2}} \, e^{-c^*/2t} \sin\left[\left(\omega_{mn}^2 - \frac{c^{*2}}{4}\right)^{1/2}(t-\tau)\right]$$
$$\cdot\, q(\xi,\eta,\tau)\,d\xi d\eta d\tau \,. \tag{5.68}$$

Assuming as before that one can justify interchanging summation and integration in (5.68), one obtains the closed form solution

$$w(x,y,t) = \int\limits_o^t \iint\limits_S \hat{G}(x,y,t,\xi,\eta,\tau)q(\xi,\eta,\tau)d\xi d\eta d\tau \ , \qquad (5.69)$$

where

$$\hat{G}(x,y,t,\xi,\eta,\tau) = \sum_{m,n} \frac{\phi_{mn}(\xi,\eta)\phi_{mn}(x,y)}{\rho h\left[\omega_{mn}^2 - \frac{c^{*2}}{4}\right]^{1/2}} e^{-c*/2t}$$

$$\cdot \sin\left[\left(\omega_{mn}^2 - \frac{c^{*2}}{4}\right)^{1/2}(t-\tau)\right] \ . \qquad (5.70)$$

For an assessment of the performance, realize that with

$$\dot{w}(x,y,t) = \psi(x,y,t)w(x,y,t) \ , \qquad (5.71)$$

where $\psi(x,y,t)$ is an appropriate function, equation (5.63) changes into

$$\rho h\ddot{w} + N\nabla^4 w + c\psi w = q \ , \qquad (5.72)$$

$$\rho h\ddot{w} + N\nabla^4 w = q - c\psi w = f \ , \qquad (5.73)$$

respectively.

Applying to (5.73) the same manipulations as to (5.46) yields

$$w(x,y,t) + c \int\limits_o^t \iint\limits_S G_0(x,y,t,\xi,\eta,\tau)\psi(\xi,\eta,\tau)w(\xi,\eta,\tau)d\xi d\eta d\tau = w_0(x,y,t) \qquad (5.74)$$

Setting the left hand side in (5.74) equal to $\hat{T}w$, one has accordingly

$$\hat{T}w = w_0 \ .$$

In order to show that $w = \varepsilon w_0$, $0 < \varepsilon < 1$, which is the required performance of w, one has then to require that this time

$$\|\hat{T}\| = \sup_{s,t \text{ o } S}\left(1 + c \max \iint \int^{t} |G_0| |\psi| d\xi d\eta d\tau\right) > 1 \ ,$$

which is obviously satisfied.

The performance of the plate controlled by the way of
equation (5.63) has again been verified by means of a numerical
calculation whose results are depicted in Figures 5.3 and 5.4.
As previously, the norms (5.61) and (5.62) have been used to
indicate the difference in behaviour between uncontrolled and controlled plate.

In order to raise another issue, let now the solution in
its form (5.44) be used. This solution may be expressed in an
abbreviated form as

$$w = Gq \ , \tag{5.75}$$

where G is the operator involved in (5.44).

One may refer to operator G in (5.75) as the "transfer
operator" relating output w with input q. In classical control
theory it is quite common to work with the "transfer" concept. It
may therefore be welcome to recognize traces of that concept also
in the context of this theory.

Following closely a suggestion by E.P. Popow [5.2], it
is assumed that the transfer operator is equivalent to the output
of the controlled system if q is replaced in (5.75) by an expression

$$q^* = \Phi(x,y)H(t) \ , \tag{5.76}$$

where $H(t)$ is the unit step function

$$H(t) = \begin{cases} 0 & \text{when} \quad t < 0 \ , \\ 1 & \text{when} \quad t \geq 0 \ , \end{cases} \tag{5.77}$$

where the specific expression

$$\frac{\partial}{\partial \tau} q^* = \Phi(x,y)\delta(t) \tag{5.78}$$

is chosen instead of a general one, and where $\delta(t)$ is the Dirac delta function defined by means of

$$\int_{-\infty}^{\infty} \delta(t)dt = 1 , \qquad \int_{-\infty}^{\infty} f(t)\delta(t)dt = f(0) . \qquad (5.79)$$

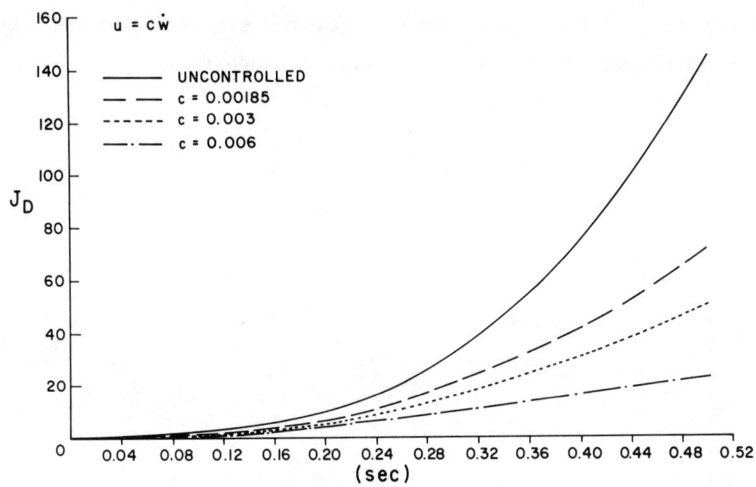

Figure 5.3 - Deflection Mean Square Response at Plate Centre

Using (5.76) in (5.44), and observing (5.77), (5.78) and (5.79), yields

$$w(x,y,t) = \int \int_S [-2\tilde{G}(x,y,t,\xi,\eta,t) + \tilde{G}(x,y,t,\xi,\eta,0)]\Phi(\xi,\eta)d\xi d\eta .$$

$$(5.80)$$

According to Popow's suggestion, the quantity on the right side of (5.80) is the transfer operator of the plate system.

Major attention is being paid to the time behaviour of the system's output. Briefly, it can be said that in the sense of a rudimentary "stability" notion, one would require that the system should guarantee a bounded output in response to a bounded input. Whether this requirement is satisfied, can in a way be evaluated

by means of the transfer operator: the bounded input is given by
(5.76). The bounded output may follow from (5.80) if the right
side is bounded as a function of time. Obviously, it is the time
behaviour of the "Green" function $\tilde{G}$ which decides on whether the
system reacts in a stable way (stable being interpreted in a rather
broad way). The fact that such an assessment of the system's behaviour is possible, is the reason for the significance and importance attributed to the system's transfer operator.

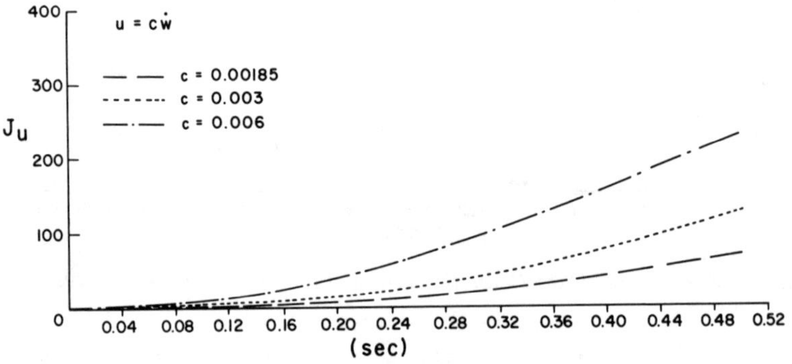

Figure 5.4 - Control Force Mean Square Response at Plate Centre

Referring back to (5.75), it becomes clear that the
operator G is only responsible for the "structural stability" of
the system. The term "structural stability" shall be explained
in more detail soon. Obviously, the input q, the "perturbation"
of the system, could also lead to instability of the output w.
Therefore, it may now be established which conditions for q are
sufficient to warrant stability of the system under the assumption

that there is already structural stability. Then, by definition

$$|\tilde{G}|_{\substack{max \\ s,t}} < \tilde{K} < \infty .$$

(5.81)

From (5.44) follows

$$|w| \leq S|\tilde{G}|_{\substack{max \\ s,t}} \left| \int_{o}^{t} \frac{\partial q}{\partial \tau} \, d\tau \right| + 2S|\tilde{G}|_{\substack{max \\ s,t}} |q|_{\substack{max \\ s,t}} .$$

(5.82)

Use the fact that according to (5.39,ii)

$$|q| < K_1 ,$$

(5.83)

and

$$\left| \int_{o}^{\infty} \frac{\partial q}{\partial \tau} \, d\tau \right| < K_2 < \infty ,$$

(5.84)

one has

$$|w| \leq S\tilde{K}(K_2+2K_1) < \infty ,$$

(5.85)

which is stability. Hence, conditions (5.39,ii) for q are sufficient for a stable output, provided that the structural stability
of the system is guaranteed by means of (5.81).

Now, attention shall be payed to the structural stability
of the problem. By virtue of (5.45), the behaviour of $\tilde{G}$ as a function of time is proper, as $\tilde{G}$ remains bounded for any time t. This
is the case because the parameter c* is a real valued constant.
Moreover, the eigenfunctions ϕ_{mn} and their corresponding eigenvalues
ω_{mn} are predetermined and unchangeable by virtue of the choice of
the auxiliary problem (5.29). Hence, the function $\cos[\omega_{mn}^2+c*]^{1/2}(t-\tau)$
remains a real valued trigonometric function for any time t and
is therefore harmless.

However, let the auxiliary problem have to be changed
from its form (5.29) into another form. Then, the quantity ω_{ij}
may become a function $\omega_{ij} = \omega_{ij}(g)$ of a "structural parameter" g.
If this parameter q is changed due to changes in the structure of

the system, it may happen that ω_{ij}^2 becomes negative or even complex
so that a situation may occur at which the cos-function in $\tilde{G}$ changes
into a hyperbolic one, rendering the controlled system unstable.
Since this loss of stability has happended by virtue of a variation
of the structural parameter g, one says that the system has lost
its "structural stability".

Let it be shown, how the form of the auxiliary equation
(5.29) can be made to vary. Assume that in (5.9) the control u
reads (5.10), i.e., instead of (5.14) and u = cw, one has now
u = Aw, A being a more general operator, for example
$A = gv(x,y(\partial^2/\partial x^2)w$, where g is a structural parameter and v a
certain appropriate function.

Setting

$$v(x,y) \frac{\partial^2}{\partial x^2} w = A^*w , \tag{5.86}$$

the basic set of equations (5.25) changes into

$$\rho h\ddot{w}+N\nabla^4 w+gA^*w = q , \quad U[w]_B = 0 , \quad w(0,x,y) = 0 , \quad \dot{w}(0,x,y) = 0 . \tag{5.87}$$

The corresponding auxiliary equation replacing (5.29) now reads

$$N\nabla^4\phi_{ij}(x,y)+gA^*\phi_{ij}(x,y) = \rho h\lambda_{ij}^2(g)\phi_{ij}(x,y) , \tag{5.88}$$

where the eigenfunctions ϕ_{ij} satisfy the boundary conditions of the
problem, i.e.,

$$U[\phi_{ij}]_B = 0 . \tag{5.89}$$

It must be emphasized that the eigenvalues λ_{ij} in (5.88) are functions of the structural parameter g, i.e.,

$$\lambda_{ij} = \lambda_{ij}(g) , \tag{5.90}$$

by virtue of which, changes in the structure, i.e., changes in g,
can lead to the loss of structural stability.

Condition (5.88) together with condition (5.89) may result in a nonselfadjoint problem, as has for example been pointed out by H. Leipholz in [5.3,5.4]. Let it be assumed for the sake of generality, that this is the case. Then, the problem which is adjoint to (5.88) and (5.89), must also be brought into play.

With

$$D = N\nabla^4 \, \cdots \, + gA^* \, \cdots \, , \tag{5.91}$$

equations (5.88) and (5.89) can be written as

$$D\phi_{ij} = \rho h\lambda_{ij}^2(g)\phi_{ij} = 0 \, , \quad U[\phi_{ij}]_B = 0 \, . \tag{5.92}$$

The adjoint problem is

$$D_a\psi_{ij} = \rho h\lambda_{ij}^2(g)\psi_{ij} = 0 \, , \quad U_a[\psi_{ij}]_B = 0 \, , \tag{5.93}$$

where the operators D_a and U_a are defined by means of the condition

$$\int\int_S \psi_{ij} D\phi_{ij} dxdy = \int\int_S \phi_{ij} D_a\psi_{ij} dxdy \, , \tag{5.94}$$

and where the eigenfunctions ϕ_{ij}, ψ_{ij} form a bi-orthonormal set, i.e.,

$$\int\int_S \phi_{ij}\psi_{k\ell} dxdy = \delta_{ik}\delta_{j\ell} \, . \tag{5.95}$$

In (5.95), the δ_{ij}, $\delta_{j\ell}$ are Kronecker symbols.

By means of Laplace transformation, the differential equation in (5.88) changes into

$$\rho h p^2 \overline{w} + (N\nabla^4 + gA^*)\overline{w} = \overline{q} \, . \tag{5.96}$$

Setting

$$\overline{w} = \sum_{m,n} A_{mn}\phi_{mn} \, , \quad \overline{q} = \sum_{m,n} B_{mn}\phi_{mn} \, , \tag{5.97}$$

and realizing that

$$B_{mn} = \int\int_S \bar{q}\psi_{mn}\,dxdy \; , \tag{5.98}$$

allows one to rewrite (5.96) as

$$\sum_{m,n} \rho h p^2 A_{mn}\phi_{mn} + \sum_{m,n} (N\nabla^4+gA^*)\phi_{mn} = \sum_{m,n} B_{mn}\phi_{mn} \; . \tag{5.99}$$

By virtue of (5.91), (5.92), equation (5.99) becomes

$$\sum_{m,n} A_{mn}\{\rho h[p^2+\lambda^2_{mn}(g)]\}\phi_{mn} = \sum_{m,n} B_{mn}\phi_{mn} \; . \tag{5.100}$$

Hence, taking (5.98) into account,

$$A_{mn} = \frac{B_{mn}}{\rho h[p^2+\lambda^2_{mn}(q)]} = \frac{\frac{1}{\rho h}\int\int_S \bar{q}\psi_{mn}\,d\xi d\eta}{p^2+\lambda^2_{mn}(q)} \; . \tag{5.101}$$

Using (5.101) in (5.96) yields

$$\bar{w}(x,y) = \sum_{m,n} \frac{1}{p^2+\lambda^2_{mn}(g)} \phi_{mn}(x,y) \int\int_S \frac{\bar{q}(\xi,\eta)}{\rho h} \psi_{mn}(\xi,\eta)\,d\xi d\eta \; . \tag{5.102}$$

Applying to (5.102) the inverse Laplace transformation results in

$$w(x,y,t) = \sum_{m,n} \int_o^t\!\!\int\!\int_S \frac{\phi_{mn}(x,y)\psi_{mn}(\xi,\eta)}{\rho h\lambda_{mn}(g)} \sin\lambda_{mn}(g)(t-\tau)$$

$$\cdot q(\xi,\eta,\tau)\,d\xi d\eta d\tau \; . \tag{5.103}$$

Assume again that in (5.103) integration and summation
can be interchanged, the solution (5.103) can be rewritten as

$$w(x,y,t) = \int_o^t\!\!\int\!\int_S G_{uns}(x,y,t,\xi,\eta,\tau)q(\xi,\eta,\tau)\,d\xi d\eta d\tau \; , \tag{5.104}$$

where

$$G_{uns}(x,y,t,\xi,\eta,\tau) = \sum_{m,n} \frac{\phi_{mn}(x,y)\psi_{mn}(\xi,\eta)}{\rho h \lambda_{mn}(g)} \sin\lambda_{mn}(g)(t-\tau) , \qquad (5.105)$$

is an *unsymmetric* Green function.

Since $\lambda_{mn}(g)$ is a function of the structural parameter g, it is possible that, by varying g, $\lambda_{mn}(g)$ can be made to assume complex values in (5.88) and (5.89). If that happens, the trigonometric function in (5.105), (5.103), respectively, becomes a hyperbolic function, rendering the response (5.103), (5.104), respectively, of the controlled system unstable. Since this instability came about by varying the structural parameter g, one is faced with structural instability. Moreover, since the instability is caused by the eigenvalue $\lambda_{ij}(g)$ of the auxiliary problem (5.88), (5.89), which involves the control $u = gA^*\phi_{ij}$, one realizes that this is *the kind of control* which can make a controlled system unstable.

It is the task of the designer to make instability through control impossible. For that purpose, he has to investigate the stability of problem (5.87), i.e., he has to find a range

$$0 \leq g < g_{crit} \qquad (5.106)$$

of the structural parameter g for which the eigenvalues $\lambda_{ij}(g)$ of the auxiliary problem (5.88) remain real valued.

5.4 A LIAPUNOV-LIKE ASSESSMENT OF A SPILLOVER-SAFE DESIGN

Referring back to the previously raised concern about spillover, it is obvious that the stability investigation should not be carried out by means of a modal approach. Rather, a Liapunov-like global assessment of stability should be carried out. If the problem is a conservative one, that is a simple problem, as the Hamiltonian can always serve as a Liapunov functional. Therefore, attention

shall be focussed here on a nonconservative, nonselfadjoint problem.
It will be shown in the following that also in such a case a
Liapunov-like approach is possible which then warrants a spillover-
safe design of the controlled system.

Let as an example a system be considered which is still
a continuous one, but, it is, as compared with the previously presented plate problems, a simpler one as it involves a flexible
cantilever rod.

This cantilever rod is supposed to be the antenna arm
of a satellite, see Figure 5.5. In unaccelerated flight of the
satellite, the cantilever can be considered to be in "equilibrium"
relative to the satellite. This equilibrium position due to the
effect of a uniformly distributed load p follows from the boundary
value problem

$$\alpha y''''(x) = p \; , \qquad\qquad (5.107)$$

$$y(0) = y'(0) = y''(\ell) = y'''(\ell) = 0 \; . \qquad\qquad (5.108)$$

Assuming the cantilever to have length ℓ and a constant bending
stiffness α, its elastic line is given by

$$y(x) = \frac{1}{24} x^4 \frac{p}{\alpha} - \frac{1}{6} x^3 \frac{p\ell}{\alpha} + \frac{1}{4} x^2 \frac{p\ell^2}{\alpha} \; , \qquad\qquad (5.109)$$

yielding the deflection

$$y(\ell) = \frac{1}{8} \frac{p\ell^2}{\alpha} \qquad\qquad (5.110)$$

at the tip of the cantilever. This can be an unacceptable large
value, if α is small.

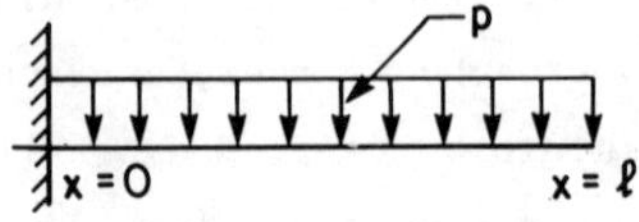

Figure 5.5 - Uniformly Loaded Cantilever

For a satellite antenna, this will most probably be the case. Hence, control of the antenna arm may be considered.

A possible control may consist in sensing the curvature of the arm and exerting distributed forces on the arm which are proportional to the measured curvature. These control forces can, for example, be implemented by means of a properly controlled magnetic field surrounding the arm.

For such control one has

$$\alpha y''''(x) = p - gu , \qquad u \equiv y''(x) , \tag{5.111}$$

where u is the control and g a factor of proportionality. Consequently, the elastic line of the controlled arm follows from the boundary value problem

$$\alpha y'''' + g y'' = p , \tag{5.112}$$

$$y(0) = y'(0) = y''(\ell) = y'''(\ell) = 0 , \tag{5.113}$$

which can easily be recognized to be nonselfadjoint.

The solution to the control problem (5.112), (5.113) is

$$y_c = \frac{\alpha p}{g^2} \left(\sin \sqrt{\tfrac{g}{\alpha}}\, \ell \right) \left(\sin \sqrt{\tfrac{g}{\alpha}}\, x \right) + \frac{\alpha p}{g^2} \left(\cos \sqrt{\tfrac{g}{\alpha}}\, \ell \right) \left(\cos \sqrt{\tfrac{g}{\alpha}}\, x \right)$$

$$- \sqrt{\tfrac{\alpha}{g}}\, \frac{p}{g} \left(\sin \sqrt{\tfrac{g}{\alpha}}\, \ell \right) x - \frac{\alpha}{g}\, \frac{p}{g} \left(\cos \sqrt{\tfrac{g}{\alpha}}\, \ell \right) + \frac{1}{2}\, \frac{p}{g}\, x^2 . \tag{5.114}$$

From (5.114) follows for the tip's deflection, assuming $g = 20\alpha/\ell^2$, the value

$$y_c(\ell) = 0.04\, \frac{p\ell^4}{\alpha} . \tag{5.115}$$

As compared with the tip's deflection of the uncontrolled antenna arm, which is given by (5.110), one has a distinct gain in "stiffness" through control, as

$$y_c(\ell) = 0.32 y(\ell) . \tag{5.116}$$

This fact encourages one to investigate further. By means of
(5.109) and (5.114), and using again

$$g = 20\alpha/\ell^2 \tag{5.117}$$

as the proportionality factor in the control law, one obtains the
following table (Table 5.1). The values in this table have been
used to plot the corresponding elastic lines in Figure 5.6.

Table 5.1

	deflection of the cantilever		ratio $|y_c(x)|/|y(x)|$
	uncontrolled $y(x)$	controlled $y_c(x)$	
at x =			
$\ell/4$	0.013	-0.003	0.226
$\ell/2$	0.044	0.011	0.242
$3\ell/4$	0.084	0.024	0.286
ℓ	0.125	0.039	0.312

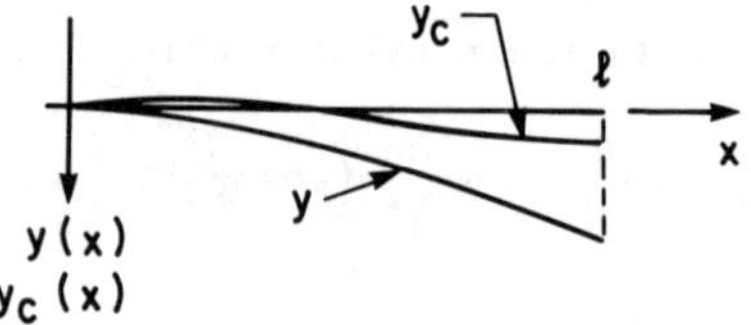

Figure 5.6 - Deflection y of Uncontrolled Cantilever
Deflection y_c of Controlled Cantilever

Taking these results into account, one can safely conclude that the behaviour of the antenna arm has advantageously been
changed by the proposed control. One could hastily propose to
increase the value of g beyond that one given by (5.117) as to
increase the stiffening effect of the control. In doing so, one
would make a decisive mistake, since, due to the nonselfadjointness
of the control problem, larger values of g could cause *flutter*
of the antenna arm. Such phenomenon would be devastating and must
be avoided. Its possible occurrence will be revealed in the

subsequent stability investigation, which also confirms that the value of g in (5.117) is the largest one that can safely be assumed if flutter is to be ruled out.

The relative equilibrium position of the antenna arm follows from (5.112), (5.113). If this position is perturbed, a small vibration about it will be initiated, which is described

$$\mu\ddot{w}(t,x) + \alpha w''''(t,x) + g w''(t,x) = p \, , \tag{5.118}$$

$$w(t,0) = w'(t,0) = w''(t,\ell) = w'''(t,\ell) = 0 \, . \tag{5.119}$$

In (5.118), (5.119) μ is the cantilever's mass per unit length and

$$w(t,x) = v(t,x) + y(x) \, . \tag{5.120}$$

Obviously, $\ddot{w}(t,x) \equiv \ddot{v}(t,x)$. Therefore, using (5.120) in (5.118), (5.119) and by virtue of (5.223), (5.113), one arrives at

$$\mu\ddot{v} + \alpha v'''' + g v'' = 0 \, , \tag{5.121}$$

$$v(t,0) = v'(t,0) = v''(t,\ell) = v'''(t,\ell) = 0 \, . \tag{5.122}$$

This is the problem which determines the behaviour of the perturbation v. If v remains bounded for any time, the cantilever will be stable. If v increases unboundedly, one has instability. As can be expected, the magnitude of the eigenvalue g will determine whether there is stability or not.

In order to solve problem (5.121), (5.122), one may set

$$v(x,t) = \psi(t)\phi(x) \, , \quad \psi(t) = e^{i\omega t} \, . \tag{5.123}$$

Then,

$$-\mu\omega^2\phi + \alpha\phi'''' + g\phi'' = 0 \, , \tag{5.124}$$

$$\phi(0) = \phi'(0) = \phi''(\ell) = \phi'''(\ell) = 0 \, , \tag{5.125}$$

results. This is a boundary value problem involving the two eigenvalues ω^2 and g. If for a certain value of g the eigenvalue ω^2
should turn out to be negative or complex, the time factor $\psi(t)$ in
(5.123) would become unbounded, thus indicating instability.

As one easily recognizes, problem (5.121), (5.122),
(problem (5.124), (5.125), respectively), is identical to the
famous problem of Beck. In Beck's problem, the factor g is interpreted as P, which is a tangential, compressive force at the tip
of the cantilever. Due to this analogy, one can use Beck's procedure to solve problems (5.124), (5.125). In this way, one arrives
at an eigenvalue curve as displayed in Figure 5.7. As one sees,
ω^2 is never negative, but for $g < g_{cr}$, ω^2 becomes complex. Due to
complex ω^2, the time factor $\psi(t)$ in (5.123) turns out to be
unbounded, which implies *flutter* of the cantilever. Hence, values
$g > g_{cr}$ must be avoided.

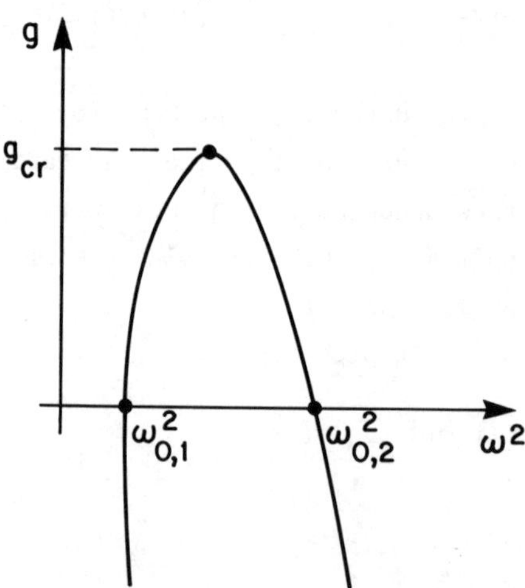

Figure 5.7 - Eigenvalue Curve of Problem (5.124), (5.125)

From Beck's calculations [5.5], or comparable solution
to the problem, which are reported in [5.6], [5.7] or [5.8], one
finds

$$g_{cr} = 20.05\alpha/\ell^2 \; . \tag{5.126}$$

Hence, it is now understandable why earlier the fact was stressed
that g just given by (5.117) should be used to evaluate the effect
of the control (or a smaller value of g), but not a larger value
of g.

In Figure 5.7, only a single branch of the eigenvalue
curve has been shown. The peak value of this branch, that is
$g_{cr} = 20.05\alpha/\ell^2$, is the first flutter load of problem (5.124),
(5.125). For $g \equiv 0$, the curve intersects the ω^2-axis in $\omega_{0,1}^2$ and
$\omega_{0,2}^2$ where $\omega_{0,i}$, $i = 1,2$, are the natural frequencies of free
vibration of the antenna arm. Yet, there are infinitely many
more branches to the actual eigenvalue curve as indicated in
Figure 5.8, yielding infinitely many more flutter loads $g_{cr,i}$,
$i = 1,2,3, \ldots$. The question is whether $g_{cr,1} < g_{cr,2} < g_{cr,3} \cdots$
holds, i.e., whether $g_{cr,1}$ is indeed the *smallest* flutter load.
Only, if this fact were secured, one could design safely against
instability, avoiding "spillover" completely. Since problem (5.124),
(5.125) is nonselfadjoint, the classical extremum properties of
conservative (selfadjoint) problems do not hold, and a statement
concerning the above question cannot be made easily. If one would
use the modal approach for the calculation of the eigenvalue curve,
one were only able to produce a few of its branches. Even if in
the case of a few branches known, the evidence were that $g_{cr,1}$ is
smaller than any other $g_{cr,i}$ so far produced, this evidence were
not conclusive, as there could be other, not considered branches
of the eigenvalue curve which might lead to a value of g_{cr} smaller
than that of $g_{cr,1}$, thus invalidating any design against flutter
based solely on $g_{cr,1}$.

In order to find a design against flutter, and to safeguard the antenna arm against "spillover", without solving the problem exactly, the following solution to the problem will be posed.

Consider the functional

$$V = \frac{1}{2} \int_0^\ell [\mu\dot{v}^2 + \alpha(v'')^2 - g(v')^2]dx + \frac{1}{2} gv(\ell,t)v'(\ell,t) . \qquad (5.127)$$

In (5.127),

$$H = \frac{1}{2} \int_0^\ell [\mu\dot{v}^2 + \alpha(v'')^2 - g(v')^2]dx \qquad (5.128)$$

is the Hamiltonian of the controlled antenna arm and

$$W = \frac{1}{2} gv(\ell,t)v'(\ell,t) \qquad (5.129)$$

is proportional to the negative work done by the non-potential force $gv'(\ell,t)$ on the deflection $v(\ell,t)$ at the tip of the antenna.

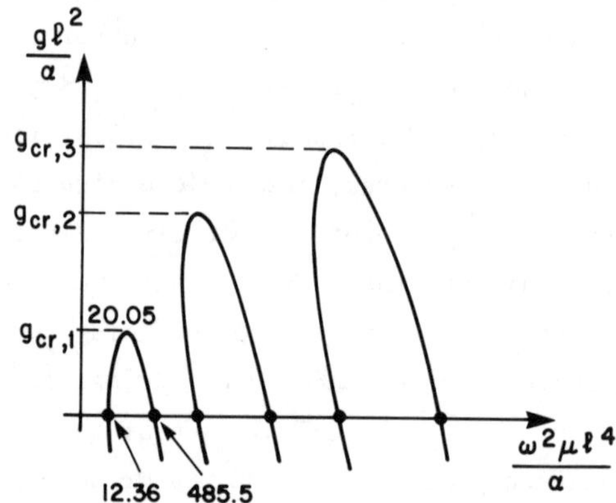

Figure 5.8 – *Additional Branches of the Eigenvalue Curve and Additional Flutter Loads $g_{cr,i}$, $i = 1,2,3,\ldots$, for Problem (5.124), (5.125)*

It is easily found that

$$\dot{H} = -gv'(\ell,t)\dot{v}(\ell,t) . \qquad (5.130)$$

For $\dot{W}$ one has

$$\dot{W} = \frac{1}{2}\,\dot{g}v(\ell,t)v'(\ell,t) \; + \; \frac{1}{2}\,gv(\ell,t)\dot{v}'(\ell,t) \;. \tag{5.131}$$

Using (5.123), one can write

$$gv(\ell,t)\dot{v}'(\ell,t) = g[\psi(t)\phi(\ell)][\dot{\psi}(t)\phi'(\ell)] \equiv g[\dot{\psi}(t)\phi(\ell)][\psi(t)\phi'(\ell)]$$

$$= g\dot{v}(\ell,t)v'(\ell,t) \;. \tag{5.132}$$

By virtue of (5.132), equation (5.131) can be transformed into

$$\dot{W} = gv'(\ell,t)\dot{v}(\ell,t) \;. \tag{5.133}$$

Also, due to (5.128) and (5.129), the functional V can
be written as

$$V = H + W \;, \tag{5.134}$$

so that

$$\dot{V} = \dot{H} + \dot{W} \;. \tag{5.135}$$

Using (5.130) and (5.133) in (5.135) yields

$$\dot{V} = 0 \;, \qquad V = \text{const} \;. \tag{5.136}$$

Hence, V can be used as a Liapunov functional as long as it is
positive. In order to establish a condition for V to be positive,
return to (5.134): The sign of V were essentially the same as
that of H, if it could be shown that W is always positive.

For the purpose of proving this, let the expression
$gv(\ell,t)v'(\ell,t)$ be investigated. By virtue of (5.123) and the
boundary conditions in (5.125), one has

$$gv(\ell,t)v'(\ell,t) = g\psi^2(t)\phi(\ell)\phi'(\ell) \equiv g\psi^2(t)\phi(x)\phi'(x)]_0^\ell \;.$$

The sign of W is therefore determined by the sign of $g\phi(x)\phi'(x)]_0^\ell$.
Obviously,

$$g\phi\phi']_0^{\ell} = g \int_0^{\ell} \frac{d}{dx} (\phi\phi')dx = q \int_0^{\ell} [\phi''\phi+(\phi')^2]dx . \qquad (5.137)$$

Using (5.124), equation (5.137) can be transformed into

$$g\phi\phi']^{\ell} = \int_0^{\ell} [\mu\omega^2\phi^2-\alpha\phi''''\phi+g(\phi')^2]dx . \qquad (5.138)$$

Applying integration by parts to (5.138) changes it into

$$g\phi\phi']_0^{\ell} = \int_0^{\ell} [\mu\omega^2\phi^2-\alpha(\phi'')^2+g(\phi')^2]dx = F . \qquad (5.139)$$

Hence, the expression $gv(\ell,t)v'(\ell,t)$, and consequently W, will be positive when the functional F is positive.

Using the norm concept

$$||f|| = \left[\int_0^{\ell} f^2 dx \right]^{1/2} ,$$

which holds for any square integrable function, one finds from (5.139) that F is positive if

$$g||\phi'||^2 > \alpha||\phi''||^2-\mu\omega^2||\phi||^2. \qquad (5.140)$$

From [5.6], p. 197, follows $||\phi||^2/||\phi'||^2 < \ell^2/2$. Therefore, (5.140) yields as a condition for a positive F the inequality

$$g > \alpha||\phi''||^2/||\phi'||^2-\mu\omega^2\ell^2/2 . \qquad (5.141)$$

Yet, ϕ can be used as an admissible function in Rayleigh's quotient for Euler's buckling load P_E of the antenna arm:

$$R = \alpha||\phi''||^2/||\phi'||^2 > P_E = \pi^2\alpha/4\ell^2 . \qquad (5.142)$$

Hence, approximately

$$g > \pi^2\alpha/4\ell^2 - \mu\omega^2\ell^2/2 .$$

Moreover,

$$\omega > 1.875^2 \alpha^{1/2} / \ell^2 \mu^{1/2} = \omega_{0,1} , \tag{5.143}$$

where $\omega_{0,1}$ is the first frequency of free vibration of the antenna arm. Therefore, again approximately,

$$g > 2.47\alpha/\ell^2 - 6.18\alpha/\ell^2 = -3.71\alpha/\ell^2 . \tag{5.144}$$

This inequality indicates that F is positive for positive g. Consequently, also W is positive for positive g.

Using this result in (5.134), one has

$$V > H \qquad \text{for} \qquad g > 0 , \tag{5.145}$$

which shows that the problem of the positiveness of V has been reduced to the problem of the positiveness of H.

From (5.128) follows

$$H > \frac{1}{2}\left[\alpha \int_o^\ell (v'')^2 dx - g \int_o^\ell (v')^2 dx\right] . \tag{5.146}$$

Yet, from Rayleigh's principle one obtains

$$\alpha \int_o^\ell (v'')^2 dx > P_E \int_o^\ell (v')^2 dx , \tag{5.147}$$

where $P_E = \pi^2 \alpha/4\ell^2$ is again Euler's buckling load of the cantilever. With (5.147) in (5.146) one has

$$H > \frac{1}{2} (P_E - g) \int_o^\ell (v')^2 dx . \tag{5.148}$$

This result indicates that H is positive, and so is V, for

$$g < P_E = \pi^2 \alpha/4\ell^2 . \tag{5.149}$$

From $\dot{V} = 0$ (see (5.136)) follows $V(t=0) \equiv V(t)$ for any t. Choosing appropriate initial conditions, one can set

$$V(t=0) < \varepsilon^2/2 \,, \quad \varepsilon > 0 \,, \text{ small} \,. \tag{5.150}$$

Then, also

$$0 < H < V = H+W = \varepsilon^2/2 \quad \text{for} \quad g < P_E \text{ at any } t \,. \tag{5.151}$$

Thus, for any time, and for $g < P_E$,

$$0 < \frac{1}{2}(P_E-g) \int_o^\ell (v')^2 dx < H < \varepsilon^2/2 \,. \tag{5.152}$$

Inequality (5.152) yields

$$\int_o^\ell (v')^2 dx < \varepsilon^2/(P_E-g) = \kappa^2 \quad \text{for} \quad g < P_E \text{ at any } t. \tag{5.153}$$

Yet,

$$\int_o^\ell (v')^2 dx = ||v'||^2 \,,$$

and from [5.6], p. 196, one has

$$v^2 \leq \ell ||v'||^2 \,.$$

Hence,

$$|v| \leq \kappa \ell^{1/2} \quad \text{for} \quad g < P_E \quad \text{at any } t. \tag{5.154}$$

Inequality (5.154) indicates that the vibration of the antenna
arm about its relative equilibrium can be kept arbitrarily small,
provides the initial perturbation of the arm is chosen to be sufficiently small.

The final conclusion is that it is sufficient for the
stability of the antenna arm's control that g is kept below the
value P_E of Euler's buckling load of the antenna arm. This fact
can be used as a design criterion for the antenna arm's control
with the certainty in mind that the control will be stable for any
modes, i.e., without spillover.

5.5 REFERENCES

[5.1] LUSTERNIK, L.A. and SOBOLEW, W.I., *Elemente der Funktional-analysis*, Akademie-Verlag, Berlin, 1955.

[5.2] POPOW, E.P., *Dynamic von Systemen der Automatischen Regelung*, Akademie-Verlag, Berlin, 1958.

[5.3] LEIPHOLZ, H.H.E., "Assessing Stability of Elastic Systems by Considering their Small Vibrations", *Ing. Archiv*, Vol. 53, 1983, pp. 345-362.

[5.4] LEIPHOLZ, H.H.E., "On Fundamental Integrals for Certain Mechanical Problems having Non-Selfadjoint Differential Equations", *Hadronic Journal*, Vol. 6, 1983, pp. 1488-1508.

[5.5] BECK, M., "Die Knicklast des einseitig eingespannten, tangential gedrückten Stabes", *ZAMP*, Vol. 3, 1952, pp. 225, 476.

[5.6] LEIPHOLZ, H.H.E., *Stability Theory*, Academic Press, New York and London, 1970.

[5.7] LEIPHOLZ, H.H.E., *Direct Variational Methods and Eigenvalue Problems in Engineering*, Noordhoff International Publishing, Leyden, 1977.

[5.8] LEIPHOLZ, H.H.E., *Stability of Elastic Systems*, Sijthoff et Noordhoff, Alphen aan den Rijn, The Netherlands, 1980.